The Role of Public Policy in K-12 Science Education

A volume in
Research in Science Education (RISE)

Series Editors:
Dennis W. Sunal and Cynthia S. Sunal, *University of Alabama*
Emmett L. Wright, *Kansas State University*

Research in Science Education

Dennis W. Sunal, Cynthia S. Sunal, and Emmett L. Wright, Series Editors

Reform in Undergraduate Science Teaching for the 21st Century (2003)
edited by Dennis W. Sunal, Emmett L. Wright, and Jeanelle Bland Day

The Impact of State and National Standards on K–12 Science Teaching (2006)
edited by Dennis W. Sunal and Emmett L. Wright

The Impact of the Laboratory and Technology on Learning and Teaching Science K–16 (2008)
edited by Dennis W. Sunal, Emmett L. Wright, and Cheryl Sundberg

Teaching Science with Hispanic ELLs in K–16 Classrooms (2010)
edited by Dennis W. Sunal Cynthia, S. Sunal, and Emmett L. Wright

The Role of Public Policy in K–12 Science Education (2011)
edited by George E. DeBoer

The Role of Public Policy in K-12 Science Education

edited by

George E. DeBoer
AAAS Project 2061

Information Age Publishing, Inc.
Charlotte, North Carolina • www.infoagepub.com

Library of Congress Cataloging-in-Publication Data

DeBoer, George E.
The role of public policy in K-12 science education / George E. DeBoer.
p. cm. -- (Research in science education)
Includes bibliographical references.
ISBN 978-1-61735-224-9 (pbk.) -- ISBN 978-1-61735-225-6 (hardcover) --
ISBN 978-1-61735-226-3 (e-book)
1. Science--Study and teaching--United States. 2. Civics--Study and teaching--United States. I. Title.
LB1585.3.D44 2010
507.1'073--dc22

2010045145

Printed in the United States of America

CONTENTS

PREFACE TO THE SERIES

Science education as a professional field has been changing rapidly over the past two decades. Scholars, administrators, practitioners, and students preparing to become teachers of science find it difficult to keep abreast of relevant and applicable knowledge concerning research, leadership, policy, curricula, teaching, and learning that improve science instruction and student science learning. The literature available reports a broad spectrum of diverse science education research, making the search for valid materials on a specific area time-consuming and tedious.

Science education professionals at all levels need to be able to access a comprehensive, timely, and valid source of knowledge about the emerging body of research, theory, policy and practice in their fields. This body of knowledge would inform researchers about emerging trends in research, research procedures, and technological assistance in key areas of science education. It would inform policy makers in need of information about specific areas in which they make key decisions. It would also help practitioners and students become aware of current research knowledge, policy, and best practice in their fields.

For these reasons, the goal of the book series, *Research in Science Education,* is to provide a comprehensive view of current and emerging knowledge, research strategies, and policy in specific professional fields of science education. This series presents currently unavailable, or difficult to gather, materials from a variety of viewpoints and sources in a usable and organized format.

Each volume in the series presents a juried, scholarly, and accessible review of research, theory, and/or policy in a specific field of science education, K–16. Topics covered in each volume are determined by current issues and trends, as well as generative themes related to up-to-date research findings and accepted theory. Published volumes will include empirical studies, policy analysis, literature reviews and positing of theoretical and conceptual bases.

PREFACE

The goal of this volume of *Research in Science Education* is to examine the relationship between science education policy and practice and the special role that science education researchers play in influencing policy. It has been suggested that the science education research community is isolated from the political process, pays little attention to policy matters, and has little influence on policy. But to influence policy, it is important to understand how policy is made and how it is implemented. This volume sheds light on the intersection between policy and practice through both theoretical discussions and practical examples.

This is not a "how to" book. It is not a book about how to develop policy, how to become a policy researcher, or how to get your own research considered in policy decisions. There is already a vast literature on education policy. The *Handbook of Education Policy Research* published in 2009 is a thorough examination of many of the most important issues in education policy research. This current volume in the *Research in Science Education* series is an introduction to the field of policy as it intersects with science education. Its primary intent is to lay out what I will call the "policy terrain" in science education, those places where policy is relevant to science education. That includes policy decisions that have affected practice in science education, including classroom instruction, curriculum development, and assessment. It also includes discussions of how science education research is used by policymakers in their deliberations about the best policies to enact.

Nor is this a book for policymakers. The book does not provide an authoritative description of the state of knowledge with respect to various policy initiatives that would be helpful to a policymaker. There are numerous review articles on the current state of research on assessment, curriculum development, instructional models, and teacher education that can be used by policymakers. And the book does not take advocacy positions, informing policymakers of the best policy positions to take, although in a few places authors do suggest how certain policies might be beneficial for improving teaching and learning.

Instead, the book is meant to be useful to science education researchers and to practitioners such as teachers and administrators because it provides information about which aspects of the science education enterprise are affected by state, local, and national policies. It also provides helpful information for researchers and practitioners who wonder how they might influence policy. In particular, it points out how the values of people who are affected by policy initiatives are critical to the implementation of those policies. When it comes to curriculum and assessment, for example, developers need to know something about the real-world setting where they want their materials to be used. No matter how carefully researchers and developers plan their studies and materials, if practitioners hold other values, it is surprisingly easy for them to ignore, work around, or otherwise undermine new policy initiatives. In the chapters that follow, a number of policy researchers, policy developers, and policy implementers discuss a wide range of issues dealing with the impact that the science education community has on policy development and the effect that these policies have on practice. The volume should be considered as a first step in thinking through what the field of policy research in science education might look like.

This book was written primarily about science education policy development in the context of the highly decentralized educational system of the United States. But, because policy development is fundamentally a social activity involving knowledge, values, and personal and community interests, there are similarities in how education policy gets enacted and implemented around the world. Therefore, two chapters focus on examples of policy development outside the United States. Each example traces the process of policy development from initial vision to final implementation.

Following an introductory chapter that briefly discusses some of the more general aspects of the policy world, the volume is organized into three parts. Part I focuses on the many influences on policy development and enactment; Part II focuses on how policy affects curriculum, instruction, and the equitable treatment of all students; and Part III focuses on policy implementation.

George E. DeBoer

ACKNOWLEDGMENTS

The people most to be thanked for this volume in the *Research in Science Education* series are the authors who gave so generously of their time to conduct the research and write the manuscripts that you see here. Policy is a relatively new area of investigation in science education, with very limited resources to draw upon. Authors were faced with the challenge of having to forge new ground and provide their own interpretations of what the intersection of policy and science education involves. As editor, I was fortunate to be able to call upon some of the most well respected science educators who were willing to share their ideas about the role and importance of policy in various aspects of our field. Thanks also go to the graduate students and others who assisted the authors in their investigations and preparation of the manuscripts.

Another group of individuals to whom a great deal of credit must go for helping make this project possible are the reviewers who offered critical feedback on the chapters. They provided the additional insights needed to help craft the contributions that make up this volume. The reviewers were Alicia Alonzo, Rolf Blank, Rodger Bybee, Joanne Carney, Daryl Chubin, Janet Coffey, Angelo Collins, Thomas Corcoran, Charlene M. Czerniak, James Donnelly, Mauricio Duque, Richard Duschl, Jacob Foster, Magnia George, Jessica Hammock, Wynne Harlen, Peter Hewson, Jane Butler Kahle, Okhee Lee, Shirley Malcom, Ron Marx, Barbara Nagle, Sharon Nelson-Barber, Mary Koppal, Barbara Olds, William Penuel, Harold Pratt, Mary Ratcliffe, Senta Raizen, Jo Ellen Roseman, Patricia Rowell, Judith Opert Sandler, Kathryn Scantlebury, Mark Schneider, Susan Snyder, John Staver, Iris Weiss, and Karen Worth.

I also wish to think my wonderful colleagues at AAAS for their generous and support during this project. And, finally, sincere appreciation to Dennis Sunal and Emmett Wright, series editors, for seeing the value of such a volume and proposing it to me several years ago, their persistence until I finally said yes, and their encouragement throughout the project.

George E. DeBoer

CHAPTER 1

INTRODUCTION TO THE POLICY TERRAIN IN SCIENCE EDUCATION

George E. DeBoer

INTRODUCTION

In general terms, policies are the rules, regulations, and established practices that guide the behavior of social groups. In the special case of science education, these policies include federal mandates to test students in science, state requirements concerning the number of science courses students must take to graduate from high school, and state requirements concerning teacher certification and licensure. Local school districts have policies that determine the courses that will be available to students, what textbooks will be used, the nature of laboratory and field experiences for students, and how large class sizes will be. All of these policies affect what students are taught in science and how well students learn. Some policy researchers study the impact of existing policies or the likely impact of proposed policies. Others generate knowledge that could become part of future policy discussions.

The Role of Public Policy in K–12 Science Education, pp. 1–9

As many have noted, there is virtually no literature in science education on how research affects policy, how policy affects practice, or how the personal values of teachers, parents, administrators, and students are relevant to policy enactment or implementation. It's not that policy researchers have not studied various aspects of education related to the teaching of science, but the tradition of doing policy research resides primarily in the report literature of the federal government, foundations, and think tanks, not in referred academic journals in science education. RAND, for example, studied the relationship between reform-oriented instruction and student achievement in science and mathematics (Le et al., 2006), and CPRE has engaged in a number of studies related to science education including a recent study of learning progressions in science (Corcoran, Mosher, & Rogat, 2009). But there is simply no tradition of policy analysis or policy research *in the field of science education*. The American Educational Research Association's recently published 1,000-page *Handbook of Education Policy Research* (Sykes, Schneider, & Plank, 2009) has just a single two-page entry on science education policy. It is part of a chapter on global education and describes how policymakers have linked low science scores of American students on international tests to faulty curriculum and then concluded that curricular reform was needed to fix the problem (Baker, 2009, p. 965). Baker also noted that policymakers have linked student scores on international tests to a country's economic performance and concluded that if the U.S. improved its curriculum it would improve its economic competitiveness internationally. Baker argues that because the curriculum has become so similar from country to country it is unlikely that the curriculum can be the explanation for the differences in performance or differences in economic performance. He says that such quick fixes deny the complexity of the problem. He concludes the section by saying:

> If the United States wishes to further enhance its mathematics and science training, these global trends suggest the nation is not handicapped by "low achieving curricula," but that other qualities of the system, such as instruction and inequalities in resources and so forth, should receive serious empirical investigation before charging off half-cocked towards curricular reform. (p. 966)

The point is not to argue the validity of Baker's critique but rather to point out that because we have a very limited commitment to policy analysis, there is a very limited research base upon which to make policy decisions. In his 2008 Keynote address at the Annual Meeting of the National Association for Research in Science Teaching in Baltimore, MD, Peter Fensham said that science education researchers have been politically naïve with respect to policy issues. He noted three areas of naïveté—curriculum development, exaggerating the generalizability of findings, and not recognizing the contested nature of science in the curriculum. Clearly, these are good reasons for science educators to become more aware of how policy interacts with science education practice.

THE POLICY AGENDA

Broadly conceived, the science education policy agenda includes all of the issues under consideration either formally or informally by the general public, the media, and the educational research community having to do with the teaching of science. These include the many issues regarding assessment, instruction, curriculum, teacher certification, graduation requirements, and so on. When viewed more narrowly, the science education policy agenda includes just those issues that are being formally considered by policymakers, such as whether there should be national standards ("common standards") or national testing in science. One place where issues of this kind have been debated are debated is in the reauthorizations of the Elementary and Secondary Education Act (McNeil, 2010), legislation that was originally enacted in 1965 and reauthorized a number of times since then including as the No Child Left Behind Act of 2001. At the national level, policy decisions are made by the U.S. Congress and federal regulatory agencies in a two-part process that first includes passage of an authorization bill, which may then be followed by an appropriations bill that provides the funds to make it possible to carry out the functions of the authorization bill. In many cases funding is not appropriated for authorized programs, as in the case of the *America Competes Act* (Robelen, 2010). At the state level, policy decisions are made by state legislatures and state education departments. At both the state and federal levels, the courts may determine the constitutionality of legislation that has been enacted.

WHO DEFINES POLICY ISSUES?

Out of the hundreds or even thousands of issues of public concern that might be of potential interest to policymakers, only a relatively small number of them ever find their way onto a policy agenda. The policy agenda—the issues that get considered and deliberated by policymakers—is influenced by many factors. Frances Fowler (2004), in her book, *Policy Studies for Educational Leaders*, says: "In the United States almost all education policy issues are defined within a loosely linked set of institutions that some call the **education policy planning and research community** (EPPRC)" (p. 171, emphasis in original). Among the key players in this loose network of institutions responsible for issue definition, Fowler lists a small number of well known foundations including Carnegie, Ford, Kellogg, Pew, and Rockefeller; a small number of think tanks such as the American Enterprise Institute, the Brookings Institution, the Committee for Economic Development, and the RAND Corporation; and the five universities that originally made up the Consortium for Policy Research in Education (CPRE), namely, Harvard University, Stanford University, the University of Michigan, the University of Pennsylvania, and the University of Wisconsin at Madison (Fowler, 2004, pp. 172–173). Since then, Northwestern University and Teachers College Columbia have also joined CPRE. To those we can add professional organizations such as the American Association for the Advancement of Science, the National Research Council of the National Academies, and the National Science Teachers Association. It's not that other institutions are not also involved in this very informal and diffuse process of issue definition, but rather that these are some of the institutions that are most involved in education policy definition and are the most familiar to us. Funding for the idea development that takes place at these institutions comes largely from foundations, corporations, and the federal government. It is also important to note that policymakers themselves typically take initiative in policy issue generation and in issue definition.

GETTING ON THE POLICY AGENDA

Fowler (2004) identifies two types of policy agendas. The first she refers to as the *systemic agenda*, by which she means all of the issues that are being discussed outside of the formal governmental process. These include the issues being discussed by professional organizations and other interest groups, the issues that the media has decided to emphasize, and the issues that the public is paying attention to (p. 182). The second agenda is the *governmental agenda*. This is "the list of subjects or problems to which governmental officials … are paying some serious attention at any given time" (Kingdon, 1995, p. 3, as cited in Fowler, 2004, p. 182). The governmental agenda is made up of issues generated by the policymakers themselves and those that reach the policymakers from the systemic agenda. What is important to understand about both the systemic and governmental policy agendas is that there are many issues that never gain the attention of professional organizations, the media, the general public, or policymakers. And those that do find their way there often take a long and circuitous route with many stops and starts along the way.

Perhaps the most significant thing to consider when thinking about how a policy issue becomes part of the policy agenda is that policymakers must be made aware of the issue and they must appreciate its importance at the national, state, or local level. This means that policymakers need to receive communications and arguments about the issue from influential people whom they trust. In most cases, these issues percolate up to policymakers through a very complex system of individuals and organizations. Just as members of professional societies, university researchers and theoreticians, and public and private foundations engage in issue definition, they also engage in communicating their ideas into the larger society through written and spoken presentations. Unfortunately, more often than not, these communications are to other specialists in their field and not tailored to policymakers and their staffs. When ideas *are* communicated effectively, they can gain traction, and they can ultimately become part of the much smaller agenda that receives formal consideration by policymakers.

A number of things are clear when considering how issues come to the attention of policymakers: First is that access to policymakers is key to getting issues on the policy agenda. One particular type of institution that has played a very significant role in this regard is the world of private and public foundations. Heads of foundations, especially heads of the largest and most well-funded foundations, often have access to policymakers that most people do not have. It is also important to note when considering whether policymakers will be interested in an issue, that when the issue is highly charged and personal values are likely to play a significant role, policymakers often find it difficult to ignore their own deeply held personal feelings about the issue. Policymakers want to create policy that is in the best interest of society, but it is difficult for them to support policies that are not consistent with their own values. But policymakers also want to make informed decisions rather than uninformed ones, so when it comes to issues that are more pragmatic in nature and less highly charged, well-reasoned arguments and supporting data can be compelling to policymakers. This provides education researchers with an opportunity to influence the policy agenda by communicating clearly the results of their own research. Researchers can also provide policymakers with data that may help policymakers rethink faulty assumptions they are making (see Osborne, Chapter 2, this volume). To assist in getting these results and arguments in front of policymakers, many professional organizations in science education, of which science education researchers are members, have a legislative office that acts as a liaison with policymakers. In this regard, it is important to note that the staffs of governmental policymakers play a critical mediating role in "translating" issues and getting them on the policymakers' radar screen. It is usually with staff members that professional organizations communicate (see Peterson, Chapter 9, this volume).

THE EFFECT OF POLICY ON PRACTICE

With regard to science education, policymakers are charged with creating the structures and processes that provide the best opportunity for students to learn science. At the state level, state education departments set policies regarding requirements for teacher certification, including alternative routes to certification; graduation requirements for students; and standards for professional development. Policymakers often are forced to make their decisions on the basis of their collective judgment, without the availability of compelling research evidence to support those decisions.

In the end, the goal of policy development in science education is to set a direction for the development of educational practices that advance students' knowledge and competence in science. Some policies do that and some do not. Unfortunately, we often do not have good empirical evidence to answer the question of whether any particular policies are or will be effective. In part this is due to a failure to systematically collect data on outcomes that could be linked to policy initiatives so that explanations and predictions could be made. But it is also due to the complexity of the task: there are both long- and short-term effects to consider as well as unforeseen and unintended consequences along with those that are intended. A policy issue that is currently being discussed at the national level at the time of this writing is the issue of common core standards (Cavanagh, 2010). Common core standards have been developed in mathematics and English language arts, and a process has begun that is likely to lead to common core standards in science as well. Common core standards have been developed because of a belief that they will lead to improved student learning. But this belief is not the result of a body of empirical evidence from carefully controlled studies. In fact, in a review of the research literature on the effects of national standards on student outcomes, McCluskey (2010) concludes that the case for common standards shared by all states is empirically weak. To emphasize the difficulty of making policy decisions on the basis of empirical evidence, Cavanagh notes that scholars in this area have pointed out that: "Determining which aspect of a nation's education system drives achievement—standards, as opposed to well-trained teachers, or a strong societal emphasis on education, or other factors—is difficult" (Cavanagh, 2010, para. 2, Lessons From Abroad section). The reason for this is that often the systems we are studying are too complex for researchers to find single factors that can be identified as causal, possibly because the outcome is overdetermined or possibly because we have not yet been able to identify the many interrelated parts of a very complex but singular cause. So, when research is available, clear, and convincing, it may be used in making policy decisions, but when it is not, policy still gets made if policymakers believe the policy will be beneficial in some way.

POLICY IMPLEMENTATION

Once educational policies have been enacted by policymakers, they still have to be implemented, turned into actual educational programs and practices. The process of implementation can be contentious. When implementing new policies, especially at the school level, it is important to understand the various positions held by the community at large, the parents of the students who are affected, the students themselves, teachers, and school administrators—and try to gain their support. As with policy formation, policy implementation takes place in a social world that is shaped by people's beliefs and values, one in which social structures, personal leadership, and political determination play a large part. In addition to those who support a new policy, there are usually individuals or groups who may have an interest in subverting or circumventing it. Often this comes from a desire on the part of groups for local control over their own affairs. This is something that policymakers and those charged with policy implementation must always be aware of. If local community members have not been adequately involved in the decision-making process, it may be difficult to gain their support.

Emerson's (1839, as cited in Brown, 1997) statement that "there is no history, only biography" may be particularly relevant in the context of policy development and implementation. The extended quote is: "The new individual must work out the whole problem of science, letters, & theology for himself, can owe his fathers nothing. There is no history; only biography" (p. 170). Policies are intimately connected to the values of the individuals within a society and must be understood in terms of the people that make up that society. In a democratic society, policies last only as long they are thought to be of value. In fact, in the United States, the legislative process *requires* that legislation be reviewed and reauthorized periodically. If not reauthorized, the legislation dies. Policy formation and implementation are organic processes, with policies continuously under review and subject to change.

THE NEED FOR POLICY RESEARCH IN SCIENCE EDUCATION

There is a great need for research in science education that is of the quality and accessibility to nonspecialists that policymakers can use in the decisions they make. This is perhaps the best way for science education researchers to influence policy, as long as their research is relevant to the real world problems that policymakers deal with. There are also opportunities for science education researchers to become actual policy researchers, that is, individuals who study the very interesting and complex processes involved in policy formation and implementation. And there are opportunities to become policy analysts, to study the effectiveness of existing policies, to use past experiences to develop predictions about the likely efficacy of alternative policy directions, and to advise policymakers by bringing evidence to them to inform the policymakers' values. All of these are ways that science education researchers can be involved with the world of education policy if they have the desire to do so.

REFERENCES

Baker, D. P. (2009). The invisible hand of world education culture: Thoughts for policy makers. In G. Sykes, B. Schneider, & D. N. Plank (Eds.) *Handbook of education policy research* (pp. 958–968) New York, NY: Routledge.

Brown, L. R. (1997). *The Emerson museum: Practical romanticism and the pursuit of the whole*. Cambridge, MA: Harvard University Press.

Cavanagh, S. (2010). U.S. common-standards push bares unsettled issues: Familiar themes emerge in resurgent debate (Published in Print as: Resurgent debate, familiar themes). *Education Week*. Published online January 14, 2010 (from http://www.edweek.org/ew/articles/2010/01/14/17overview.h29.html)

Corcoran, T., Mosher, F. A., & Rogat, A. (2009). Learning progressions in science: An evidence-based approach to reform. CPRE Research Report # RR-63. Retrieved from http://www.cpre.org/images/stories /cpre_pdfs/lp_science_rr63.pdf

Fensham, P. (2008, April). *The link between policy and practice in science education: Where does research fit in?* Paper presented at the annual meeting of the National Association for Research in Science Teaching, Baltimore, MD. Retrieved from http://www.narst.org/annualconference/postconference08/ peterfenshamintrolecture.pdf

Fowler, F. C. (2004). *Policy studies for educational leaders*. Upper Saddle River, NJ: Pearson.

Kingdon, J. W. (1995). *Agendas, alternatives, and public policies* (2nd Ed.). New York, NY: HarperCollins.

Le,Vi -N., Stecher, B. M., Lockwood, J. R., Hamilton, L. S., Robyn, A., Williams, et al. (2006) *Improving Mathematics and Science Education: A Longitudinal Investigation of the Relationship between Reform-Oriented Instruction and Student Achievement*. Santa Monica, CA: RAND.

McCluskey, N. (2010). Behind the curtain: Assessing the case for national curriculum standards. *Policy Analysis, 661*. Retrieved February 17, 2010, from http:// www.cato.org/pub_display.php?pub_id=11217

McNeil, M. (2010). House committee to hold hearings on new ESEA: House hearings mark start of reauthorization process. *Education Week*. Published online

February 22, 2010, Published online February 22, 2010 (www.edweek.org/ew/articles/2010/24/22esea.h29htm)

Robelen, E. W. (2010). Many authorized STEM projects fail to get funding. *Education Week*. February 23, 2010. Retrieved online from http://www.edweek.org/ew/articles/2010/02/24/22stem_ep.h29.html. (Published in print February 24, 2010 as Talk Bigger Than Federal Funding for STEM Projects.)

Sykes, G., Schneider, B., & Plank, D. N. (2009). *Handbook of education policy research*. New York, NY: Routledge.

PART I

MULTIPLE INFLUENCES ON POLICY DEVELOPMENT AND ENACTMENT

Part I of the volume focuses on a wide range of influences on policy development and enactment from a variety of perspectives. The section begins with two chapters on the influence that science education researchers have on policy development. Jonathan Osborne uses examples from Great Britain and the European Union to argue that for science educators to have an influence on science education policy, they must step out of their scholarly worlds and engage directly with the practical world of policymakers. Jane Butler Kahle and Sarah Beth Woodruff point out that there is very little research available for policymakers who are setting teacher education policy and how badly needed that research is. This focus on the role of researchers is followed by two chapters that describe the role that funders play in shaping policy. Dennis Cheek and Margo Quiriconi demonstrate through historical analysis how important foundations have been in influencing education policy, especially when federal spending was much more limited, and Janice Earle points out how federal funders, especially the National Science Foundation, have become more and more important in shaping the policy agenda. Then, Jean-Pierre Sarmant, Pierre Léna, and Edith Saltiel use the case of *La main à la pâte* to show how a nongovernmental entity, the French Academy of Sciences, was able to change science education policy at the elementary level in France and is continuing to spread its influence worldwide. Cheek and Quiriconi follow their first chapter with a chapter on the influence of state education departments in science education policy development, pointing out how little research has been done on the role that state education departments play in the national policy context. Rodger Bybee focuses on the intricacies of policy development and enactment—at the international, national, and state level—in the context of student assessment. Finally, Jodi Peterson provides some practical advice to individuals, including science education researchers, who wish to approach policymakers with ideas for the policy agenda and how organizations such as the National Science Teachers Association can help science educators become involved in policy advocacy.

CHAPTER 2

SCIENCE EDUCATION POLICY AND ITS RELATIONSHIP WITH RESEARCH AND PRACTICE

Lessons From Europe and the United Kingdom

Jonathan Osborne

INTRODUCTION

The intersection between research, policy, and practice is complex. In this chapter, I explore the nature of that relationship as it has developed in Europe, but more particularly in the United Kingdom in the context of science education. The primary role of research is to contribute to our knowledge of the world, in the case of education, to understand a particular aspect of the social world. Policy and practice, in contrast, are dominated by debates about what *should* be done and, in the case of science education, by issues such as the best way to improve students' engagement with science, whether the level of student understanding of science is as high as we think it should be, and whether the numbers of

The Role of Public Policy in K–12 Science Education, pp. 13–46

students pursuing full-time study of science should be increased. In other words, researchers answer fundamentally different kinds of questions than policymakers do: research deals with questions of what is and why, whereas policy deals with questions of value, judgment, and action. This chapter offers an analysis of the interrelationship between research, policy, and practice. It is intended to illuminate the lessons that might be learned by those who wish their research to influence policy and ultimately affect practice.

The first half of the chapter examines how research intersects with policy and practice. It begins by exploring the contributions research can make to the decision-making process policymakers and practitioners must undertake. First, research supplies empirical evidence of cause and effect relationships and provides insights into certain forms of behavior and action. In addition, because many decisions—such as whether the emphasis of a curriculum should be on science for future scientists or on science for citizenship—are ultimately issues of values, evidence provided by researchers cannot definitively answer such questions. Rather, evidence from research can inform the debate by providing reliable and valid empirical evidence of what is known or by highlighting the values underlying the arguments in the debate. As Willard (1985) points out: "values emanate from practice and become sanctified with time. The more they recede into the background, the more taken for granted they become" (p. 444). The scholarship of researchers can help, therefore, to bring to light the tacit assumptions that relevant actors may hold and the consequences of those assumptions for practice. The first half of the chapter concludes by looking briefly at the ways in which researchers might communicate with their audiences. The second half of the chapter then draws on specific examples from science education to explore the themes of the first half of the chapter.

THE INTERSECTION OF POLICY AND PRACTICE

Science Education Policy in the European Context

Any discussion of the relationship between the research community and policy communities in Europe must start with the recognition that the diversity of systems and structures within Europe is much larger than that within any single country. Even the United Kingdom (U.K.) consists of four semiautonomous entities, England, Wales, Scotland, and Northern Ireland. Each of these countries sets its own curriculum standards and runs separate systems of assessment. Likewise, Germany consists of 16 states (Länder) and is more akin to the United States in that each state sets its own education standards. Switzerland, too, has its own regional areas as does Spain. Other countries, such as the Netherlands, Norway, and France are more unified in the way their school curricula are specified. Of these, it is the French system that is possibly the most highly centralized, where the curriculum has a status akin to that of a law and is an official bulletin of the National Ministry of Education (Bulletin officiel du ministère de l'éducation nationale). As such, all French schools are required to follow this curriculum and any changes come from the National Ministry. This means that individual variations between regions (Département) are not permitted. (See Sarmant, Saltiel, and Léna, Chapter 6, this volume, for a further discussion of the French educational system.) Many of the countries of Europe belong to the European Union (EU), but the European Commission (the executive branch of the EU) has limited executive authority. For example, it does not set policy in areas such as taxation, social welfare, or education. Policy in these areas is the prerogative of the individual countries. The work of the EU in the domain of education is restricted to supporting research and engaging in coordination activities. For example, one of its current major programs, a consequence of an agreement signed in Bologna in 1999 with the eponymous title of "the Bologna agreement," is an attempt to harmonize the meaning of educational qualifications across Europe by setting equivalent periods of study for all academic degrees across the EU. This action was necessary because, in some countries, an undergraduate degree was a 3-year program and in others a 5-year program. As the free movement of people in Europe depends on having common educational standards, some degree of harmonization and standardization was thought to be essential.

The outcome of such diversity within the EU, and the lack of authority over education by the European Commission, is that education researchers who have an interest in influencing policy and practice will look within their own countries to where the levers of power exist and the means by which policymakers are influenced, rather than looking to Europe as a whole. Despite differences in their educational systems, Europeans share a set of common values that shape and enable a common discourse. For example, Europe consists of advanced societies that are heavily dependent on science and technology for maintaining and improving their industrial and knowledge-based economies as well as the infrastructure of buildings, roads, rail, air travel, and communications on which the society is built. The advanced technological nature of European societies is itself a product of a shared participation in the Renaissance and the Enlightenment. As a consequence, a common thread is the commitment to empirical evidence, sound argument, and rationality as the basis of belief. This belief in the value of evidence provides education researchers with excellent opportunities to influence policy and practice throughout all of Europe, even if the starting point is their own country.

One matter of significant concern and challenge to policymakers in Europe, that has relevance to science education researchers, is their desire to sustain Europe's economic position in a competitive global environment. It is this concern, in particular, that has had an impact on policy issues associated with science education across the European Union. The Lisbon Strategy, also known as the Lisbon Agenda, for example, established an action and development plan for the European Union in the year 2000. Its aim was, and still is, to make the EU "the most dynamic and competitive knowledge-based economy in the world capable of sustainable economic growth with more and better jobs and greater social cohesion, and respect for the environment by 2010" (European Council, 2000). This goal led the EU to establish a small group of scientists and science educators to examine the future supply of scientists and technologists. The outcome was the report *Europe Needs More Scientists* (European Commission, 2004). The publication was important in identifying the nature and extent of the problem and what it would take to achieve the primary objective of the Lisbon agenda, pointing to the fact that:

> In 2001, the number of researchers per 1000 of the workforce (in full-time equivalent, FTE) was 5.7 for the EU.... These figures should be compared to a value of 9.14 researchers per 1000 of the workforce (FTE) for Japan and 8.08 for the USA. Only some countries in Europe (Finland, Sweden, Norway) reach that ratio and the most populated ones show much lower figures (Germany: 6.55, U.K.: 5.49, France: 6.55). (p. iii)

Similar points have been made in reports written in the United Kingdom. In 2001, for example, the government commissioned one of its chief scientific advisors, Sir Gareth Roberts, to examine the future supply of scientists and technologists. In the report Roberts (2002) identified:

> a "disconnect" between this strengthening demand for graduates (particularly in highly numerate subjects) on the one hand, and the declining numbers of mathematics, engineering and physical science graduates on the other. (p. 2)

Roberts (2002) argued that the issue of supply was in need of urgent attention if the United Kingdom was to remain competitive in the global economy. Education was portrayed as one means of remediating this problem, and a whole chapter of the report was devoted to the state of U.K. science education. Six years later, a similar report was written by Lord Sainsbury (Lord Sainsbury of Turville, 2007), who served as Minister for Science in the U.K. government from 1997–2005. Here, the issue was framed bluntly in its title: "The Race to the Top." His principal argument was that, in an era of globalization, the only way to sustain the competitive edge of the United Kingdom was to move to "high-value goods, services and industries" and that this objective was only achievable through an improvement in the process of knowledge transfer and research. Sustaining and developing such an economy would depend on reversing the 20-year decline in the number of students choosing to pursue Science, Technology, Engineering, and Mathematics (STEM) related careers. The means of achieving this objective was a campaign to improve the quality and quantity of STEM teachers and a government drive to increase the number of students choosing to study physics, chemistry, and biology. It should be noted that such a policy concern is not unique to Europe, as similar concerns have been articulated in the United States (National Academy of Sciences: Committee on Science Engineering and Public Policy, 2005) and in Australia (Tytler et al., 2008). Indeed, the stimulus funds provided in 2009 for education within the US have an identical title to that of the Sainsbury report.

The concern of EU policymakers about the supply of STEM professionals has led to a request for evidence about the nature of the problem and its causes. Policymakers have turned to science education researchers, among others, to inform that debate with relevant information about supply and demand and the feasibility of certain proposed policy initiatives. But, this specific policy concern also offers challenges because a focus on increasing the number of scientists may conflict with other goals of science educators that they believe to be more important for the society to pursue. For instance, all of these reports are based on the assumption that the role of science education is to act as a "pipeline" that will supply the next generation of scientists, engineers and technologists. But that assumption has been contested (Jenkins, 1999; Millar & Osborne, 1998; National Academy of Science, 1995; Schwab, 1962). A number of these writers have argued that although it may be legitimate for a nation to provide incentives to students to study science in order to increase the number of scientists and engineers, there are other rationales for science education that are more important. For example, there is a need to educate young people to be critical consumers of scientific knowledge, given that many of the contemporary political and moral dilemmas are posed by science, a point made by the European Commission in their White paper on Education and Training (European Commission, 1995), where they argued that:

> Democracy functions by majority decision on major issues which, because of their complexity, require an increasing amount of background knowledge. For example, environmental and ethical issues cannot be the subject of

> informed debate unless young people possess certain scientific awareness. At the moment, decisions in this area are all too often based on subjective and emotional criteria, the majority lacking the general knowledge to make an informed choice. Clearly this does not mean turning everyone into a scientific expert, but enabling them to fulfill an enlightened role in making choices which affect their environment and to understand in broad terms the social implications of debates between experts. There is similarly a need to make everyone capable of making considered decisions as consumers. (p. 28)

Framed as a "pipeline," science education is reduced to a form of preprofessional training, where the curriculum is defined by what are the most appropriate knowledge and skills necessary to attract and prepare future STEM professionals. More importantly viewing science education this way neglects the needs of the vast majority of students whose study of science will terminate when they leave high school, if not before. From a liberal education perspective, science education is valued because it represents important ideas that are part of our cultural heritage. Translated into specific goals, this requires an education in science that not only introduces students to the major explanatory theories that science has to offer about the material world, but also explains how we know what we know and why it matters. In addition, it should attempt to develop students' higher order thinking skills and assist them to engage critically with science in their daily lives. Goldacre (2008) argues passionately that the growth of pseudo-scientific "rubbish" requires a public able to discriminate good from bad science. But the process of "interpreting evidence isn't taught in schools, nor are the basics of evidence-based medicine and epidemiology, yet these are the scientific issues which are on most people's minds" (Goldacre, p. 334).

The issue of the goals of the curriculum is, therefore, fundamentally an issue of values, and the concerns of the policy community now focus on the needs of a specific subset of students. What interests should research in science education serve? If, for example, high-level policymakers are operating from the assumption that the most important need of the society is to increase the number of technical personnel while science educators think that an informed citizenry is the most important need of the society, what can science education researchers do? Herein lies the first point required to understand the practical realities regarding the intersection of policy and research: Framing of the policy discourse is set by high-level policymakers.

Failure by researchers to recognize the specific policy agenda of policymakers simply leads to a situation where the ideas and thinking of science education researchers are ignored.

So, then, how can science education researchers reconcile their own agendas for reform with those of the policymakers? Should they work within the constraints policymakers create and try to find the space in which they can influence practice, or should they provide arguments against the agendas of the policymakers? In part, the answer is both. On the one hand, it is the policymakers who decide the relative importance of any given topic for the political agenda. And, as such, their political power to set and control the discourse is substantial. On the other hand, it is the researchers whose work, writing, and scholarship can influence the importance that is given to a topic by policymakers. Researchers have the data and theoretical insights that policymakers do not have, and policymakers would rather make informed than uninformed judgments whenever possible. In the context of the policy concern about the supply of future scientists, for example, there is a body of empirical work (Butz et al., 2003; Lowell, Salzman, Bernstein, & Henderson, 2009; Lynn & Salzman, 2006; National Research Council, 2008) and scholarship (Hill, 2008) that questions the premises of the debate. For example, within the past 40 years, the United Kingdom, along with many European countries, has gone from an economy largely dependent on a manufacturing base to one where it has been largely subsumed by other elements such as financial services, tourism, and employment in the public sector. Thus, basing policy decisions on the questionable assumption that forms of employment will be very similar in 10 or 20 years time is to risk educating or training young people in a body of skills and knowledge that may be of little utility when they complete their formal education. This example is offered not so much to argue for the validity of one side of the debate or the other or to resolve the dilemma. Rather, it illustrates how researchers who can provide empirical data and arguments that are salient to a policy issue can engage with the debate and offer an informed and critical commentary on issues of policy.

The Evidence-Based Policy and Practice Movement

So far, this chapter has simply attempted to provide the reader with an appreciation for the kinds of policy issues that are of central concern in Europe and to show how issues of value and evidence permeate this debate. In this section, the question of what contribution research can make to policy is considered in more detail. What exactly is it that educational researchers offer? One change in the landscape regarding the nature of public policymaking is the growing belief that social practice, and the policies guiding that practice, should be based on evidence of what works. It is becoming more and more widely accepted that just as evidence is a central epistemic justification for belief in science, the discourse surrounding policy also requires, as far as is possible, an evidential base on which to frame discussions and for policymakers to justify their decisions. Consequently, in the United Kingdom, but also more broadly across Europe, there has been a return to an instrumental view that research should provide a body of empirical evidence on which practice should be based. This applied focus for education in the United Kingdom was most succinctly expressed by Hargreaves (1996) who argued:

> In education, there is simply not enough evidence on the effects and effectiveness of what teachers do in classrooms to provide an evidence-based corpus of knowledge. The failure of educational researchers, with a few exceptions, to create a substantial body of knowledge equivalent to evidence-based medicine means that teaching is not—and never will be—a research based profession unless there is major change in the kind of research that is done in education. (p. xx)

Drawing on his knowledge of research in medicine, Hargreaves argued for education to provide "a type of applied research which gathers evidence about what works in what circumstances" (p. 2). Similar arguments have been made by Stokes about the importance of "use-inspired" scientific research in the influential book *Pasteur's Quadrant* (Stokes, 1997).

This vision of the role of research can be seen as part of a wider movement for evidence-based policy and practice across many areas of professional work (Davies, Nutley, & Smith, 2000). Arguably, it has led to the development of higher standards for reporting research, which enhances both its validity and reliability (APA Publications and Communications Board Working Group, 2009). In the context of medicine, the idea that practice should be evidence-based—in the sense that it should take into account the results from systematic research as well as the experience of the individual clinician—became well established in the early 1990s (Davies, Nutley, & Smith, 2000), and its influence has subsequently extended to other areas of policy both in Europe and elsewhere. The U.S. Department for Education, for example, has established the "What Works Clearing House" (http://ies.ed.gov/ncee/wwc/), which aims to be a "central and trusted source of scientific evidence for what works in education." Whether that is achievable is, perhaps not surprisingly, controversial (Millar, Osborne, & Springer, 2009; Shavelson & Towne, 2002; Slavin, 2002). Nevertheless, good arguments have been generated supporting a need to be more systematic and rigorous in accumulating evidence to support various policies and practices, *systematic* in the sense that we need to establish procedures for identifying salient or relevant research, and *rigorous* in the sense that we need to establish commonly agreed upon criteria as to what kind of research provides worthwhile evidence to justify those policies and practices. Such an approach would have the benefit of revealing gaps in our knowledge and would avoid unneeded replication of already well-established findings. Hence, the Ministry responsible for education in the U.K. funded the creation of the "Center for Evidence-Informed Education Policy and Practice" (the EPPI Center), which has been influential in conducting research reviews and also in developing the methodology to improve the quality of such reviews. Likewise, in Europe, the role of evidence in governance has also been recognized in a European Commission White paper (European Commission, 2001).

Hence first, the change in the political zeitgeist about the role of evidence offers a significant opportunity for researchers to influence the policy debate, as they are the primary source of the evidence policymakers need. Researchers can also help policymakers answer questions such as: "What will be the costs of implementing such policies?" "Is the proposed solution likely to make a significant improvement in current practice?" "How certain can we be of the validity of the research evidence in support of this policy or practice?" The policymaking process is also influenced, therefore, by articulate and well-constructed arguments (Majone, 1989) about the quality of the research evidence, the cost of implementation, and the likelihood of its success. A second point in comprehending the intersection between policy and research is that providing evidence of what works and developing coherent arguments that are grounded in scholarship are the primary means by which researchers can engage with policy.

Taking Account of Values

Researchers must acknowledge, however, that evidence of what works is not the sole determinant of policy or practice. Many times, issues of policy require value-based judgments. In the case of science education, much of the value-based debate involves the issue of goals. For example, to what extent should the curriculum teach not just the content of science but also explore how science works? The case for paying greater attention to the nature of science and the scientific enterprise has come out of a large body of theoretical work conducted in the domain of science studies. See Fuller (1997) or Irwin (1995) for examples.

In science education, decisions about forms of curriculum, assessment, and national and local standards are overwhelmingly made on the basis of values. For example, although we can show empirically that a consensus has been building that the nature of science should be a feature of the compulsory school curriculum (DeBoer, Chapter 10, this volume; Osborne, Ratcliffe, Collins, Millar, & Duschl, 2003), or that modifying a curriculum in certain ways can result in enhanced engagement with science by girls (Haussler & Hoffmann, 2002), arguments about whether one curriculum is "better" than another depend on the availability of a valid and reliable metric and an agreement about the salient outcomes. Disagreements about the relative values of differing goals mean that decisions about curriculum, therefore, are largely made on the basis of competing values rather than empirical evidence of outcomes. Moreover, these values are often highly contested (Black & Atkin, 1996; Perks, 2006). Hence: The third point needed to understand the intersection between policy and research is that the thoughtful recognition and articulation of values is a significant opportunity for researchers to influence policy.

In addition, researchers need to know that teachers typically rely heavily on a body of tacit knowledge that is highly situated in the context of their own classrooms and the students with whom they work rather than on scholarly research. Decisions about goals and the means to achieve them within the classroom, therefore, are most often either value-based or dominated by professional judgment rather than empirical evidence. Unlike medicine, where the goal is clearly to improve the health of the patient, goals in education are not so self-evident. In the science classroom, any given activity may aim to improve students' skills, their active engagement in the learning process, their knowledge, or a combination of all three. Claims of the relative importance of these educational outcomes become matters of personal value and professional judgment made by the individual teacher.

A corollary of the reliance by teachers on their own personal experience regarding the effectiveness of an educational innovation is their frequent reaction to empirical evidence for innovative approaches to learning. Teachers will commonly argue, and did in our work (Millar, Leach, Osborne, & Ratcliffe, 2006), that the context of a research study with which they were presented was exceptional, that the same intervention would not work in their school or with their students, or that the teachers in the reported study were specially chosen. In other words, the teachers believed that similar effects were unlikely to be attained in classrooms like their own, which had not been involved in the research study. Such responses have similarities with the responses of students observed in Chinn and Brewer's (1993) study who were presented with unexpected data. In that study, seven distinct forms of student responses were observed, only one of which was to accept the data and change their theories and their actions. The other six responses involved discounting the data in various ways in order to protect the preexisting theory.

Why do people tend to ignore empirical evidence and choose instead to hold onto their personal beliefs? In part it is born of a natural reluctance to change, the unwillingness of individuals to accept that their standard practice can be improved sufficiently to justify the effort and the demands of change. Change, as Claxton (1988) has pointed out, involves threats to an individual's sense of competence (as new techniques are unfamiliar and untested); sense of control (as the outcomes and reactions of the students are uncertain); sense of confidence (as there is no base of previous experience on which to rely); and sense of comfort (as the emotions associated with the three prior concepts are unsettling).

Moreover, teachers, like any other individual, will change their professional practice only if they have some level of dissatisfaction with existing practice. To borrow from the model of conceptual change advanced by Posner, Strike, Hewson, and Gerzog (1982), if there is no dissatisfaction with existing practice, or if there are doubts either about the plausibility of the research evidence, the potential value of the new suggested practice, or any combination of these, then there is little incentive to change (Fullan, 2001). Moreover, change in practice is rarely reducible to a single action but, rather, requires a set of actions for which teachers need an understanding of their theoretical rationale and a course of training in their use. Such training can rarely be achieved quickly (Joyce & Showers, 2002; Loucks-Horsley, Hewson, Love, & Stiles, 1998; Lovering, 1995) and is often not available.

The reluctance of individuals to change practice when evidence conflicts with their strongly held beliefs and values is important for researchers to consider when framing their recommendations, and for policymakers to consider when deciding their actions. Even policymakers find it difficult to be objective and unbiased. Kay (2008), for example, points to the fact that "evidence based policy is sought by government but mostly the result is policy-based evidence. Only facts and arguments that support the desired policy are admitted, so the analytic basis of decision-making is eroded not enhanced" (p. 11). Put another way, only evidence that accords with pre-existing beliefs is recognized, and what might be presented as an evidence-based decision is, in reality, value-based. The implication for researchers in science education is that the study of teachers' values and their tacit knowledge is needed to engage teachers and affect practice. Thus, a fourth observation we can make about the intersection of policy and research is that it is difficult for people to change their beliefs, especially those beliefs that have been reinforced through personal experience. Educational researchers need to keep this in mind if they are to have any impact on education practice.

Although teachers' values and their beliefs are a major determinant of their own practice, their influence on broader policy is, in contrast, relatively minimal, given that they rarely present a common voice. Indeed, as individuals, their influence is usually felt only when there is strong collective dissatisfaction with the practices that policymakers may have implemented. Even then, it is more likely that there will be an adaptation or subversion of the required changes so that the intentions of the new policy are subsumed within the previous goals of the teacher. A good example of this is the manner in which the requirements for investigative work in school science required by the National Curriculum in England and Wales were simply ritualized so that students would score highly when assessed (Donnelly, Buchan, Jenkins, Laws, & Welford, 1996), actions that effectively subverted the pedagogic intent of the new requirements.

In summary, the science education researcher needs to know that values are a significant determinant of policy and action. For the researcher, empirical evidence remains a tool that can be used to challenge deeply held beliefs. Although the practice of science might be secured by its commitment to evidence, however, the practice of science education often falls well short of such standards of rationality. Where evidence is thin or questionable, values will predominate. And even where evidence does exist, both policymakers and practitioners will be selective in choosing the evidence they will attend to. Communicating and influencing practitioners and policymakers therefore requires an understanding of their values and of how deeply they are held.

Communicating Research Findings to the Public

Another challenge to education researchers is communicating research findings to policymakers and practitioners. We know that policymakers rarely read academic journals and, even if they did, how could they possibly keep up with everything that is published? The world is awash with data and information. The challenge to researchers is to distill that information so that it is salient to policymakers' central concerns and easy to understand. The common view is that to get a Minister to read what a researcher has to say, it must be summarized on one page; to get his or her advisor to read it, it must be summarized on three pages; and, if you give them a report of 25 pages, do not expect either one to read it.

Although this might seem overly skeptical, what it does point to is the importance of executive summaries to fulfill this function. It also explains why quantitative results often predominate in such reports. Quantitative findings can be presented in a chart or table where the principal relationships are often self-evident. For example, Figure 2.1 summarizes a large body of data that tells a very clear story about declining student engagement with the study of science relative to other subjects in countries of the developed world.

The presentation of data in this chart immediately raises the question "Why?" and it quickly generates possible explanatory hypotheses. Indeed, there is a 0.93 negative correlation between the student

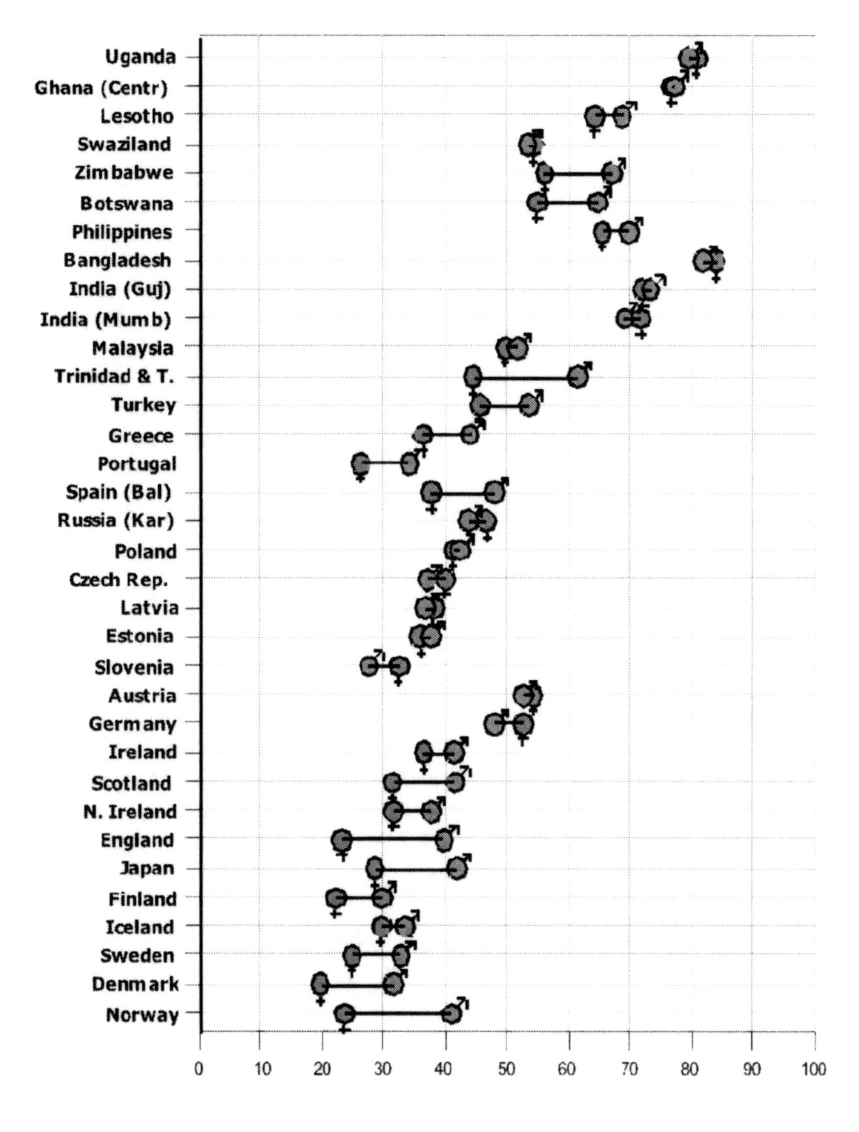

Source: Sjøberg and Schreiner (2005).

Figure 2.1. Percentage answering "Agree" or "Strongly Agree" by gender to the question "I like school science better than most other school subjects."

responses to this question and a country's UN index of Human Development, which seems to suggest that the student responses reflect something deeply cultural.

In contrast to communicating with quantitative data, communicating the evidential base of *qualitative* research studies is much harder. Researchers are reduced to using quotations to communicate their points in an articulate and convincing form. One such example comes from a study of English students' views about their school science. Twenty focus groups in a variety of locations in the U.K. participated in the study. The data set for this study was large, and the research was published in a 200-page report as well as in an article in an academic journal (Osborne & Collins, 2001). In conveying the findings to wider audiences, the researchers extracted salient quotations that communicated major findings from the research. In this case, the explanation for student disengagement with science was captured succinctly in the statement:

> The blast furnace, so when are you going to use a blast furnace? I mean, why do you need to know about it? You're not going to come across it ever. I mean look at the technology today, we've gone onto cloning. I mean it's a bit away off from the blast furnace now, so why do you need to know it? (p. 449)

The selective extraction of powerfully expressed quotations might be questionable in academic contexts because the practice raises questions about how representative the quotation is. However, reports for practitioners and policymakers must address their audience in language that is clear and, most important, convincing. Direct quotations are a good way to achieve this goal. Reports of this nature (e.g., Osborne & Dillon, 2008) will typically also have a stronger focus on the implications of the study, which are presented as a set of recommendations. Communicating research findings can also be aided by attracting the attention of the media, by presenting work in public forums such as at meetings of scientific societies, writing short pieces for quality newspapers, or sending summaries directly to policymakers.

Researchers who are not prepared to engage in this kind of writing should not be surprised if their influence extends only within the academic community. Policy is continually wrestling with important contemporary problems, "Why are students choosing not to pursue further study of science?" Research that might inform the policy debate will be of interest to policymakers if it is relevant to a policy concern and if it is communicated clearly, succinctly, and convincingly. Research that is concerned with the competing merits of two different methodological approaches to conducting research, on the other hand, will rarely be of interest to policymakers whose focus is on "what works."

Knowledge Intermediaries as Communicators of Research Findings

Another way in which the results of research influence policy and practice is through individuals who act as intermediaries between researchers, on the one hand, and the policymakers and teachers of science on the other. These "knowledge intermediaries" include curriculum and textbook writers, local education authority advisors, and those engaged in teacher professional development or other initiatives in the field of science education. These individuals often have to find an empirical or theoretical base for the practices that they advocate and will often look to scholarly research for the source of that knowledge, either through reading research publications or by attending conferences where researchers present their work. Thus, these intermediaries can act as an important conduit by which research results assist in two ways. First, they can identify the kind of evidence or arguments that are salient to the issues of the day. Second, many of them are skilled at distilling the findings of research into a form that is more easily read and understood. These are skills researchers themselves sometimes lack as they are often detached by the nature of their work from their audience, their values, and what matters to them. Lacking an understanding of their audience means that sometimes they fail to communicate their work effectively. Thus, knowledge intermediaries fill an important role in the communication of research results to policymakers and practitioners.

EXAMPLES OF THE INTERSECTION BETWEEN RESEARCH, POLICY, AND PRACTICE

Having discussed some of the general issues involved in the intersection of research, policy, and practice, I now seek to illustrate them with three specific examples drawn from the field of science education. The first example is the body of research on student attitudes toward science, which informs the debate about the supply and demand of scientists that I discussed earlier. This is a body of research that policymakers have paid particular attention to. The second is the development of a new National Curriculum in England, which has placed more emphasis on teaching about the nature of science. The third is a comprehensive body of evidence in support of a particular educational practice that research has shown to be highly successful but that has failed to significantly influence policy or classroom practice.

1. The Future Supply of Scientists

The concern about the future supply of scientists discussed earlier is informed by a body of research on students' attitudes toward the study of science. The research of Sjøberg and Schreiner (2005) in Norway; Osborne and Collins (2001) in the United Kingdom; and Lindahl (2007) in Sweden have all helped to identify a declining interest in the study of science. This research is supported by data from the Program for International Student Assessment (PISA) (Organization for Economic Cooperation and Development, 2007) and Trends in International Mathematics and Science Study (TIMSS) (Martin, Mullis, & Foy, 2008), both of which measure student attitudes toward science. Although these data have identified that there is a potential problem, additional information is beginning to accumulate about when student interest first forms, which may suggest to policymakers what kinds of action to take. For example, the work of Tai, Qi Liu, Maltese, and Fan (2006) is exceptional both in its use of longitudinal data drawn from the U.S. National Assessment of Educational Progress (NAEP) gathered between 1988 and 2000 and the clarity of its findings. Its major finding was that the majority of students make a tacit, if not explicit, decision about the broad focus of their future career trajectory before age 14. Additional evidence to support this conclusion has been reported by Bandura, Barbaranelli, Caprara, and Pastorelli (2001); Lindahl (2007); Ormerod and Duckworth (1975); and The Royal Society (2006). What the research does not do, but which is of even more interest to policymakers, is address how the students come to such decisions. Surveys of student attitudes toward science have led researchers to conclude that the reason many students are turning away from science is that they find either the content of science or how it is taught unengaging. See, for example, the work of Haste (2004), Schreiner (2006), and Osborne and Collins (2001).

The longitudinal nature of Tai et al.'s (2006) data and its elegant summarization in a graphical form has enabled researchers to readily communicate the results to policymakers and teachers and to explore the implications of the findings in reports such as *Science Education in Europe: Critical Reflections* (Osborne & Dillon, 2008) and in other publications and presentations. In the case of the United Kingdom, these observations have led the government to make science education a major element of an initiative on Science, Technology, Engineering, and Mathematics (HM Treasury, 2006) and to establish a $6 million research program whose primary objective is to answer the following questions:

1. What are the key factors that shape patterns of engagement and achievement in science and/or mathematics education by children and young people?

2. What can we learn from the effectiveness of past and current interventions, initiatives, and practice to inform the design and development of more effective future interventions?
3. How can research-informed approaches help to address some of the key challenges in enhancing engagement and achievement in science and mathematics identified by recent research and reports?
4. What specific new interventions offer the greatest potential to improve engagement and learning in science and mathematics and how could their potential effectiveness and feasibility be assessed more fully?

The complementary aspect of this debate about supply and demand is the question of what the future demand for STEM educated professionals and individuals might be. The issue of demand often does not gain the same attention as that of supply. Concerns about supply were raised in the United Kingdom more than 40 years ago (Dainton, 1968), yet the doom-laden scenarios envisaged then never materialized. Furthermore, the existing data suggest that in many domains of science, particularly the life-sciences, there is now an over-supply of graduates (Butz et al., 2003; Lynn & Salzman, 2006; Teitelbaum, 2007) and that the global production of doctoral students in STEM subjects is fairly healthy (Jagger, 2007). Indeed, at a seminar organized by the U.S. National Research Council on "Research on Future Skill Demands," one expert argued that employers' greatest concern was not the supply but the quality of training and the ability of individuals to work collaboratively (National Research Council, 2008). These examples illustrate how empirical information can be used to identify important features of a phenomenon—the uptake of science by young people—or to criticize the basic assumptions behind various policy proposals. One person to argue against the basic assumptions of policymakers on this issue is Hill (2008) who thinks that we are moving to a postscientific society where one of the key features of that society will be an emphasis on "innovation leading to wealth generation" where:

> productivity growth will be based principally not on world leadership in fundamental research in the natural sciences and engineering, but on world-leading mastery of the creative powers of, and the basic sciences of, individual human beings, their societies, and their cultures. (p. 78)

The point is not whether Hill's (2008) prediction is correct but more that his argument, and that of others (Gilbert, 2005; Tapscott, 2009), offers a critique of the background assumptions and values on which government policy regarding the supply and demand of scientists is framed. Indeed, another critique developed by scholarship within the science studies community argues that all of these concerns about the future supply of scientists are simply an effort to sell science and its value to policymakers. Hence, although the claims are advanced with a degree of apparent objectivity, they are actually based on carefully selected evidence to buttress the political claims of scientists for financial support (Fjaestad, 2007; Nelkin, 1995). What is important to recognize here is that there are no neutral participants in this (or any) policy debate. All arrive at the table with their own values and beliefs. The function of good research and scholarship is to use data to identify flawed or self-serving arguments and to provide arguments that are capable of surviving critical scrutiny themselves.

2. Changing the English Science Curriculum: 1996–2006

Arguably, the major influence on science education policy in the United Kingdom in the past decade has been the report *Beyond 2000: Science Education for the Future* (Millar & Osborne, 1998). The idea for this report was conceived as an attempt to articulate a vision of what school science should be by the late Professor Rosalind Driver and myself. At the time, we were concerned that the extant curriculum embodied in the current version of the National Curriculum simply failed to offer an appropriate education in science for the vast majority of students who would relinquish its study at the end of compulsory schooling at age 16. In addition, there was a rising tide of concern that school science needed to develop the skills required to critically evaluate the social applications and implications of science (Solomon & Aikenhead, 1994). The notion that more needed to be done to develop the public's understanding of science and its implications was an idea that had been initiated by the Bodmer (1985) report. This report characterized the public as lacking a knowledge of science, and it initiated a debate about exactly what kind of understanding of science the public should have (Irwin, 1995; Irwin & Wynne, 1996). This debate also permeated the field of school-based science education where a discussion began about what kind of science education was most appropriate for those who would be citizens and not practicing scientists. In a brief but influential article Millar (1996), a science education researcher, argued that the existing science curriculum had not been designed to meet the needs of contemporary youth. Rather, the form of the curriculum had evolved from its antecedents, but its basic focus was largely unchanged: The curriculum for 14–16 year olds was intended to prepare students for the curriculum of 16–18 year olds, which only a minority chose to study. This, in turn, was intended to meet the needs of an even smaller minority who chose to study science at University. Nowhere had any consideration been given to the needs of the educated citizen. Notably, this article was written in a professional journal—the *School Science Review*—where it was more widely read than if it had been published in an academic journal. The article thus raised questions about the appropriateness of the common form of science education. The arguments formed the basis of a proposal to the Nuffield Foundation that it was time to conduct a review of the aims and goals of science education by consulting with those stakeholders who had an interest in both its practice and outcomes, including teachers, curriculum developers, and researchers. The approach was somewhat unusual in the field of science education because researchers were joining with policymakers to discuss the formulation of a definitive, values-based policy statement that included the collective wisdom of the research community and its scholarship.

The Nuffield Foundation is the U.K.'s equivalent of the Ford Foundation in the United States, a philanthropic organization formed from the profits made by Morris Motors (before the industry fell into demise in the 1980s). (See Cheek & Quiriconi, Chapter 4, in this volume, for a discussion of the role of foundations in influencing education policy.) Part of its mission has been to influence education policy and curriculum, and it takes a largely instrumental view of the research it funds by asking how the outcomes of the research can affect policy or improve practice. This proposal was, therefore, a natural fit with the foundation's goals. The report (Millar & Osborne, 1998) was drafted by its principal authors and revised by a subgroup of interested researchers. Its argument was presented as series of brief chapters and a set of 10 recommendations, with the major one being that the compulsory school science curriculum should be framed in terms of the needs of the majority for whom each science course might be the last one they took, rather than the needs of the minority who would continue with the study of science. Such courses should place an emphasis on an education for scientific literacy, enabling individuals to become critical consumers of scientific knowledge rather than potential producers of scientific knowledge—the tacit if not explicit goal of traditional syllabi. Nuffield made the decision that the report was important enough to merit its professional production. This lent the report an air of authority that helped its reception with policy audiences. Moreover, the report was launched at an invited conference that the President of the Royal Society chaired. These points are made to demonstrate that constructing policy documents that will be read requires considerable work and professionalism.

Even when reports are generated carefully and professionally, however, the large number of reports produced means that many of them will fail to gain attention. It is difficult to explain the success of this particular document in influencing the thinking about science education. In part, it can be explained by the political zeitgeist of the time. Similar influential documents had been written in the North American context, notably *Science for All Americans* (Rutherford & Ahlgren, 1989); in addition, there was an emerging recognition that the existing curriculum was not fit for its purpose; and there was a rising concern about the relationship between science and society. The positive reception to the report gave Nuffield the confidence to support the development of a syllabus for 17-year-olds, entitled "Science for Public Understanding" (Hunt & Millar, 2000), which was the first attempt to produce a curriculum that embodied the goals and recommendations of the report. The aims of the new course were to develop students' understanding of both the major explanatory theories of science and the nature of contemporary scientific practice by teaching explicitly a set of "ideas-about-science." The intention was that ideas-about-science and science explanations should be taught and learned within the context of relevant science topics that have social importance (e.g., air pollution, hereditary diseases, life in the universe). The textbook was positively reviewed, and an evaluation of the course, although critical of the lack of teacher training and pedagogy, found that the course was largely successful in addressing its aims and engaging students (Osborne, Duschl, & Fairbrother, 2002).

By 2002, there was an increasing momentum to develop a course for the final stages of compulsory schooling, an initiative that was successful in attracting large-scale funding from both the Nuffield and Wellcome Foundations. In keeping with the recommendations of the original report, *Beyond 2000: Science Education for the Future* (Millar & Osborne, 1998), the initial materials for this new curriculum development project titled *Twenty-First Century Science* were pilot tested with a group of 70 schools. This pilot curriculum was evaluated (Ratcliffe & Millar, 2009), and the feedback was used to produce the revised materials and programs of professional training.

Simultaneously, in 2004 those responsible for revising the National Curriculum recognized that the existing program of study needed to give more attention to teaching about the nature of science, what is commonly termed "how science works." The *Beyond 2000* report had made a strong argument that school science should more explicitly teach a set of "ideas-about-science." Empirical evidence to support this argument emerged from a U.K. research-council-funded project, which had conducted a Delphi study of the key elements that stakeholders in the science curriculum thought should be taught about science (Osborne, Ratcliffe, Collins, Millar, & Duschl, 2003). Emerging from this work was a consensus around nine themes about the nature of science that should form an essential element of any school science education. In addition, empirical evidence from the evaluation of the new *Science for Public Understanding* course

(Osborne, Duschl, & Fairbrother, 2002) had shown that it was possible to incorporate many of these elements into a curriculum and teach them in an effective manner. In short, the view that school science should address not only what we know but also how science works was an argument that had gained significant ground with those responsible for drafting the National Curriculum.

As a consequence, the requirement to teach about "how science works" was increased and the detailed specification of the content was reduced. In addition, the new National Curriculum framework (Qualifications and Curriculum Authority, 2005) was designed to be much more skeletal so that many different curricular offerings could be offered rather than one standard course for all. The origins of the rationale for this change cannot be attributed to any one cause. However, the *Beyond 2000* report had made a recommendation that there should be an element of choice in the curriculum. Having one standard curriculum for the complete diversity of students at age 14–16 was increasingly seen as no longer justifiable. In addition, the development of new courses such as *Twenty First Century Science* (Millar, 2006) was possible only if the national specification for the curriculum moved more toward defining a required set of outcomes and specifying in much less detail the route by which these were to be attained. Hence, this policy change must be seen as a response to demands for a more flexible curriculum framework.

The work of developing the *Twenty First Century Science* curriculum was considerable because a whole new form of examination had to be tested and approved by the English Qualification and Curriculum Authority, and a program of professional training had to be offered to all teachers. The final product was one of several new curricula offered to schools that matched the new National Curriculum specification, and it was offered nationally to those schools that wished to adopt it in 2006. Its launch was, however, not without controversy, much of the critique coming from practicing teachers who saw it as undermining academic standards and intellectual rigor (Perks, 2006). Others were more positive (Reynolds, 2008):

> There are two things that have happened this year that have not happened before. The first thing is that lessons have been completely taken over by *students* asking the questions. The second, and by far the best, is that not once, not on a single occasion, not ever this year have I had to answer the question "why do we have to learn this?" (p. 8)

The findings of the evaluation conducted of the initial pilot suggest that the course has been generally successful in its aims. Teachers found the course more enjoyable to teach and more topical than previous courses, but had some difficulty adapting their pedagogy to some of the more interactive techniques the course required (Ratcliffe, Hanley, & Osborne, 2007).

Ten years after the initial publication of the *Beyond 2000* report, examination data show that approximately a quarter of all 16-year old students were taking the *Twenty First Century* science course as their compulsory science course. What this story demonstrates is that for researchers to have an impact on practice, a community of researchers who perceive advocacy to be a significant element of their work is required. Research alone is unlikely to impact practice significantly; rather it requires extended political engagement of the kind outlined here. It also calls for dialogue between researchers and policymakers. That is undoubtedly easier to achieve in a context where policy is generated centrally rather than as a dispersed responsibility as in the United States or Germany. In the U.K. context, dialogue is aided by the fact that many policymakers in science education were once teachers of science themselves, as are many of the researchers. Their common backgrounds enhance dialogue and understanding, making policymakers more receptive to the arguments that researchers might advance. Undoubtedly, however, there is an element of happenstance in the positive reception in that policymakers saw the lack of engagement of young people with science as a significant problem. *Beyond 2000* and its implications were seen as offering a solution to this problem, even though this was not its basic intention.

The Extension to Europe

The Nuffield Foundation regarded the funding of the report *Beyond 2000: Science Education for the Future* as one of its most influential projects. Therefore, in 2006, the director of the foundation—notably not researchers—initiated a conversation with the authors of *Beyond 2000* and several others about the possibility of engaging a group of European science educators in producing a report that would examine the issues and dilemmas confronting science education in Europe. Seeing this as an opportunity to make a more far-reaching statement about the nature of science education and the challenges it faced—in essence to sketch out an evidence-based set of recommendations for EU policy—Justin Dillon and I accepted this invitation.

The report, *Science Education in Europe: Critical Reflections* (Osborne & Dillon, 2008), was the product of two invited seminars where the invitees, a mix of researchers and EU policymakers, presented papers on the nature of science education in their own country and on other related issues. The final, 25-page report was written principally by the two authors, drawing on these submissions and other sources of information. Again, a considerable effort was put into its professional presentation and to disseminating it through various networks of contacts.

Essentially, this was an attempt to apply the previous success in the U.K. policy context to the larger setting of the EU. Although individual countries are always interested in their neighbor's practices, for any report to affect the policies of such a diverse group of countries, the arguments had to be clear and convincing. Member states will naturally focus on the soundness of the arguments, the quality of the data that is presented, and the values and existing practices of the people who will be affected by any policy initiative. How influential this particular report will be remains to be seen, but it has served the function of offering one clear vision of what needs to be done to improve the quality of science education in Europe.

3. Cognitive Acceleration in Science Education (CASE): The Impact of Evidence on Teachers' Practice

In this final example, I look at a case in which the research has been sound but the response of the majority of practitioners and policymakers has been *not* to change policy or practice based on those research findings. Cognitive Acceleration in Science Education (CASE) is a set of Piagetian-based teaching interventions designed to develop children's cognitive capabilities. It is the outcome of a program of research that has been conducted over a 30-year period of time, a feature that makes it exceptional compared to most research conducted in education. In the mid-1970s, Shayer and his colleagues first established the "levels of thinking" of U.K. students using a representative sample of students from ages 11 to 16 (Shayer, Küchemann, & Wylam, 1976) based on a Piagetian schema, and they developed instruments for measuring the students' level of scientific reasoning.

Arguing that the imperative for research was to develop interventions that would "accelerate" children's thinking to higher levels, they then developed an extensive 2-year intervention program for 11-14 year old children consisting of a detailed and tightly specified set of activities that addressed logico-mathematical reasoning skills, to be used by the teacher at 2-week intervals (Adey, Shayer, & Yates, 1989). Schools were specifically recruited for this project, and the findings were based on a sample of 11 schools that participated and 16 control schools. The study used a quasi-experimental design in which the researchers worked with teachers who were specifically trained in the use of the materials as well as in the underlying theoretical rationale and aims of the project. The intervention was found to have had significant positive effects on student learning in science (Adey & Shayer, 1990), and those gains increased one year later (Shayer & Adey, 1992a). Two years later, the intervention had a significant positive effect on the performance of the participating students in their national examinations in science (Figure 2.2). Even more surprisingly, similar improvements were found in the students' performance in English and mathematics (but not for girls in the first cohort studied) (Shayer & Adey, 1992b), although in the next cohort girls did show similar gains as well (Shayer & Adey, 1993). The findings of this study have since been replicated on an annual basis.

In summary, this is a research study that consists of: (a) a tightly defined and scripted intervention whose materials were later made available by a commercial publisher; (b) an extensive program of professional development to assist teachers in understanding the nature of the program and its implementation; (c) a clearly defined set of outcome measures (national tests) that are commonly used to assess student performance; and (d) a set of instruments, whose reliability has been tested, developed from a previous program of research for measuring students' ability to reason. Moreover, translating the research into a usable form required the professional production of the instructional materials

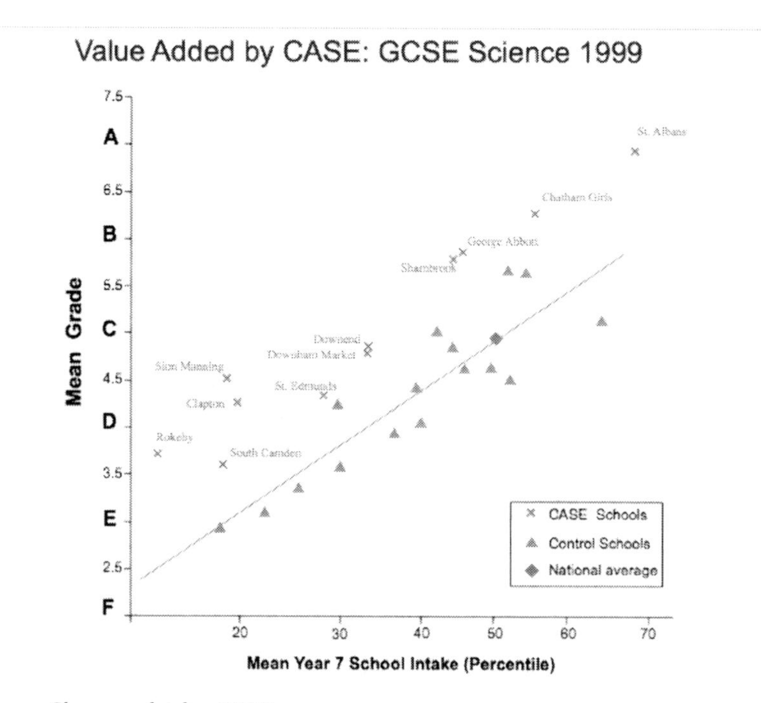

Source: Shayer and Adey (1993).

Figure 2.2. Results for students taking national science examinations (GCSE) at age 16 for CASE schools and control schools.

and the development of a program of professional training to support teachers who wanted to adopt the program (Adey, Landau, Hewitt, & Hewitt, 2003).

Overall, this work provides some of the most conclusive evidence that exists for the effectiveness of an intervention program within research in science education, a program that represents a considerable investment both intellectually and financially. Given the quality of the evidence in support of the intervention's effectiveness, and particularly its effect on raising student achievement in national examinations, one might have expected the uptake of such a program to have been near universal, especially as the findings of the project received considerable national publicity. Yet this has not happened, and the program is pursued by only a small minority of schools. Why?

Answers to the question must be sought in the professional values of teachers. As with all experimental interventions, questions can be raised about the causal mechanism underlying the obtained outcome (Leo & Galloway, 1996) or the validity of the instruments. However, would additional evidence have helped? The best explanation of the skepticism surrounding the reception of this work lies in the view that teaching is a highly situated activity where judgments are made on the basis of detailed knowledge of students that is acquired over time. In each class, each individual student is unique, and the students' actions are contextually specific. In evaluating the validity of research findings, teachers are engaging in the process of relating generalized conclusions obtained from a range of unfamiliar domains to the individual students in their own classroom. Given only the materials that were the product of the research and development, a summary of the findings from several research studies, and little understanding of the program's underlying rationale, teachers are naturally skeptical about whether such findings would be valid for their practice. It is also possible that accelerating students' cognitive development in this way was inconsistent with the teachers' ideas about the purpose of teaching science.

This example is given to point out once again that research alone is usually insufficient to influence practice or the policies that determine practice. This is true both when the link between research and practice is direct as in the CASE example, and also when information obtained from research is used by policymakers to inform their decisions. For, although research findings play a role in the deliberations of policymakers, policymakers must also consider the cost of the policy intervention, their own values, and the values of the people affected. In the CASE example, the teachers saw no reason to modify their teaching behavior in order to accelerate the cognitive development of their students, even though researchers had provided them with the tools to do so and demonstrated convincingly that it was possible to improve student performance. For reasons that are not well understood, they simply doubted the evidence or felt that the program did not match their conception of what it means to teach science (Ratcliffe, Leach, Millar, & Osborne, 2005). A policy that directed them to do so would probably have failed in the face of passive resistance.

THE INFLUENCE OF POLICY ON PRACTICE

This brings us finally to a brief discussion on the relationship between policy and practice. As we noted at the end of the last section, a policy that would have required teachers to adopt the CASE approach probably would not have been accepted by the teachers. But, in another example of the interface between policy and practice, teachers were required to make changes in their practice. This was the requirement of the English and Welsh (Department for Education and Science, 1991) national curriculum for students to conduct open-ended investigative laboratory tasks, and the requirement that such work should be formally assessed. In the opinion of science education experts, however, the vision of science offered and codified within the documents was "narrow, over-dependent on the notion of variables and prone to algorithmic interpretation" (Donnelly, Buchan, Jenkins, Laws, & Welford, 1996, p. 47). Given that the assessment of student performance made a substantive contribution to student's final examination grade, the outcome was perceived by both teachers and their students to be "high-stakes." As a consequence, teachers manipulated their practice and limited students' work to four practical investigations that fit the model (Watson, Goldsworthy, & Wood Robinson, 1999). The outcome was a narrowing of the curriculum and an impoverished educational experience for students. A similar finding emerged from Au's (2007) meta-analysis of studies of the overall effect of high-stakes assessment on schools where Au found that the overwhelming outcome was a contraction of content, a more fragmented curriculum and an increasingly teacher-centered pedagogy.

Ball and Bowe, with Gold (1992), drawing on the work of Barthes (1972), offer another way of looking at curricular change. They have argued for a conception of text as being "writerly" or "readerly." "Readerly" texts signify an authoritarian relationship with minimal space for critical interpretation by the reader. Attempts to make policy interventions "teacher-proof" and an insistence that the program be implemented with fidelity take this view of texts. "Writerly" texts leave space for critical interpretation and the possibility for acts of resistance or creative re-interpretations. Although the authors of the science national curriculum attempted to make the texts "readerly," the response of teachers to the statutory requirements of the national curriculum for investigative work was "writerly." That is, teachers reinterpreted the requirements so as to advance the attainment of the students rather than to faithfully address the intent of the curriculum documents. This brief example illustrates how well-intentioned policies often have unintended consequences, especially when the ideas and values of those charged with implementing the policies have not been thoroughly investigated in advance.

This point leads to one final reality that educational researchers need to be aware of. This is the distinction between "evidence-based practice," of which there is little in education, and "evidence-informed practice," the domain in which research makes most of its potential contribution (Millar, Leach, Osborne, & Ratcliffe, 2006). Researchers should recognize that the goal of most research is to offer teachers and policymakers the best available evidence to inform their judgments and to establish a culture where these groups routinely look to researchers to contribute what they know to assist their professional judgment. But, whether and how such evidence is used will always be unpredictable (Ogborn, 2002). Teachers may well take ownership of any innovation, but in doing so the innovation inevitably undergoes a degree of transformation, and the fidelity it has with the original practice is often lost. Sometimes the original intent is maintained even if the form of the innovation is changed, but in other cases the changes are inconsistent with the original intent, making it questionable whether the practice can be called evidence-based at that point. This is simply something that educational researchers must face as they attempt to get their ideas implemented in the real world of school practice.

CONCLUSIONS AND SUMMARY

For the researcher, the translation of research findings into practice is a substantial task. It requires a thoughtful and considered account of the relevance of the research to practice. And then it requires integrating the research findings with practitioner knowledge to produce teaching materials and guidelines that can be used by the teachers. In addition, there is usually a need for an extensive program of professional development to help teachers understand the relationship between the activities of the program and the program's rational. Both the research and professional development require significant time and financial support. The materials for CASE were developed and turned into a publication *Thinking Science* (Adey, Shayer, & Yates, 1989), now in the third edition. Schools interested in using the materials are also encouraged to enroll in an extensive (and expensive) program of professional development which is supported by the researchers. Likewise, my own research on the teaching of ideas, evidence, and argument in school science (Osborne, Erduran, & Simon, 2004a) required a year of work to turn these ideas into something that could support the professional development of teachers with DVD-based materials (Osborne, Erduran, & Simon, 2004b). The message that comes from these examples is that for research to have any impact on practice, researchers have to engage in a process of translating their findings into a form that is usable and comprehensible to a practitioner or policy audience. Perhaps, one of the best examples of this process is offered by the team of people led by Paul Black that is studying formative assessment. After publishing a long academic review of the role and nature of formative assessment (Black & Wiliam, 1998a), Black and Wiliam produced a short but readable summary of the implications for practice and policy in a short 16-page pamphlet, *Inside the Black Box* (Black & Wiliam, 1998b). This format made it readily readable and, to date, this work has been enormously influential in the field of educational assessment.

This chapter has explored how research may contribute and affect policy in the context of science education in Europe drawing predominantly on examples from the United Kingdom. Three examples were chosen to explore how research can influence practice and the nature and limitations of that relationship. In the U.K. context as well as in a number of other European countries, the relatively centralized nature of the policy development process supports such a dialogue between the research and policy community. For academics considering this kind of engagement with policymakers themselves, there are many factors to weigh. First, there is the issue that efforts to directly influence policy rarely earn the support of academic institutions, which tend to value most highly basic research and publications in leading academic journals. That is not to say that academic institutions do not value work that contributes to the public domain and influences public policy, especially when it valorizes the institution but, rather, that that kind of work typically will contribute less to advancing the career of the individual researcher.

There will be those who do not see it as their responsibility to disseminate their findings beyond publication and presentations at academic conferences. Their attitude is that if the policymakers can find their work and make use of it, so much the better, but in their view, it is not the responsibility of the researcher to convince the policymakers of its value. To those who would see scholarly reporting of basic research as their primary mission, it is worth remembering that the world is awash with information. The challenge of contributing to society goes beyond the production of new knowledge to one of gaining the attention of a relevant audience. Simon (1971) pointed presciently to the fact that information "consumes the attention of its recipients. Hence a wealth of information creates a poverty of attention and a need to allocate attention efficiently among the overabundance of information sources that might consume it" (p. 40). Thus, as Goldhaber (1997) argues, we live increasingly in an economy where the commodity that is most scarce is attention or time.

Europe is not an island in the international context. Many of its governments have been concerned by the poor performance of their country in tests such as PISA and TIMSS. They see the effective functioning of their educational systems as a significant means of sustaining the international competitiveness of their economies. And, given their advanced technological base, education in STEM subjects and their improvement are of particular significance. Policymakers are, therefore, predisposed to listen. Research of significance or value, however, is unlikely to be attended to unless researchers are prepared to engage in a scholarly process of sifting, summarizing, and translating that information and evidence into a form that is readily assimilable. Those who choose not to engage in such a process—that is, who chose not to place their scholarly efforts into the hands of those who have to make decisions about policy and practice—should not be surprised at the lack of public recognition or use of their work.

REFERENCES

Adey, P., Landau, N., Hewitt, G., & Hewitt, J. (2003). *The professional development of teachers: Practice and theory.* Dordrecht, The Netherlands: Kluwer.

Adey, P., & Shayer, M. (1990). Accelerating the development of formal thinking in middle and high school students. *Journal of Research in Science Teaching, 27*(3), 267–285.

Adey, P., Shayer, M., & Yates, C. (1989). *Thinking Science: The Curriculum Materials of the CASE project.* London: Thomas Nelson and Sons.

APA Publications and Communications Board Working Group. (2009). Reporting Standards for Research in Psychology. *American Psychologist, 63*(9), 839–851.

Au, W. (2007). High stakes testing and curricular control: A qualitative metasynthesis. *Educational Researcher, 36*(5), 258–267.

Ball, S., & Bowe, R., (with Gold, A.) (1992). *Reforming education and changing Schools.* London: Routledge.

Bandura, A., Barbaranelli, C., Caprara, G. V., & Pastorelli, C. (2001). Self-efficacy beliefs as shapers of children's aspirations and career trajectories. *Child Development, 72*(1), 187–206.

Barthes, R. (1972). *Mythologies* (A. Lavers, Trans.). New York: Hill and Wang.

Black, P., & Atkin, J. M. (Eds.). (1996). *Changing the subject: Innovations in science, mathematics and technology education.* London: Routledge.

Black, P., & Wiliam, D. (1998a). Assessment and classroom learning. *Assessment in Education, 5*(1), 7–74.

Black, P., & Wiliam, D. (1998b). *Inside the Black Box: Raising standards through classroom assessment*. London: King's College.

Bodmer, W. F. (1985). *The public understanding of science*. London: The Royal Society.

Butz, W. P., Bloom, G. A., Gross, M. E., Kelly, T. K., Kofner, A., & Rippen, H. E. (2003). *Is there a shortage of scientists and engineers? How would we know?* Santa Monica, CA.: RAND.

Chinn, C. A., & Brewer, W. F. (1993). The role of anomalous data in knowledge acquisition. A theoretical framework and implications for science instruction. *Review of Educational Research, 63*, 1–49.

Claxton, G. (1988). *Live and Learn: An introduction to the psychology of growth and change in everyday life*. Milton Keynes: Open University Press.

Dainton, F. S. (1968). *The Dainton Report: An Inquiry into the flow of candidates into science and technology*. London: HMSO.

Davies, H. T. O., Nutley, S. M., & Smith, P. C. (Eds.). (2000). *What works? Evidence-based policy and practice in public services*. Bristol, England: The Policy Press.

Department for Education and Science. (1991). *Science in the National Curriculum*. London: HMSO.

Donnelly, J., Buchan, A., Jenkins, E., Laws, P., & Welford, G. (1996). *Investigations by order: Policy, curriculum and science teachers' work under the Education Reform Act*. Nafferton: Studies in Science Education.

European Commission. (1995). *White paper on education and training: Teaching and learning—Towards the learning society* (White paper). Luxembourg: Office for Official Publications in European Countries.

European Commission. (2001). *Governance in the EU: A White Paper*. Brussels, Belgium: European Commission.

European Commission. (2004). *Europe needs More Scientists: Report by the High Level Group on Increasing Human Resources for Science and Technology*. Brussels: European Commission.

European Council. (2000). Presidency conclusions [Electronic Version]. Retrieved from http://www.europarl.europa.eu/summits/lis1_en.htm

Fjaestad, B. (2007). Why journalists report science as they do. In M. W. Bauer & M. Bucchi (Eds.), *Journalism, science and society*. (pp. 123–132). New York, NY: Routledge.

Fullan, M. (2001). *The new meaning of educational change* (2nd Ed.). London: Cassell.

Fuller, S. (1997). *Science*. Buckingham, England: Open University Press.

Gilbert, J. (2005). *Catching the knowledge wave? The Knowledge Society and the Future of Education*. Wellington, New Zealand: NZCER Press.

Goldacre, B. (2008). *Bad Science*. London: Harper Collins.

Goldhaber, M. (1997). The attention economy and the net. *First Monday, 2*(4). Retrieved October 15, 2010, from http://firstmonday.org/htbin/cgiwrap/bin/ojs/index.php/fm/article/view/519/440

Hargreaves, D. (1996a). *Teaching as a research based profession: Possibilities and prospects* (The Teacher Training Agency Annual Lecture, 1996). London: The Teacher Training Agency.

Hargreaves, D. H. (1996b). *Teaching as a research based profession: Possibilities and prospects* (The Teacher Training Agency Annual Lecture). London: The Teacher Training Agency.

Haste, H. (2004). *Science in my future: A study of the values and beliefs in relation to science and technology amongst 11–21 year olds*. London: Nestlé Social Research Programme.

Haussler, P., & Hoffmann, L. (2002). An intervention study to enhance girls' interest, self-concept, and achievement in physics classes. *Journal of Research in Science Teaching, 39*(9), 870–888.

Hill, C. (2008). The Post-scientific Society. *Issues in Science and Technology, 24*(1), 78–84.

HM Treasury. (2006). *Science and Innovation investment Framework: next steps*. London: HMSO.

Hunt, A., & Millar, R. (Eds.). (2000). *Science for Public Understanding*. London: Heinemann Educational.

Irwin, A. (1995). *Citizen science*. London: Routledge.

Irwin, A., & Wynne, B. (Eds.). (1996). *Misunderstanding science: The public reconstruction of science and technology*. Cambridge, England: Cambridge University Press.

Jagger, N. (2007, Sept 17–18). *Internationalising doctoral careers.* Paper presented at the Conference on The National Value of Science Education. University of York.

Jenkins, E. (1999). School science, citizenship and the public understanding of science. *International Journal of Science Education, 21*(7), 703–710.

Joyce, B., & Showers, B. (2002). *Student Achievement Through Staff Development* (3rd ed.). White Plains, NY: Longman.

Kay, J. (2008). Darwin's marriage and the war in Iraq: A missing link. *Financial Times*, p. 11.

Leo, E. L., & Galloway, D. (1996). Conceptual links between Cognitive Acceleration through Science Education and Motivational Style: a critique of Adey and Shayer. *International Journal of Science Education, 18*(1), 35–49.

Lindahl, B. (2007). *A longitudinal study of students' attitudes towards science and choice of career.* Paper presented at the 80th NARST International Conference New Orleans, Louisiana.

Lord Sainsbury of Turville. (2007). *The Race to the Top: A Review of Government's Science and Innovation Policies*. London: HM Treasury.

Loucks-Horsley, S., Hewson, P., Love, N., & Stiles, K. E. (1998). *Designing Professional Development for Teachers of Science and Mathematics*. Thousand Oaks, CA: Corwin Press Inc.

Lovering. (1995). *A study of the social and spatial processes determining the development of the European scientific labour market.* Hull, England: Department of Earth Sciences, University of Hull.

Lowell, B. L., Salzman, H., Bernstein, H., & Henderson, E. (2009). *Steady as she goes? Three generations of students through the science and engineering pipeline*. New Jersey, US: Rutgers University.

Lynn, L., & Salzman, H. (2006, Winter). Collaborative advantage: New horizons for a flat world. *Issues in Science and Technology*, , 74–81.

Majone, G. (1989). *Evidence, argument and persuasion in the policy process*. New Haven, CT: Yale University Press.

Martin, M. O., Mullis, I. V. S., & Foy, P. (2008). *TIMSS 2007 International Science Report: Findings from IEA's Trends in International Mathematics and Science Study at the Fourth and Eighth Grades* Boston, MA: TIMSS & PIRLS International Study Center, Lynch School of Education.

Millar, R. (1996). Towards a science curriculum for public understanding. *School Science Review*, 77(280), 7–18.

Millar, R. (2006). Twenty-first century science: Insights from the design and implementation of a scientific literacy approach in school science. *International Journal of Science Education, 28*(13), 1499–1521.

Millar, R., Leach, J., Osborne, J. F., & Ratcliffe, M. (2006). *Improving Subject Teaching: Lessons from Research in Science Education*. Abingdon, Oxon, England: Routledge.

Millar, R., & Osborne, J. F. (Eds.). (1998). *Beyond 2000: Science Education for the Future*. London: King's College London.

Millar, R., & Osborne, J. F. (2009). Research and Practice: A Complex Relationship? In M. C. Shelley II, L. Yore, & B. Hand (Eds.), *Quality Research in Literacy and Science Education*. Springer.

National Academy of Science. (1995). *National Science Education Standards*. Washington, D.C.: National Academy Press.

National Academy of Sciences: Committee on Science Engineering and Public Policy. (2005). *Rising Above the Gathering Storm: Energizing and Employing America for a Brighter Economic Future*. Washington, DC: National Academy Sciences.

National Research Council. (2008). *Research on Future Skill Demands*. Washington DC: National Academy Press.

Nelkin, D. (1995). *Selling science: How the press covers science and technology*. New York, NY: Freeman.

Ogborn, J. (2002). Ownership and transformation: Teachers using Curriculum Innovation. *Physics Education, 37*(2), 142–146.

Organization for Economic Co-operation and Development. (2007). *PISA 2006: Science Competencies for Tomorrow's World: Volume 1: Analysis*. Paris, France: Author.

Ormerod, M. B., & Duckworth, D. (1975). *Pupils' attitudes to science*. Slough, England: NFER.

Osborne, J. F., & Collins, S. (2001). Pupils' views of the role and value of the science curriculum: a focus-group study. *International Journal of Science Education, 23*(5), 441–468.

Osborne, J. F., & Dillon, J. (2008a). *Science education in Europe: Critical reflections*. London: Nuffield Foundation.

Osborne, J. F., Duschl, R., & Fairbrother, R. (2002). Breaking the mould: Teaching science for public understanding. Retrieved from http://www.kcl.ac.uk/education/hpages/jopubs.html

Osborne, J. F., Erduran, S., & Simon, S. (2004a). Enhancing the quality of argument in school science. *Journal of Research in Science Teaching, 41*(10), 994–1020.

Osborne, J. F., Erduran, S., & Simon, S. (2004b). *Ideas, evidence & argument in science education: A CPD Pack*. London: King's College London.

Osborne, J. F., Ratcliffe, M., Collins, S., Millar, R., & Duschl, R. (2003). What "ideas-about-science" should be taught in school science? A Delphi study of the "expert" community. *Journal of Research in Science Teaching, 40*(7), 692–720.

Perks, D. (2006). *What is science education for.* London: Institute of Ideas.

Posner, G. J., Strike, K. A., Hewson, P. W., & Gerzog, W. A. (1982). Accommodation of a scientific conception: Toward a theory of conceptual change. *Science Education, 66*, 211–227.

Qualifications and Curriculum Authority. (2005). *Programme of study for KS4 from 2006*. London: Author.

Ratcliffe, M., Hanley, P., & Osborne, J. F. (2007). *Evaluation of twenty-first century science GCSE strand 3: The teaching of twenty-first century science GCSE, and teachers' and students' views of the course*. Southampton, England: University of Southampton.

Ratcliffe, M., Leach, J., Millar, R., & Osborne, J. F. (2005). Evidence-based practice in science education: The researcher-user interface. *Research Papers in Education, 20*(2), 1470–1146.

Ratcliffe, M., & Millar, R. (2009). Teaching for understanding of science in context: Evidence from the pilot trials of the <I>twenty-first century science</I> courses. *Journal of Research in Science Teaching, 46*(8), 945–959.

Reynolds, H. (2008). Some positive thoughts on the new KS4 curriculum. *Institute of Physics Education Group Newsletter,* 5.

Roberts, G. (2002). *SET for success: The supply of people with science, technology, engineering and mathematics skills.* London: HM Treasury.

Rutherford, F. J., & Ahlgren, A. (1989). *Science for All Americans: A project 2061 report*. Washington DC: AAAS.

Schreiner, C. (2006). *EXPLORING A ROSE-GARDEN: Norwegian youth's orientations towards science—seen as signs of late modern identities.* Thesis submitted for Doctor Scientarium. Oslo, Norway: Faculty of Education.

Schwab, J. J. (1962). *The teaching of science as enquiry.* Cambridge, MA: Harvard University Press.

Shavelson, R. J., & Towne, L. (2002). *Scientific research in education*. Washington, DC: National Academy Press.

Shayer, M., & Adey, P. (1992a). Accelerating the development of formal thinking in middle and high school students II: Postproject effects on science achievement. *Journal of Research in Science Teaching, 29*(1), 81–92.

Shayer, M., & Adey, P. (1992b). Accelerating the development of formal thinking in middle and high school students III: Testing the permanency of effects. *Journal of Research in Science Teaching, 29*(10), 1101–1115.

Shayer, M., & Adey, P. (1993). Accelerating the development of formal thinking in middle and high school students IV: Three years after a two-year intervention. *Journal of Research in Science Teaching, 30*(4), 351–366.

Shayer, M., Küchemann, D. E., & Wylam, H. (1976). The distribution of Piagetian stages of thinking in the British middle and secondary school children. *British Journal of Educational Psychology, 46*, 164–173.

Simon, H. (1971). Designing organizations for an information-rich world. In M. Greenberger (Ed.), *Computers, communications and the public interest*. Baltimore, MD: John Hopkins University Press.

Sjøberg, S., & Schreiner, C. (2005). How do learners in different cultures relate to science and technology? Results and perspectives from the project ROSE. *Asia Pacific Forum on Science Learning and Teaching, 6*(2), 1–16.

Slavin, R. E. (2002). Evidence-based educational policies: Transforming educational practice and research. *Educational Researcher, 31*(7), 15–21.

Solomon, J., & Aikenhead, G. (Eds.). (1994). *STS education: International perspectives on reform*. New York, NY: Teachers College Press.

Stokes, D. E. (1997). *Pasteur's quadrant*. Washington, DC: Brookings Institution Press.

Tai, R. H., Qi Liu, C., Maltese, A. V., & Fan, X. (2006). Planning early for careers in science. *Science, 312*, 1143–1145.

Tapscott, D. (2009). *Grown up digital: How the net generation is changing your world*. New York: McGraw-Hill.

Teitelbaum, M. (2007, Sept 17-18). *Do we need more scientists and engineers?* Paper presented at the Conference on The National Value of Science Education, University of York, England.

The Royal Society. (2006). *Taking a leading role*. London: Author.

Tytler, R., Osborne, J. F., Williams, G., Tytler, K., Clark, J. C., & Tomei, A. (2008). *Opening up pathways: Engagement in STEM across the primary-secondary school transition. A review of the literature concerning supports and barriers to science, technology, engineering and mathematics engagement at primary-secondary transition. Commissioned by the Australian Department of Education, Employment and Workplace Relations*. Melbourne: Deakin University.

Watson, R., Goldsworthy, A., & Wood Robinson, C. (1999). What is not fair with investigations? *School Science Review, 80*(292), 101–106.

Willard, C. (1985). The science of values and the values of science. In J. Cox, M. Sillars & G. Walker (Eds.), *Argument and Social Practice: The Fourth SCA/AFA Summer Conference on Argumentation* (pp. 435–444). Annadale, VA: Speech Communication Association.

CHAPTER 3

SCIENCE TEACHER EDUCATION RESEARCH AND POLICY

Are They Connected?

Jane Butler Kahle and Sarah Beth Woodruff

INTRODUCTION

> In many of the most important contemporary debates about teacher quality and teacher preparation, the central focus ... is research itself, particularly on the fundamental question of whether there is a research basis for teacher education and, if so, what that research base suggests. (Cochran-Smith & Fries, 2005, p. 69)

In this chapter we examine the relationship between science teacher education policy and research. Is there a connection? We are primarily interested in whether or not there is a research base that drives policy, both for preservice teacher education and in-service professional development, but also are there broader state and federal governmental policies that affect the nature of research in this area?

The Role of Public Policy in K–12 Science Education, pp. 47–75

Historical trends suggest that the role of federal and state governmental authorities in education policymaking is increasing. While there is no federal obligation to intervene in education, history is marked by dramatic and increasingly frequent interventions by the federal government. The reason for this intervention is dissatisfaction on the part of national leaders with the educational system, in particular the poor performance of students in science in this country. While there may be some consensus on the dismal state of student achievement in science, there is no clear understanding or agreement concerning the appropriate policy response or on the role that research should play in informing or directing education policy.

In the 1990s, the question of *what* science should be taught was addressed mutually by scientists and science educators and published in *Science for All Americans* (Rutherford & Ahlgren, 1990), *Benchmarks for Science Literacy* (American Association for the Advancement of Science, 1993), and the *National Science Education Standards* (National Research Council, 1996). During that time, researchers focused on how science should be taught (Schroeder, Scott, Tolson, Huang, & Lee, 2007) and how science teachers should be educated (National Research Council, 1997, 2000). *Science for All Americans* and *NSES* discussed many of these principles of good teaching, and to some extent helped to create a consensus on what good science teaching is that has helped focus teacher education in science. The documents, developed in the 1990s, stressed updated scientific knowledge as well as inquiry as the preferred instructional strategy. The intent was for students to learn, or discover, science as scientists did. Further, the National Research Council publications recommended that science teachers learn science through inquiry, especially in their undergraduate science courses. In this chapter, we will explore whether or not the consensus is strong enough to determine what science teachers should learn in terms of both content and pedagogy.

As the twentieth century drew to a close, public confidence in teacher education was in steady decline, driven in part by the 1983 report of the National Commission on Excellence in Education (NCEE), *A Nation at Risk,* and similar critiques (Carnegie Forum on Education and the Economy, 1986). In response, federal and state governments' involvement in education increased. As an example, in 1998, Congress reauthorized the Eisenhower Professional Development Program, Title II of the Elementary and Secondary Education Act (ESEA) and mandated that each state annually report the percentage of teaching candidates who successfully passed state certification tests. However, schools of business, law, medicine, and nursing were not being asked by the government to report on the percentage of candidates who were passing professional certification exams in those fields (Wilson & Youngs, 2005). Implementation of this policy implied that teacher education was less effective than preparation for other professions, although there was little research indicating that this was the case. Further, it focused research on the effectiveness of teacher education and on a single measure of effectiveness, teacher content knowledge.

This Congressional action led to increased state requirements for teacher education programs and state certification. The 2005 Higher Education Act, Title II (HEA), required states to apply explicit criteria to assess the performance of teacher preparation programs. According to the *Secretary's Fifth Annual Report on Teacher Quality* (U.S. Department of Education [USDOE], 2006), in response to that mandate, 44 states have adopted or integrated National Council for the Accreditation of Teacher Education (NCATE), Teacher Education Accreditation Council (TEAC), or Interstate New Teacher Assessment and Support Consortium (INTASC) standards as criteria for this purpose. State criteria for evaluating teacher preparation program performance include: (a) indicators of teachers' knowledge and skills in 50 states; (b) passing rates on state certification tests in 34 states; and (c) qualifications of program staff, quality of clinical experiences, or teacher rehire and retention rates in 28 states. In 2005, 17 programs in 11 states were deemed "at-risk" or "low-performing" (USDOE, 2006).

The emphasis on quality teacher education and teaching is also central to the No Child Left Behind Act of 2001 (NCLB), which called for the reform of education using research-based methods to produce excellence. NCLB specifically included science as a core content area. It required states to establish academic standards in science as well as in reading/language arts and mathematics and to administer standards-based assessments to determine student achievement in those subjects. Further, NCLB mandated that states ensure that all students are taught by *highly-qualified* teachers and that only evidence-based professional development programs receive federal funding. Ingersoll (2007) noted that NCLB defines a *highly-qualified* teacher as one who (a) has a bachelor's degree, (b) holds a regular or full state-approved teaching certificate or license, and (c) is competent in each of the academic subjects s/he teaches.

Whether or not someone has met the first two requirements can be objectively determined. One either does or does not have a bachelor's degree and a state approved teacher certificate. Each state, however, determines what constitutes "competency" in a subject. These differences in definition lead to different estimates of the numbers of teachers who are *highly qualified* in each state. For example, in 2005, Tennessee and Alabama reported that only one-third of classes were taught by highly-qualified teachers, while Georgia and North Carolina reported that 90% were taught by highly-qualified teachers (Cochran-Smith & Zeichner,

2005). But, no matter how states define content competence, they cannot ignore it. Federal policy (e.g., NCLB) and government agencies (e.g., USDOE) have chosen to emphasize content knowledge as the key measure of teacher quality and have given little attention to teacher instructional practice despite policy language that stresses improving both the subject matter knowledge of teachers and how effectively they teach. The theory of change driving federal education policy implementation draws a direct connection between teacher quality, measured in large part by teacher content knowledge, and student achievement. This chapter explores whether this theory of change is supported by evidence from research on teacher education and student achievement.

PREPARING TO BE A SCIENCE TEACHER

> The research on teacher preparation that is most needed in the contemporary context is that which attempts to link the training, learning, and policy aspects of teacher preparation to one another. (Cochran-Smith & Fries, 2005, p. 80)

A Policy Focus on Science Content Knowledge

More than 1,000 teacher education programs in the U.S. prepare the nation's science teachers. Teacher education programs vary significantly within and across states. Although national-level, normative standards (e.g., NCATE and TEAC) exist, teacher professional preparation is embedded in complex policy issues largely at the state level. However, increasingly it is also impacted directly by federal policy initiatives. And although states generally control entry to the teaching profession by requiring and setting policies regarding teacher certification, the specific requirements and activities of teacher preparation programs are under the purview of colleges and universities. State oversight of university teacher education ranges from rigorous external evaluation of programs in a handful of states (e.g., North Carolina and Mississippi) to blanket approval of programs in most others. For example, blanket approval would exist if a teacher education program has been approved by its college or university, the state accepts that approval without further review. Additionally, states collect little data on graduates of teacher education programs in order to inform policy decisions.

Over a half century ago, Vannevar Bush's (1945) report to President Truman, *Science: The Endless Frontier,* indicated an urgent need to improve science and mathematics education for national security purposes. Bush recommended the establishment of an organization at the federal level that would support both scientific research and the improvement of science education. That organization, founded in 1950, was the National Science Foundation (NSF). The establishment of NSF represented a significant entry of the federal government into science education. First, because there was a growing discrepancy between what science teachers knew and the science they needed to know to be effective teachers, NSF developed and funded science and mathematics teacher training institutes that focused on content knowledge and that led to the attainment of advanced degrees. Then, because a similar discrepancy was found between the information in science textbooks and the current state of scientific knowledge, NSF began to fund the development of new textbooks in the various science disciplines. These curricula, commonly referred to as the alphabet soup curricula (e.g., CBA, BSCS, PSSC, etc.), included texts in biology, physics, chemistry, earth science, and elementary science. These texts included not only cutting-edge science, but they also promulgated new ways of teaching, primarily emphasizing laboratory experiences. During this period, only a few studies documented the effects of the teacher training institutes; however, there was considerable research concerning the efficacy of the NSF-supported curriculum materials (Kahle, 2008).

With the passage of the NCLB, the federal government again, as it did in the 1950s and 60s, developed policy aimed at improving science teachers and teaching. One critical assumption underlying the NCLB policies was that subject matter preparation was a key aspect of teacher quality. The assumption made in the 1950s, and again in 2001, was that teacher content knowledge would be directly related to student achievement in science, but that assumption was not based on findings from research. In a review of research on teacher preparation, commissioned by the U.S. Department of Education, Wilson, Floden, and Ferrini-Mundy (2001) found an absence of research connecting teacher knowledge to student learning. Their review included 57, out of 313 identified, studies that met stringent criteria. They concluded that no research directly assessed prospective teachers' subject matter knowledge and evaluated the relationship between teacher subject matter preparation and student learning. Despite the focus on improving teacher quality by increasing teacher subject matter knowledge, policymakers did not provide funds and directives for conducting research that might provide evidence to support this policy direction. Although associations between teacher content knowledge and improved student learning may be tenuous, some research (e.g., Magnusson, Borko, Krajcik, & Layman, 1992; Sanders, Borko, & Lockard, 1993) has shown that differences in teachers' content knowledge influence their instructional practices, and these practices influence student learning. In other words, the effect of content knowledge operates indirectly through teacher instructional practice. This finding was confirmed in a report of the Mathematics and Science Partnership Knowledge Management and Dissemination Project (MSP-Knowledge Management and Dissemination, 2007). It found that how much teachers know about their subject affects how they engage students with subject matter, how they evaluate and use instructional materials, and what content they deliver to students. These findings are more frequent for mathematics than for science, but this may be a result of the larger number of studies in mathematics, compared to those in science. Given that there is a strong assumption among teacher education policy makers that there is a link between teacher content knowledge and student learning and that this assumption is driving teacher education policy at all levels of the system, in the next section we examine more closely whether that assumption has adequate research support.

Research on Subject Matter Knowledge

Some research on teacher subject matter knowledge looks for a link between gross measures of teacher knowledge and teacher quality. For example, according to Wilson, Floden, and Ferrini-Mundy (2001),

> Researchers conducting large-scale studies have relied on proxies for subject-matter knowledge, such as majors or coursework.... The conclusions of these few studies are provocative because they undermine the certainty often expressed about the strong link between college study of a subject matter area and teacher quality. (p. 6)

Further, it is difficult to generalize across such studies because different definitions of teacher subject matter knowledge (e.g., discipline major, number of undergraduate courses, scores on standardized tests, etc.) as well as different measures of student science achievement are used. Indeed, as Monk (1994) concluded, "Gross measures of teacher preparation (such as degree levels, undifferentiated credit counts, or years of teacher experience) offer little useful information for those interested in improving pupil performance" (p. 142). These results are not surprising because such studies do not evaluate the relevance of courses for the preparation of science teachers, the prospective teachers' comprehension of what was taught, nor the teachers' ability to communicate the knowledge to students.

We found only a few studies that investigated the relationship between teacher subject matter knowledge and student learning in science. Lawrenz's (1975) study specifically focused on science teachers. She surveyed a random sample of 236 science teachers (84 biology, 111 chemistry, and 41 physics) and one randomly selected class of students for each teacher in order to assess different aspects of science teacher education and science teaching. One finding pertains directly to teacher content knowledge; that is, teachers' subject matter knowledge, as measured by the National Teacher Examination (NTE), made no significant contribution to student achievement, as measured by the Test on Achievement in Science (TAS).

More recently, two studies involved both science and mathematics teachers and large national databases. Using Longitudinal Study of American Youth (LSAY) data from fall 1987 to spring 1990, Monk (1994) reported on the subject area preparation of middle and high school science and mathematics teachers and student achievement. Although there are more positive findings for mathematics, compared to science, several science findings are pertinent. He found a significant positive relationship between teacher coursework in the physical sciences and Grade 10 and 11 students' achievement gains. However, he reported no impact on student achievement with teacher undergraduate coursework in life sciences. He also found positive effects between science education courses taken at the graduate level and student achievement. Monk concluded, "While the results using the LSAY data are positive and encouraging, we need to keep in mind that many of the estimated relationships between teacher course preparation and pupil gains are by no means large in magnitude" (p. 143). Indeed, Monk is very cautious about his findings and possible uses of them in setting policy, stating that there may be more cost-effective ways of improving the subject matter knowledge of prospective teachers than by requiring, for example, undergraduate majors in a discipline followed by a fifth year of professional education. Monk proposed instead that in-service teacher education efforts aimed at preventing the obsolescence of teacher content knowledge obtained in preservice education may be more cost-effective.

Another study by Goldhaber and Brewer (2000) also used a large national database, the 1998 National Education Longitudinal Survey (NELS). Their primary findings concern teacher certification and student achievement; however, concerning subject matter knowledge, they found that there was no significant relationship between a prospective teacher's subject matter major and student achievement in science. Furthermore, although students of teachers with degrees in mathematics, compared to students of teachers without mathematics degrees, had higher test scores, there was no similar effect in science. One possible explanation for the difference between mathematics and science may be the relative coherence of mathematics as a discipline, compared to the multi-disciplinary aspect of science. For example, it is questionable whether teacher subject matter knowledge in a particular discipline of science (e.g., biology) would affect student achievement in a different science discipline (e.g., chemistry). Another possibility is that although a teacher may have a major in one field of science (e.g., chemistry), s/he may be assigned to teach in another field of science (e.g., physics).

Because the amount of science that a preservice teacher studies is largely determined by whether the prospective teacher majors in a discipline or in education, we also examined the research concerning 4-year and 5-year science teacher education programs. Typically, a 4-year program requires fewer science courses because those degree programs also include pedagogical courses as well as clinical experiences. Beginning with the Holmes' Group initiative, there has been a push for prospective teachers to major in a discipline and to spend a fifth year in pedagogical courses and clinical experiences. The Holmes Group was founded at Michigan State University in 1986 and included 100 research universities. Primarily because of the cost to both students and institutions, the Holmes' Group initiative involved mostly well-resourced flagship universities, while former teacher colleges and smaller institutions continued their 4-year programs. By 1995, the number of universities involved in the group had dropped to 74 (Germann, Dumas, & Barrow, 1995). Typically, fifth-year programs are characterized by a year-long internship. Though the number of universities offering 5-year programs has declined, in 2005, 34 states required that all teacher certification candidates hold a content-area specific bachelor's degree to receive initial certification (USDOE, 2006).

In general, the research concerning the effectiveness of 4-year and 5-year teacher education programs is not specific to science. For example, in their comprehensive review, Wilson, Floden, and Ferrini-Mundy (2001) found only two studies that strongly supported the extended internship of the fifth-year model, and neither study included science. The studies are: Andrews (1990) and Grisham, Laguardia, and Brink (2000). Monk's (1994) conclusion that there are more cost-effective ways than a fifth year of college to convey subject matter knowledge to teachers (e.g., content-focused in-service professional development) is relevant. Based on the research that has been reported in the literature, there is little support for placing teacher content knowledge at the center of policies to improve science teacher quality.

Research on Pedagogical Knowledge and Teaching Strategies

Another line of research has examined the relationship between teachers' pedagogical knowledge and how teachers teach. These studies ask: Does what teachers learn in their pedagogical preparation affect their teaching behavior? As mentioned, teacher preparation programs are divided among discipline courses, pedagogical courses, and clinical experiences. According to Wilson, Floden, and Ferrini-Mundy (2001),

"There is no research that directly assesses what teachers learn in their pedagogical preparation and then evaluates the relationship of that pedagogical knowledge to student learning or teacher behavior" (p. 12). But unlike the lack of research demonstrating a relationship between teacher content knowledge and student learning in science, there is a growing body of evidence that *how* science is taught does affect student achievement.

Researchers in the 1980s and 1990s examined the effect of specific instructional strategies on student achievement. In 1996, Wise used a list of alternative strategies (*questioning, focusing, manipulation, enhanced materials, testing, inquiry, enhanced context,* and *instructional media*) to examine teaching. He then contrasted the effectiveness of alternative and traditional (e.g., textbooks, lectures, worksheets) teaching strategies. He concluded that the alternative strategies, which were fundamentally inquiry-oriented, were more effective than traditional approaches in enhancing student achievement in science (Wise, 1996; Wise as cited in Schroeder, Scott, Tolson, Huang, & Lee, 2007). In a 1997 study of teaching strategies and student achievement, Wright, Horn, and Sanders concluded, "Differences in teaching strategies were found to be the dominant factor affecting student academic gain" (as cited in Schroeder et al., 2007, p. 1439).

A meta-analysis of U.S. research (1980 to 2004) done by Schroeder et al. (2007) studied the effect of specific science teaching strategies on student achievement. In the 61 studies that met the researchers' criteria for inclusion in the analysis, they identified eight alternative instructional strategies in science. The authors updated the list developed by Wise (1996), however, several strategies are the same. Using student performance as the dependent variable and the eight strategies as independent variables, they calculated effect sizes for each strategy. The strategies and their effect sizes were: *Questioning Strategies (0.74), Enhanced Material Strategies (0.29), Assessment Strategies (0.51), Inquiry Strategies (0.65), Enhanced Context Strategies (1.48), Instructional Technology (IT) Strategies (0.48),* and *Collaborative Learning Strategies (0.95).* In all cases, the strategies are related to recommendations in the *NSES*. That is, *questioning* involves open-ended questions, *materials* include computers and electronic probes, *and assessment* includes performance-based assessments. While all effect sizes were judged to be significant, the largest one was found for *Enhanced Context Strategies*, which the authors defined as making science relevant to students by presenting topics in the context of real-world examples and problems. The authors concluded: "The major implication of this research is that we have generated empirical evidence supporting the effectiveness of alternative teaching strategies in science" (Schroeder, p. 1436). The significance of this finding for science teacher education is that it provides at least some direction to teacher educators in the pedagogical preparation of future teachers.

Research on Clinical Experiences

Of the three parts of a teacher education program—content, pedagogy, and clinical experience—the least amount of research has been done on clinical experiences, especially in specific subject areas. We use the term *clinical experiences* to encompass any K–12 classroom experiences during teacher preparation. These include, but are not limited to, classroom observations, tutoring, early field experiences, and student teaching. Wilson, Floden, and Ferrini-Mundy (2001) found only 10 studies in their review of the literature, and none of the studies included prospective science teachers. All but one of those studies had samples ranging from 1 to 18 teachers; the one exception was a study that involved 93 student teachers in two different teacher education programs. The authors found that most of these studies focused on attitude shifts rather than on changes in subject matter knowledge or pedagogical skills. In conclusion, they stated, "What we know about the typical clinical experience is sobering. The research demonstrates that traditional field experiences are often disconnected from coursework, focus on a narrow range of teaching skills, and reinforce the status quo" (p. 22). One result of the lack of research-based evidence has been the lack of attention to the development of coherent policies about clinical experiences. The lack of research on the effectiveness of clinical experiences does not mean that policies regarding clinical experiences are not being made. The National Council for the Accreditation of Teacher Education (NCATE), for example, has established a separate standard and indicators for measuring the effectiveness of pre-service field and clinical experiences. The NCATE standard is descriptive; it does not specify expectations for duration or intensity, nor make recommendations regarding how a specific content area, such as science, might be addressed during pre-service clinical experience. Ten states have no policies describing expectations for teacher pre-service field experience, and 16 others have policies that place responsibility for defining the pre-service experience on the teacher education programs of colleges and universities. The remaining 24 states have policies specifying the minimum duration of teacher pre-service field experience, but they vary widely from 30 hours to 180 days. According to Education Commission of the States (ECS) data, the typical field experience of pre-service teachers is between 10 and 12 weeks and is dictated by their teacher preparation programs, as outlined by the NCATE standards (ECS, 2008a).

TEACHER CERTIFICATION: TRADITIONAL AND ALTERNATIVE

> The research literature on teacher education and accreditation of teacher-education programs provides only limited guidance for policymakers. (Corcoran, 2007, p. 3)

Traditional Teacher Certification

Another area in which state education departments make policy decisions is in the minimum requirements they establish for teacher certification. Traditional teacher preparation usually takes place at an institution of higher education having a state accredited undergraduate teacher education program, followed by the individual receiving state approved teaching certification. With the exception of Arizona, all states require some form of state approval for teacher-preparation programs (Corcoran, 2007). Wilson and Youngs (2005) discuss current policies governing teacher certification, indicating that most states require a teacher candidate to have a bachelor's degree either in education or a discipline. Further, most require some sort of supervised student teaching experience, although the length of that experience varies widely across states. Currently 48 states require teacher testing, although the types of tests and passing scores vary (Goldhaber, 2008). Due to multiple areas of certification, over 1,100 unique tests are used by states (USDOE, 2006) to assess teacher content knowledge, pedagogy, and basic skills. In 2002, 37 states required passage of a basic skills test; 33 required passage of a test of subject matter knowledge; and 26 mandated passage of tests of pedagogical knowledge (Young, Odden, & Porter, 2003).

All states have developed standards that prospective teachers must meet in order to attain initial certification. Some states have established policies that link teacher certification requirements (i.e., expectations for subject-specific content and pedagogical knowledge) with state content standards for students. While 44 states have done so for the arts, only 25 have set content-specific teacher standards that are linked to state content standards in science (USDOE, 2006).

Unlike other professions, national accreditation is not required for teacher education programs. Over half of the 1,300 teacher education programs in the U.S. are regionally accredited, while less than 40% are accredited nationally by either NCATE or by TEAC. States approve programs and make accreditation policies. This decentralized system has been described as a "national non-system" (Angus, 2001, p. 597). There is at least one current attempt to rectify this situation and create a national-level certification system. The American Board for Certification of

Teacher Excellence (ABCTE), a private, nonprofit organization dedicated to the improvement of teaching, is developing a national test battery that will provide a more common standard for states and portable licenses for teachers in states that voluntarily adopt it (Constantine, Player, Silva, Hallgren, Grider, & Deke, 2009). The ABCTE uses the following requirements for certification: Candidates must (a) hold a bachelor's degree in any subject area from an approved college or university; (b) pass a background check; (c) pass the ABCTE Professional Teacher Knowledge Exam; and (d) pass the ABCTE subject area exam in the area of his/her choice. However, it should be noted that the main purpose of developing this test is to replace the usual path to teacher certification with a test-based system.

Alternative Teacher Certification

Traditional teacher education frequently is the target of criticism both at the undergraduate program level and at the state certification level. Many critics maintain that the "best and brightest" are discouraged from teaching due to lackluster undergraduate teacher education programs and/or the onerous task of completing state certification requirements. Becker, Kennedy, and Hundersmarck (2003) proposed that there is an alternative hypothesis regarding the necessary characteristics of teachers—the bright well-educated person hypothesis—that presupposes a different type of preparation, certification, and licensure. We have used *certification* to include licensure; both are processes by which teachers are approved by the state to teach a certain grade level or specific subject. This hypothesis has led to the multiplication of policies delineating alternative paths to certification. Feistritzer (2006) identified 10 different ways to receive alternative certification, and 47 states have sanctioned at least one of these alternative route programs (USDOE, 2006). The 10 ways are: (1) program leads to full certification in all subject areas; (2) program addresses shortages, and/or specific grade levels, and/or subject areas; (3) program reviews individual's academic/professional background and is designed by the state and/or local school district; (4) program reviews individual's academic/professional background and is designed by an institution of higher education; (5) post-baccalaureate program at an institution of higher education; (6) program is basically an emergency route; (7) program for persons who have few requirements to fulfill before becoming certified through a traditional route; (8) program enables a person who has some "special" qualifications to teach certain subjects; (9) program eliminates emergency route and prepares individuals who do not meet basic requirements to enter an alternate or traditional program route; and (10) program accommodates specific populations for teaching, for example, Teach for America.

New teachers from alternative route programs rose by 15% between 2004 and 2005, and the percentage of traditionally prepared new teachers fell from 85% of those entering the workforce in 2000, to 81% by 2004. Across the states, about half of the 110 alternative teacher education programs are administered by colleges and universities. However, little comparability exists across the states because, typically, states define alternative routes with little policy guidance at the national level. Some of the common characteristics of alternative route programs include: (a) a focus on recruitment, preparation, and licensing of those who already hold bachelor degrees; (b) credit for field-based experiences in lieu of coursework; (c) coursework or equivalent experiences taken while teaching; (d) follow-up monitoring during the teacher's early career; (e) a rigorous screening process for candidates; and (f) high performance (e.g., testing) standards (USDOE, 2006).

Research findings concerning differences between traditionally certified and alternatively certified teachers are difficult to interpret because the term, "alternative teacher certification," has been used to refer to everything from emergency certification in which the teacher is allowed to teacher in an emergency situation without having met all of the state's certification requirements to, more recently, state-approved pathways outside the traditional preparation program that professionally prepare individuals who had previously earned baccalaureate degrees. For example, state policies developed to address NCLB *highly-qualified* teacher requirements typically have substituted state-approved alternative pathways to certification for the previous practice of issuing certification waivers or emergency certificates. Nationally, the number of science teachers practicing without full certification (e.g., with a waiver) has declined since NCLB to 3.0% in 2005, with a slightly higher percentage of not fully certified teachers working in high-poverty districts. Teachers participating in alternative routes who meet the requirements of a *highly-qualified* teacher are not counted as teaching on a waiver (USDOE, 2006). When possible, a distinction should be made when conducting and reporting research between those teaching on waivers (i.e., a temporary, provisional, or emergency permit) and alternatively certified teachers, as defined by Feistritzer (2006).

Further confounding comparisons between traditional and alternative programs are differences in program foci. A fundamental assumption of most alternative route programs is that enrolled students already possess subject matter mastery, so the programs tend to focus on pedagogy, not content knowledge (Feistritzer, 2006; USDOE, 2006). Another caveat is revealed in a study of teacher grade point averages and type of teacher certification in Los Angeles. In general, alternatively certified teachers in the Los Angeles Unified School District had grade point averages that met or surpassed national averages of traditionally prepared teachers. However, alternatively certified teachers' grade point averages were lower than those of traditionally prepared teachers in mathematics and science (Wilson, Floden, & Ferrini-Mundy, 2001). Additionally, it is difficult to compare student achievement by type of teacher certification because higher percentages of alternatively certified teachers teach in urban settings and work with minority children (Wilson, Floden, & Ferrini-Mundy, 2001). Because so many confounding variables are involved, one must be cautious when interpreting claims about any differences between traditional and alternative routes to certification and the comparative impact of those two types of programs on teacher quality or student achievement.

Alternative and traditional paths to certification have been compared on various factors such as retention, competency, subject matter knowledge, and student learning. Although some studies comparing the effectiveness of traditional and alternative teacher certification included science teachers, not all of them reported findings by subject area. For example, a study of alternatively and traditionally prepared teachers in Georgia (Guyton, Fox, & Sisk, 1991) involved mathematics, science, and foreign language teachers. At the end of the first year of teaching, traditionally prepared compared to alternatively prepared teachers were more positive about continuing to teach. However, the findings were not reported by subject area. Jelmberg (1996) compared graduates of teacher education programs (including science) at various colleges in New Hampshire and graduates of a state-sponsored alternative program in which teachers become teachers of record prior to any formal preparation. Alternative program teachers had 3 years in which to complete a professional development plan and demonstrate proficiency on 14 competencies identified by the state. He surveyed both the teachers and their principals. He reported statistically significant differences between traditionally prepared teachers' ratings of their preparation programs and those of alternatively certified teachers. Traditionally prepared teachers rated their professional courses, practicum supervision, and overall preparation much higher. Further, principals rated traditional, compared to alternative, teachers higher on 6 of the 14 state competencies.

A few studies were found that focused exclusively on science and mathematics teachers, and their findings are discussed next. A study by Hawk and Schmidt (1989) addressed which certification route resulted in teachers with more subject matter knowledge. They compared National Teacher Examination (NTE) scores of 16 alternatively certified teachers and 18 teacher education graduates from the same institution. No significant difference in subject matter knowledge was found between the two groups. Conflicting findings were reported from two studies of Dallas teachers, who were traditionally trained and certified or who had participated in an alternative route to certification. The researchers found that only in math and physical science did higher percentages of traditionally, compared to alternatively, certified teachers pass the state content exams. However, the mean scores of the alternatively trained teachers were higher than those of traditionally prepared teachers (Hutton, Lutz, & Williamson, 1990; Lutz & Hutton, 1989).

The debate concerning traditional and alternative teacher preparation and certification is a struggle between a *professional* view of teaching, which maintains that teaching is work that requires significant preparation, support, and standards and a *deregulationist* view, which sees teaching as something that most intelligent individuals can do. The *professional* view emphasizes the need for rigorous standards and prescriptive policies to guide preparation and to control entry of persons into the field. According to the *deregulationist* view, licensing is unnecessary and costly, needed skills can be learned on the job, and alternative certification is needed to expand the pool of potential teachers (Corcoran, 2007). Arguments for deregulation typically are based on four premises: (a) there is a current shortage of mathematics and science teachers; (b) traditional teacher preparation has not addressed (or cannot address) increasing accountability requirements and mandates; (c) states are inappropriately influential and restrictive in the governance of entry to the teaching profession; and (d) teaching can be done well by most educated persons. Persistent critique of teacher education programs has lent support to the deregulationist view and has promoted the proliferation of alternative route programs. So at the same time that states are adopting more rigorous standards for teacher education programs and entry to the profession, they are also increasing opportunities for individuals to become alternatively certified. In the *2003 Second Annual Report on Teacher Quality* (USDOE, 2004) then Education Secretary Rod Paige focused on two policy approaches for improving teacher quality—requiring teachers to pass standardized tests of content knowledge and lowering barriers of entry to the profession. It should be noted that both of these policies are incorporated into the requirements of the American Board for Certification of Teacher Excellence, discussed earlier. These two views were expressed in a different way by Levine (2009), who discussed whether teaching is a *craft* or a *profession*. He maintained that states, with their emphasis on increasing access through alternative certification, were moving teaching toward being seen as a *craft*.

PROFESSIONAL DEVELOPMENT

> The consensus view is that job-embedded professional development that is led by, designed by, and provided by teachers is the best model. But the research findings ... [favor] intensive, extended, curriculum-based training that is usually provided outside the workplace combined with on-site, job-embedded implementation support. (Corcoran, 2007, p. 6)

Recently, states have begun to systematically address teacher professional development by implementing policies that define professional development (32 states) and specify requirements for it (33 states). Interestingly, some states set requirements for in-service teacher professional development but do not provide a state-approved definition. Sixteen states do not have any policies regarding teacher professional development (Education Commission of the States [ECS], 2008b). State policies generally either describe or reference NCLB requirements for teacher professional development, and no state has policies that address specific subject areas, such as science or mathematics, though some do have policies regarding the professional development of teachers of special populations (e.g., students with disabilities, English language learners). Typically, state policies are broad and place the burden of defining, directing, and monitoring teacher professional development on school districts so that they may accommodate local constraints and contexts. The consensus view is based on standards for quality professional development developed by professional organizations and states. A summary of the consensus view may be found in Elmore (2002).

Partially due to recent NSF policies requiring research and/or evaluation of projects funded by the Foundation, research is available concerning the effectiveness of science teacher professional development from the 1990s. Indeed, beginning with the teacher training institutes of the 1950s and 60s, previously discussed, the Education and Human Resource Directorate (EHR) of NSF has focused on improving science teaching and learning primarily through funding, and requiring the rigorous evaluation of, professional development projects throughout the country. In 1991, EHR fundamentally changed its policies concerning funding. With its Statewide Systemic Initiative (SSI) program, NSF specifically targeted policies concerning teacher education at the state level. It also changed standards for the amount of funding, the length of projects, and the type of contract. For the first time, education projects could be supported for 5 years rather than limited to 3 years, and SSI projects could be funded for up to 10 million dollars. Further, support was provided through cooperative agreements, rather than through grants, allowing for continuous oversight by NSF. In the 1990s, three cycles of SSI projects were funded in 24 states and the Commonwealth of Puerto Rico. Although states were free to determine the focus of their respective SSIs, the majority, 20 states, focused on professional development of mathematics and science teachers. Another distinction of the SSI program was that evaluations of both the individual projects and the overall program were funded. As a consequence, studies were available during the course of the SSI projects concerning the effectiveness of different types of professional development for mathematics and science teachers. Further, the program's effectiveness in changing science and mathematics teaching and learning as well as state and local policies was analyzed by three major evaluation groups. SRI International, Horizon Research, Inc., and Abt Associates each were awarded a contract to assess different aspects of the SSI program and projects.

Laguarda (1998) reviewed the findings of seven SSIs that were deemed likely to generate credible evidence of change in student achievement as an outcome of the SSI. According to his review, four of the seven SSIs (Louisiana, LaSIP; Montana, SIMMS; Ohio, Discovery; and Puerto Rico, PR-SSI) had credible evidence of an impact of teacher professional development on student achievement, and two (Ohio and Louisiana) produced evidence of a disproportional impact for minority students. Laguarda concluded that student achievement gains were more likely to occur when an SSI focused its reform activities at the school, classroom, and teacher level as opposed to reforming or aligning state level policies (as New Mexico, SIMSE, and Vermont, VISMT did) or building state or local level capacity (as Kentucky, PRISM, and Vermont did). An example of reforming or aligning state policies was changing teacher education program requirements to match new areas or levels of certification. State and district capacities were built primarily through incentives directed at both levels.

NSF eventually expanded the SSI model to include cities (Urban Systemic Initiatives—USI) and rural areas (Rural Systemic Initiatives—RSI), and near the end of the decade NSF funded RAND to conduct a study of teaching practices and student achievement in six states and urban areas that had participated in the systemic initiatives (Kahle, 2008). Although there were major design issues with the study, the researchers found a positive, but weak, relationship between the frequency of use of teaching practices promulgated by the systemic reform initiatives (inquiry and problem solving) and student achievement (Klein, Hamilton, McCaffrey, Stecher, Robyn, & Burroughs, 2000).

In a smaller, well-controlled study, Kahle, Meece, and Scantlebury (2000) analyzed student achievement in eight urban middle schools whose teachers had participated in Ohio's SSI. Because of the low numbers of European American students in the schools, results were reported only for African American students. Using hierarchical linear modeling, they found that students of teachers participating in the SSI professional development, compared to students of nonparticipating teachers in the same schools, (a) scored higher on a valid and reliable measure of achievement and (b) reported that their teachers more frequently used standards-based teaching strategies that were part of the SSI's professional development efforts. The *Discovery* Inquiry Test was developed and field-tested for reliability; a group of scientists and science education experts established validity. After a MANOVA revealed three significant multivariate effects, a follow-up univariate test, adjusted for multiple comparisons, indicated that students in classes taught by the SSI teachers, compared to students in classes of match teachers, scored significantly higher on the *Discover* Inquiry Test ($F = 50.29$, $MSE = 51.9$, $df = 1{,}605$, $p < .001$; $\eta^2 = .08$) (Kahle, Meece, & Scantlebury, 2000, p. 1029). Indeed, frequency of use of standards-based teaching practices, not participation in the SSI's professional development, was a significant predictor of student achievement (Kahle et al., 2000). Another research study on the Ohio SSI found that its highly intensive (160 hours), inquiry-based professional development changed (a) teachers' attitudes toward reform, (b) their preparation to use standards-based teaching strategies, and (c) their use of inquiry instruction (Supovitz, Mayer, & Kahle, 2000), although it did not measure the impact of the program on student learning.

Later, NSF developed and funded the Local Systemic Change (LSC) initiatives for districts that were not eligible for the urban or rural initiatives. Because most of the SSIs, USIs, and RSIs were aimed at change in middle and high schools, NSF decided to have the LSCs focus on Grades K-8 teachers. The LSCs were administered separately from the other three initiatives and were evaluated separately by Horizon Research, Inc. All of the LSC projects used teacher professional development as the primary agent of change. Supovitz and Turner (2000) employed hierarchical linear modeling "to examine the relationship between professional development and the reformers' vision of teaching practice" (p. 963). They noted that during the 1990s, educators, researchers, and policymakers had reached a consensus on what constitutes high-quality professional development. The six agreed upon characteristics of high-quality professional development are: (a) *models inquiry teaching,* (b) *intensive and sustained,* (c) *engages teachers with concrete tasks and based upon teachers' experience with student,* (d) *focuses on content,* (e) *connects teaching practices to student achievement,* and (f) *connects to other aspects of school change* (Supovitz & Turner, 2000). The authors reported two relevant findings: (a) the quantity of professional development in which teachers participate that is consistent with the consensus model is strongly linked with both inquiry-based teaching practice and an investigative classroom culture, and (b) teachers' content preparation has a powerful influence on teaching practice and classroom culture. For example, related to the finding about the importance of the quantity of professional development received, with only 40–79 hours of professional development, the LSC teachers' investigative classroom culture was above that of the average teacher, and change was substantial after 160 hours of professional development. Supovitz and Turner (2000) describe an *investigative classroom culture* as one where teachers frequently "arrange seating to facilitate student discussion, require students to provide evidence to support claims, encourage students to explain concepts to one another, and have students work in cooperative groups" (p. 969). Concerning inquiry teaching, LSC teachers, compared to others, significantly increased their use of inquiry approaches after 80 hours of professional development. Supovitz and Turner's second major finding regarding the importance of content preparation of individual teachers is supported by findings from earlier studies of mathematics teachers (Cohen & Hill, 1998).

In addition to evaluations of the NSF-funded professional development programs, other major studies of PD effectiveness have been conducted. For example, Desimone, Porter, Garet, Yoon, and Birman (2002) used a purposeful sample of 207 teachers in 30 schools, in 10 districts and in 5 states to examine the characteristics of teacher professional development and the effect of professional development on teaching strategies in mathematics and science from 1996 to 1999. The study was conducted in the context of an evaluation of the Eisenhower Professional Development Program—Title II of the Elementary and Secondary Education Act (ESEA). The researchers identified six key features of professional development. Three were structural features: (a) reform type (e.g., who provides the professional development), (b) duration, and (c) collective participation (e.g., teachers from same grade, school, or district). The other three features characterized the substance of the professional development. They were: (a) active learning, (b) coherence, and (c) content focus. Using hierarchical linear modeling, they examined two levels of outcomes: the structural level and the substantive level.

Desimone et al. (2002) reported their findings in relation to the six key features. Briefly, for structural features, both reform type and collective participation led to changes in teaching practice. For the substantive features, both active learning and coherence led to changes in teaching practice. It should be noted that this study used teacher self-report of teaching practices rather than independent observation of classrooms. Unlike Supovitz et al. (2000), they found no effects for duration or for content focus. In conclusion, Desimone and colleagues stated that "professional development intended to increase a specific instructional practice must focus squarely on that specific practice. Transfer, if present at all, is not strong" (p. 91). Given that findings regarding some components or features of professional development are inconclusive, additional research is needed to investigate how interactions among features (e.g., duration, intensity, content), rather than how singular features, impact teaching and learning outcomes.

More recently, the federal government, under the auspices of NSF and USDOE, has focused attention on teacher acquisition of content knowledge through the Mathematics and Science Partnership Title IIB Program (MSP). MSP focuses on increasing the academic achievement of students by enhancing the content knowledge and teaching skills of teachers through partnerships between science, technology, engineering, and mathematics (STEM) faculty in colleges and universities and science and mathematics teachers at high-need schools. Funded projects are required to conduct rigorous evaluations, using research designs that provide empirical evidence regarding the improvement of outcomes for teachers and their students. Government Performance and Results Act (GPRA) measures applied to the MSP program include the extent to which the partnerships (a) significantly increase teacher content knowledge, (b) increase the percentage of students who meet basic and proficient standards on state assessments, and (c) successfully utilize experimental or quasi-experimental evaluation designs that may lead to scientifically valid results. Early data suggested that MSP projects were meeting annual targets for increasing teacher content knowledge and improving student achievement (using very broad achievement measures), but those studies often did not use rigorous methodologies (USDOE, 2008). In fact, projects have struggled to comply with directives to implement evaluation designs that provide credible evidence regarding project outcomes (USDOE, 2008). For example, all 15 Ohio USDOE MSP projects claimed to have made measurable and valid improvements in teachers' content knowledge (using locally developed assessments). Similarly, all projects reported having achieved measurable improvements in pedagogy, but for most projects this claim was not substantiated by classroom observation or by triangulation with other data. Four of the 15 projects reported measurable improvements in student performance on state assessments, with two reporting that these changes were statistically significant, but none reported student data for a comparison or control group (Zorn, Seabrook, Marks, Chappell-Young, Hung, & Marx, 2009).

Weiss and Pasley (2008) noted similar limitations in the collection of findings from NSF-funded MSP projects. While their MSP-Knowledge Management and Dissemination (KMD) project aimed to identify both research-based findings and practice-based insights on effective approaches to teacher professional development, the use of project evaluation plans, rather than research designs, hampered attempts to attribute outcomes to specific project features. They suggested that a synthesis of research findings and practitioner insights across multiple projects might produce the information needed to (a) build an initial, testable theory, (b) provide guidance for current practice, and (c) frame research to contribute to more robust theories. In essence, they proposed a rigorous and systematic practice-based approach to research on professional development that might better address the complexity and diversity of education systems and settings that often make experimental or quasi-experimental studies impossible.

IS THERE A CONNECTION BETWEEN RESEARCH AND POLICY?

> There is consensus that the quality of teachers and teaching matter—and undoubtedly are among the most important factors shaping the learning and growth of students. Moreover, there is consensus that serious problems exist with the quality of teachers and teaching in the United States. Beyond that, however, there appears to be little consensus and much disagreement—especially over what teacher quality entails and what the sources of, and solutions to, the problem might be. (Ingersoll, 2007, p. 1)

We hoped to answer one broad question regarding the relationship between education research and policy decisions about the education and professional development of science teachers: What is the relationship between research and policy in pre-service and in-service science teacher education?

Obstacles to responding to this question were many. First, the paucity of literature that provides empirical evidence regarding science teacher education and its relationship to teacher quality and student achievement is troubling. Generally, studies of teacher quality focus on outcomes in reading and mathematics, rather than on science teachers and students (Bolyard & Moyer-Packenham, 2008). When empirical studies were found, differences in samples, populations, definitions, and measures of achievement made it difficult to synthesize the findings. Second, education policy seldom prescribes expectations for specific subject matter or content areas such as science. Particularly at the federal level, policies tend to be broad rather than specific, providing great flexibility to states and to practitioners charged with policy implementation. While NCLB lays out several mandates, it provides little guidance for policy implementers who struggle to interpret, contextualize, and ground the law in practitioner realities.

The greatest challenge to making sense of the association between science teacher education research and policy is found in incongruent assumptions about what quality teaching entails. Although the research knowledge base is predicated on a normative view of teaching—that is, a shared understanding that what teachers do is to help students make sense of the physical world—the views of practitioners, policymakers, and other education stakeholders sometimes differ widely from this normative view (Wilson & Youngs, 2005). Findings from the limited body of research focusing on science teacher education suggest that the underlying theory of change driving policy initiatives aimed at improving science teachers and science teaching may be flawed. That is, the assumption that increasing teacher content knowledge alone, without any consideration of how that content can be made clear to students, will result in high-quality teachers, better teaching, and improved student achievement is just that, an unsubstantiated assumption.

Research studies and the policies designed to enhance science teacher education have focused on improved teacher quality as the outcome. While researchers have explored a range of variables that may impact teacher quality, including aspects of their preservice preparation and in-service professional development discussed here, most policies have tended to focus on increasing teacher content knowledge and, to a lesser extent, on pedagogical knowledge or skills as critical measures of teacher quality. Findings of research that link teacher content knowledge and pedagogical skill directly with improved student achievement in science are mixed at best.

So have policymakers been "barking up the wrong tree?" Yes and no. We did find some evidence that science teacher content knowledge is an important variable in enhancing student learning and achievement, although most of that evidence was based on changes in instructional practice, not on direct evidence of impact on student learning. The small number of studies that do provide empirical evidence linking some measure of teacher quality to student outcomes suggest that how a teacher engages students is influenced by teacher knowledge of science content and/or understanding of the nature of science, and that in turn is related to what students learn.

When all is said and done, did we find a link between research and policy concerning science teacher education from preservice through in-service? Again, the answer is both yes and no. We found growing evidence that policies—whether prescribing the amount of content studied in preparing to teach, the tests required for certification, or the length of professional development activities—were studied for effectiveness largely *after* they were implemented. While policy evaluation is a critical component of policy development and implementation, it does not substitute for research conducted prior to policy development. We found little evidence that policies were based upon, or modified in response to, research findings. In other words, we seem to be in a situation where "the cart is before the horse."

A particularly troubling example of the lack of coordination between policy and research is evidenced in the failure to recognize that research does not provide support for the proposition that teacher content knowledge translates directly to student achievement. Policy that focuses primarily or exclusively on improving teacher content knowledge ignores important aspects of schooling that mediate student learning such as school environment and teacher instructional practice.

Perhaps a more positive lens through which to view this situation is to focus on the potential for policy to contribute significantly to research and the development of the knowledge base. If, in fact, the cart is before the horse, it is quite possible for the horse to nudge the cart along in the correct direction. Education is deeply embedded in political, social, and economic realities that make it unlikely and, perhaps, unreasonable to hope that it can be sufficiently isolated from such forces to develop and test theories in the absence of contextual pressures and preferences.

In fact, some evidence suggests that research is nudging policy. Funding agencies, notably NSF, USDOE, and National Institutes of Health (NIH), have begun to include rigorous requirements for evaluation and research in their requests for proposals. While expectations that projects will use experimental or quasi-experimental research designs may be somewhat unrealistic, the collective knowledge emerging from implementation of well-conceived science teacher pre-service and in-service education projects will increase our knowledge of what works, and of the local conditions that might facilitate effective practices. The Knowledge Management and Dissemination project, funded by the NSF and USDOE MSP programs, and led by Horizon Research, Inc., is an important example of how education researchers and policymakers can collaborate effectively. As the research base grows, more empirical evidence will be available for policymakers. This empirical evidence will be augmented with rich practitioner insights gleaned from, and validated across, multiple projects and contexts. However, the research must be easily accessible to people outside the research community for maximum effectiveness in framing policy.

One example of an effective way to reach policymakers was found in the SSI projects. In addition to papers in scholarly journals, Ohio's SSI project, *Discovery,* produced succinct reports, called *Pocket Panoramas,* modeled after the U.S. Department of Education's *The Condition of Education*. Easy to read with many tables and figures, the *Pocket Panoramas* were distributed annually to all district superintendents, state legislators, congressional representatives and senators. Changes in state policies, including more rigorous requirements for mathematics and science teacher certification and state agency support for long-term, in-service teacher professional development, reflected findings from the SSI project. Its dissemination effort also was influential in promoting the continued support of *Discovery* with state funding. This effort suggests that the perpetual swinging of the policy pendulum may be moderated by findings from research and evaluation, if they are packaged and disseminated in ways that reach practitioners and policymakers.

Assessments of the impact of policy on science teacher education differ based upon the perspective of the observer, and recent policy development suggests that federal policymakers are not yet confident that current policy will adequately address the teacher quality issue. For example, according to *The Secretary's Fifth Annual Report on Teacher Quality* (USDOE, 2006), NCLB has increased the number of *highly qualified* teachers in U.S. classrooms. Yet, the federal government recently has provided incentives through the Higher Education Reconciliation Act of 2005 aimed at improving the quality of teachers of mathematics, science, and special education in high-poverty districts; and the American Competitiveness Act of 2006 is aimed at rigorous teacher preparation and in-service training through specific strategies, including strengthening teacher preparation for Advanced Placement (AP) and International Baccalaureate (IB) programs, and providing Improving Teacher Quality (ITQ) State Grants (USDOE). The outcomes of these recent improvement efforts and the extent to which they support NCLB implementation are unclear.

Generally, practitioner views regarding the success of NCLB and similar policy initiatives differ from those of government officials and policymakers. There is some consensus among practitioners and education stakeholders that "With few exceptions, NCLB has had little effect on the quality of the teacher pool as a whole, district assignment policies, or strategies to recruit and retain teachers" (Emerick, Hirsch, & Berry, 2004, p. 2). Critics point to state policies that are either not supportive of, or in conflict with, NCLB requirements. These include: (a) the lack of coherent and accessible data systems to track student outcomes, (b) unclear requirements for implementation of NCLB *highly-qualified* teacher assurances, and (c) minimal guidance regarding how scientifically-based professional development can be identified and implemented with limited resources (Emerick et al., 2004).

In 2005, the American Educational Research Association (AERA) published the report of its panel on research and teacher education (Cochran-Smith & Zeichner, 2005). Many of the findings and conclusions are pertinent to this chapter. For example, both the chapter authors and the editors note that existing databases are inadequate for establishing links between teacher content preparation and student achievement gains. Further, the

AERA report urged more longitudinal studies to examine the effects of teacher preparation over time (Zeichner, 2005). Goldhaber (2008) agrees that there is an urgent need for data systems that provide the ability to match teachers with schools and students in order to study the efficacy of policies designed to address teacher quality issues, including the equitable distribution of quality teachers. He further observes that any new policies should be implemented simultaneously with plans to study the effects longitudinally. It is anticipated that new federal regulations will produce appropriate databases that will enable research to better focus on policy level questions.

In summary, we agree with Zeichner's (2005) conclusion that:

> There is widespread agreement that teacher education research has had very little influence on policymaking and on practice in teacher education programs. Both teacher educators and policymakers go about their work in designing and implementing policies and programs without much regard to ... research. (p. 756)

The realities that drive policymaking (e.g., pragmatism, responsiveness, political will) are unlikely to change, but recent efforts by federal agencies to promote education evaluation and research (e.g., MSP-Knowledge Management and Dissemination and the What Works Clearinghouse) are promising. Further, the Government Performance and Results Act of 1993 (GPRA) shifts the focus of federal agencies from accountability having to do with process to accountability having to do with results. Because GPRA is results oriented, it requires a shift toward evidence-based accountability, and it increases the likelihood of a stronger role for research in policymaking. (See also Osborne, Chapter 2, this volume, for a further discussion of the evidence-based policy and practice movement.) In order to foster the intersection of policy and research, science education researchers and policymakers must acknowledge both the responsibility and the opportunity to enhance student achievement in science through their research efforts.

REFERENCES

American Association for the Advancement of Science. (1993). *Benchmarks for science literacy.* New York, NY: Oxford University Press.

Andrew, M. D. (1990). Differences between graduates of 4-year and 5-year teacher preparation programs. *Journal of Teacher Education, 41*(2), 45–51.

Angus, D. L. (2001). *Professionalism and public good: A brief history of teacher certification.* Washington, DC: The Fordham Foundation.

Becker, B. J., Kennedy, M. M., & Hundersmarck, S. (2003, April). *Communities of scholars, research and debates about teacher quality.* Paper presented at the annual meeting of the American Educational Research Association, Chicago, IL.

Bolyard, J. J., & Moyer-Packenham, P. S. (2008). A review of the literature on mathematics and science teacher quality. *Peabody Journal of Education, 83*(4), 509-535.

Bush, V. (1945, July). *Science: The endless frontier.* Washington, DC: U.S. Government Printing Office.

Carnegie Forum on Education and the Economy, Task Force on Teaching as a Profession. (1986). *A nation prepared: Teachers for the 21st century.* New York, NY: Author.

Cochran-Smith, M., & Fries, K. (2005). Researching teacher education in changing times: Politics and paradigms. In M. Cochran-Smith & K. M. Zeichner (Eds.), *Studying teacher education: The report of the AERA panel on research and teacher education* (pp. 69–109). Mahwah, NJ: Erlbaum.

Cochran-Smith, M., & Zeichner, K. M. (Eds.). (2005). *Studying teacher education: The report of the AERA panel on research and teacher education.* Mahwah, NJ: Erlbaum.

Cohen, D. K., & Hill, H. C. (1998). *State policy and classroom performance: The mathematics reform in California* (CPRE Policy Brief No. RB-23). Philadelphia, PA: University of Pennsylvania, Consortium for Policy Research in Education.

Constantine, J., Player D., Silva, T., Hallgren, K., Grider, M., & Deke, J. (2009). *An evaluation of teachers trained through different routes to certification, final report* (NCEE 2009-4043). Washington, DC: National Center for Education Evaluation and Regional Assistance, Institute of Education Sciences, U.S. Department of Education.

Corcoran, T. B. (2007, February). *Teaching matters: How state and local policymakers can improve the quality of teachers and teaching* (CPRE Policy Brief No. RB-48). Philadelphia: University of Pennsylvania, Consortium for Policy Research in Education.

Desimone, L. M., Porter, A. C., Garet, M. S., Yoon, K. S., & Birman, B. F. (2002). Does professional development change teachers' instruction? Results from a 3-year study. *Educational Evaluation and Policy Analysis, 24*(2), 81–112.

Education Commission of the States. (2008a). *Teacher preparation state policy database—undergraduate programs.* Retrieved June 8, 2009, from http://www.tqsource.org/prep/ index.asp

Education Commission of the States. (2008b). *Teacher professional development state policy database.* Retrieved June 8, 2009, from http://www.mb2.ecs.org/reports

Elmore, R. F. (2002). *Bridging the gap between standards and achievement: The imperative for professional development in education.* Washington, DC: Albert Shanker Institute.

Emerick, S., Hirsch, E., & Berry, B. (2004, November). Does highly qualified mean high-quality? *Association for Supervision and Curriculum Development Infobrief, 39.* Retrieved March 23, 2009, from http://www.ascd.org/publications/newsletters/infobrief/nov04/num39 /toc.aspx

Feistritzer, C. E. (2006). *Alternative teacher certification: A state-by-state analysis.* Washington, DC: National Center for Education Information.

Germann, P., Duman, W., & Barrow, L. H. (1995). Preparation of middle school science teachers in major state research universities. *Journal of Science Teacher Education, 6*(3), 143–145.

Goldhaber, D. (2008). *Addressing the teacher qualification gap.* Washington, DC: Center for American Progress.

Goldhaber, D. D., & Brewer, D. J. (2000). Does teacher certification matter? High school teacher certification status and student achievement. *Educational Evaluation and Policy Analysis, 22*(2), 129–145.

Grisham, D. L., Laguardia, A., & Brink, B. (2000). Partners in professionalism: Creating a quality field experience for preservice teachers. *Action in Teacher Education, 21*(4), 27–40.

Guyton, E., Fox, M. C., & Sisk, K. A. (1991). Comparison of teacher attitudes, teacher efficacy, and teacher performance of first-year teachers prepared by alternative and traditional teacher education programs. *Action in Teacher Education, 13*(2), 1–9.

Hawk, P., & Schmidt, M. W. (1989). Teacher preparation: A comparison of traditional and alternative programs. *Journal of Teacher Education, 40*(5), 53–58.

Hutton, J. B., Lutz, F. W., & Williamson, J. L. (1990). Characteristics, attitudes, and performance of alternative certification interns. *Educational Research Quarterly, 14*(1), 38–48.

Ingersoll, R. M. (2007, February). *Misdiagnosing the teacher quality problem* (CPRE Policy Brief No. RB-49). Philadelphia: University of Pennsylvania, Consortium for Policy Research in Education.

Jelmberg, J. (1996). College-based teacher education versus state-sponsored alternative programs. *Journal of Teacher Education, 47*(1), 60–66.

Kahle, J. B. (2008). Systemic reform: Research, vision, and politics. In S. K. Abell & N. G. Lederman (Eds.), *The handbook of research on science education.* Mahwah, NJ: Earlbaum.

Kahle, J. B., Meece, J., & Scantlebury, K. (2000). Urban African-American middle school science students: Does standards-based teaching make a difference? *Journal of Research in Science Teaching, 37*(9), 1019–1041.

Klein, S., Hamilton, L., McCaffrey, D., Stecher, B., Robyn, A., & Burroughs, D. (2000). *Teaching practices and student achievment.* Santa Monica, CA: RAND.

Laguarda, K. (1998). *Assessing the SSI's impacts on student achievement: An imperfect science.* Menlo Park, CA: SRI International.

Lawrenz, F. (1975). The relationship between teacher characteristics and student achievement and attitude. *Journal of Research in Science Teaching, 12*(4), 433–437.

Levine, A. (2009, January). *How do we attain better quality math and science teachers?* Keynote address at the Mathematics and Science Partnership Learning Network Conference, Washington, DC.

Lutz, F. W., & Hutton, J. B. (1989). Alternative teacher certification: Its policy implications for classroom and personnel practice. *Educational Evaluation and Policy Analysis, 11*(3), 237–254.

Magnusson, S., Borko, H., Krajcik, J. S., & Layman, J. W. (1992, March). *The relationship between teacher content and pedagogical content knowledge and student content knowledge of heat energy and temperature.* Paper presented at the annual meeting of the National Association for Research in Science Teaching, Boston, MA.

Monk, D. H. (1994). Subject area preparation of secondary mathematics and science teachers and student achievement. *Economics of Education Review, 13*(2), 125–145.

MSP-Knowledge Management and Dissemination. (2007). *Why teacher content knowledge matters: Research on the relationship between teachers' mathematics/science content knowledge and their instructional practice and students' achievement.* Retrieved March 3, 2009, from http://www.mspkd.net/index.php?page=22_2b

National Commission on Excellence in Education. (1983). *A nation at risk: The imperative for educational reform.* Washington, DC: U.S. Government Printing Office.

National Research Council. (1996). *National science education standards.* Washington, DC: National Academy Press.

National Research Council. (1997). *Science teaching reconsidered: A handbook.* Washington, DC: National Academy Press.

National Research Council. (2000). *Educating teachers of science, mathematics, and technology: New practices for the new millennium.* Washington, DC: National Academy Press.

Rutherford, F. J., & Ahlgren, A. (1990). *Science for all Americans.* Washington, DC: National Academy Press.

Sanders, L. R., Borko, H., & Lockard, J. D. (1993). Secondary science teachers' knowledge base when teaching science courses in and out of their area of certification. *Journal of Research in Science Teaching, 30*(7), 723–736.

Schroeder, C. M., Scott, T. P., Tolson, H., Huang, T. Y., & Lee, Y. H. (2007). A meta-analysis of national research: Effects of teaching strategies on student achievement in science in the United States. *Journal of Research in Science Teaching, 44*(10), 1436–1460.

Supovitz, J., Mayer, D., & Kahle, J. B. (2000). Promoting inquiry-based instructional practice: The longitudinal impact of professional development in the context of systemic reform. *Educational Policy, 14*(3), 331–356.

Supovitz, J. A., & Turner, H. M. (2000). The effects of professional development on science teaching practices and classroom culture. *Journal of Research in Science Teaching, 37*(9), 963–980.

U.S. Department of Education. (2004). *The secretary's second annual report on teacher quality.* Washington, DC: Author.

U.S. Department of Education. (2006). *The secretary's fifth annual report on teacher quality.* Washington, DC: Author.

U.S. Department of Education. (2008, September). *ESEA: Mathematics and science partnerships (OESE), FY 2008 program performance report.* Washington, DC: Author.

Weiss, I. R., & Pasley, J. D. (2008, March). *Using research findings and practice-based insights: Guidance for policy, practice, and future research.* Paper presented at the annual meeting of the American Educational Research Association, New York, NY.

Wilson, S., Floden, R., & Ferrini-Mundy, J. (2001). *Teacher preparation research: Current knowledge, gaps, and recommendations.* Seattle, WA: Center for the Study of Teaching and Policy.

Wilson, S. M., & Youngs, P. (2005). Research on accountability processes in teacher education. In M. Cochran-Smith & K. M. Zeichner (Eds.), *Studying teacher education: The report of the AERA panel on research and teacher education* (pp. 591–643). Mahwah, NJ: Erlbaum.

Wise, K. C. (1996). Strategies for teaching science: What works? *Clearing House, 69*(6), 337–338.

Young, P., Odden, A., & Porter, A. C. (2003). State policy related to teacher licensure. *Educational Policy, 17*(2), 217–236.

Zeichner, K. M. (2005). A research agenda for teacher education. In M. Cochran-Smith & K. M. Zeichner (Eds.), *Studying teacher education: The report of the AERA panel on research and teacher education* (pp. 737–759). Mahwah, NJ: Erlbaum.

Zorn, D., Seabrook, L., Marks, J., Chappell-Young, J., Hung, S., & Marx, M. (2009). *The Ohio mathematics and science partnership program external evaluation: Year 2 report.* Oxford, OH: Miami University, Ohio's Evaluation & Assessment Center for Science and Mathematics Education.

CHAPTER 4

HOW DO FOUNDATIONS INFLUENCE SCIENCE EDUCATION POLICY?

Dennis W. Cheek and Margo Quiriconi

INTRODUCTION

Perhaps one of the most overlooked institutions in our society when considering which has had the greatest impact on science education in the United States is private philanthropy. Philanthropists have sought to influence the quality and extent of education in the United States since colonial times. In fact, education has been a primary beneficiary of philanthropy in the United States throughout most of its history. Table 4.1 shows the breakdown of $307.65 billion in total charitable giving in 2008 (Giving USA Foundation, 2009).

Religion annually commands vastly more overall charitable support from all sources, including members of religious organizations. Education, however, has topped every list of *foundation* support since the 1920s when such reporting began. In 2007, private foundations gave an estimated $21.6 billion in grant awards of which $4.9 billion (22.8%) was committed to education (The Foundation Center, 2009b, p. 2). By con-

The Role of Public Policy in K–12 Science Education, pp. 77–116

Table 4.1. Breakdown of Charitable Giving in 2008

Sector	*Amount (in Billions of Dollars)*	*Percentage of Total Giving*
Religion	106.9	35
Education	40.9	13
Social Services	25.9	9
United Way and Community Causes	23.9	8
Health	13.3	4
Arts and Culture	12.8	4

trast, grants for basic research and development in science and technology that year totaled only $636 million. Although the investment by foundations in education accounted for less than 3% of K–12 education revenue in 2007 (estimated by extrapolation from National Center for Education Statistics, 2007, Table 163), foundation grants represent a considerable proportion of discretionary dollars available to improve educational practices or to support new programs.

Much of this funding has gone to support school-based programs, activities, and institutions that enhance the public's understanding and engagement in science. But, despite the very significant contributions foundations have made to science education and despite a thriving science education research community, there has been little if any attention paid to the role of foundations within that science education research community. For example, a reading of the recent National Association of Research in Science Teaching (NARST) sponsored summary of research on science education (Abell & Lederman, 2007), an electronic search of the major educational research databases, and a search of the papers given at the 2008 NARST Annual Conference to find any discussion of the role of foundations in science education came up empty. Publications coming from the science education research community seem to be completely devoid of any mention of the role of foundations in science education practice or policy. This is a lacuna in the literature that is surprising given the history and involvement of foundations in Science, Technology, Engineering, and Mathematics (STEM) education and science education in particular. A special interest group (SIG) on Education and Philanthropy was launched within the American Educational Research Association in 2001, but this SIG has, so far, only organized a series of presentations at the annual meeting and has not stimulated any research specific to foundations and STEM education policy. The Association for Research on Nonprofit Organization and Voluntary Action

(ARNOVA) through its *Nonprofit and Voluntary Sector Quarterly* and its annual meetings, promotes research on the nonprofit sector as a whole, including foundations but, once again, to date this has not resulted in research on the role of foundations in national science education policy. Finally, the National Science Teachers Association (NSTA) has been the recipient of numerous foundation grants over many years including the Toyota Tapestry Grants for the past 20 years from Toyota Motor Sales (United States), the annual Shell Science Teaching Award, and the Mickelson Exxon Mobil Teachers Academy for Grades 3–5 teachers. Yet, NSTA has never produced any documentation highlighting the role foundations play to influence national K–12 science education policy.

This chapter is not intended to review the meager body of research on the role of foundations in science education policy development. Rather, it is hoped that the chapter can make a contribution to science education researchers' understanding of the role of foundations in science education by summarizing some general trends and highlighting selected developments in foundation funding over the decades that relate to science education policy development. The chapter will concentrate on the role that three large and extant foundations (Ford, Rockefeller, and Carnegie) and their subsidiaries have played in the past several decades in impacting science education policies and practices in K–12 schools. This choice has been dictated largely by the fact that these foundations have a reasonable supply of historical analyses associated with their activities, coupled with the very significant impact these three organizations made in prior decades when there were not as many large foundations and when the federal role was comparatively less than it is now in size and impact (e.g., Buss, 1980; Fosdick, 1952; Lagemann, 1989, 1999; Richardson, 2005; Woodring, 1970). But, before describing some of the specific contributions that foundations have made to science education policy in the United States, we provide some basic background information on the nature and role of foundations in American civil society.

OVERVIEW OF FOUNDATIONS

The Internal Revenue Code (IRS) (Title 26, Sec. 501) recognizes two types of charitable organizations: public charities and private foundations. Both types are accorded the status of 501(c) (3) tax-exempt entities under the tax code. The IRS explains the difference between them as follows:

> Generally, organizations that are classified as public charities are those that (i) are churches, hospitals, qualified medical research organizations affiliated

> with hospitals, schools, colleges and universities, (ii) have an active program of fundraising and receive contributions from many sources, including the general public, governmental agencies, corporations, private foundations or other public charities, (iii) receive income from the conduct of activities in furtherance of the organization's exempt purposes, or (iv) actively function in a supporting relationship to one or more existing public charities. Private foundations, in contrast, typically have a single major source of funding (usually gifts from one family or corporation rather than funding from many sources) and most have as their primary activity the making of grants to other charitable organizations and to individuals, rather than the direct operation of charitable programs. (IRS, 2009)

The process of distinguishing between public charities and private foundations has been called by the judges of the federal Tax Court, "almost frighteningly complex and difficult" (Friends of the Society of Servants of God, 75. T.C. 209, 213, 1980), so, in this chapter we will ignore the highly convoluted nature of these distinctions and their case-by-case resolution. Examples of public charities within the U.S. science, technology, engineering, and mathematics (STEM) arena include the National Academies Corporation, the funding arm of the National Academies, which includes the National Academy of Science, the National Academy of Engineering, the Institute of Medicine, and the National Research Council; the Consortium for Mathematics and its Applications (COMAP, Inc.), a nonprofit organization whose mission is to improve mathematics education; and the Challenger Center for Space Science Education (which collects and distributes money to the Challenger Learning Centers across the nation). By the end of 2008, the IRS had recognized 956,760 public charities, not counting individual churches, synagogues, mosques, and temples (National Center for Charitable Statistics, 2008). Charities within the STEM arena collect and distribute monies from many sources to advance STEM education and research across the nation, and they are present in virtually every community. A significant number of larger charities are community foundations, which hold in trust and/or distribute funds collected from many different sources within a community to causes identified either by their respective donors or by the foundation's board, following guidelines that have been self-established. Within many locales, these community foundations are significant supporters of local STEM programming run by museums, parks, community organizations, schools, colleges, and universities. The Houston Endowment, Tulsa Community Foundation, Seattle Foundation, Rhode Island Foundation and the Greater Kansas City Community Foundation, are all large community charities that have provided substantial funding for STEM programs over the years.

Private foundations include individual, family, and corporate foundations. Most of these more than 100,000 foundations are small, family foundations with assets of less than $1 million. There were a total of 112,959 IRS-registered private foundations in 2008, and 86,135 of them filed form 990, which the IRS requires of active foundations (National Center for Charitable Statistics, 2008). Some of the private foundations that are widely known to the public include Ford, Rockefeller, Gates, Annenberg, Dell, Hewlett, Packard, Lilly Endowment, Carnegie, Broad, Wallace, Kauffman, Sloan, Casey, and MacArthur. Virtually all large corporate foundations also are private foundations including Wal-Mart, GE, Intel, Eli Lilly, IBM, and Merck. For complex tax reasons, many of these corporations also give under their corporate banner in addition to what they give through their foundations. Only 750 private foundations have assets greater than $100 million, and the $367 billion total assets of this group of 750 equals 65% of the assets held by all private foundations (National Center for Charitable Statistics, 2008).

A further distinction often is drawn within the philanthropic sector between "operating foundations" and "grantmaking foundations." Public charities or private foundations can be found in either of these two categories. The key distinction is that operating foundations exist for the sole purpose of running a particular institution or a particular set of programs rather than making grants to a wide variety of organizations working on many different programs or efforts. Many operating foundations in the STEM arena are focused on a particular institution such as a local science museum (e.g., the Boston Museum of Science or the Exploratorium in San Francisco) or a specialized library (e.g., Linda Hall Library in Kansas City, MO) or a single venture such as U.S. FIRST Robotics or Project Lead the Way. Some run very large national programs targeted to a particular cause, such as the Woodrow Wilson National Fellowship Foundation (Weisbuch, 2007) whose primary mission has been to provide fellowships to gifted graduate students in support of their doctoral training in preparation for a career in higher education with money received from grantmaking foundations such as Carnegie, Gates, Ford, and Annenberg. Some large foundations also may directly run one or more programs for a short time just to get them launched.

In order for private foundations to maintain tax exempt status, they must pay out at least 5% of their assets each year, although foundations may accumulate credits for years in which they exceed the 5%, and then apply them in future years. The largest private foundation in the United States is the Bill & Melinda Gates Foundation, with assets totaling $38.7 billion in 2008 (National Center for Charitable Statistics, 2008). Some foundations, such as Rockefeller, are set up to last in perpetuity, and others, such as the Atlantic Philanthropies, are set up to completely go out of existence (sunset) within a specified number of years. As economies prosper, the number of foundations tends to rise, with a tripling of the number of foundations in the United States between 1974 and 2004 (Stannard-Stockton, 2007). But when economic times are bad, some foundations go out of business, and others find their investment portfolios substantially reduced, which affects their level of giving.

In 2008, foundations in the United States contributed a combined total of $45.6 billion. Independent foundations represented $33 billion, corporate foundations $4.4 billion, community foundations $4.6 billion, and operating foundations $3.6 billion (Lawrence & Mukai, 2009). When we move more specifically to the realm of elementary and secondary education, the 50 largest private foundations contributed approximately $861 million in 2007 via 3,078 grants, with the Bill & Melinda Gates Foundation holding the number one spot with $193 million contributed, the Walton Family Foundation at number two with grants of $58 million, W. K. Kellogg at number three with $40 million, the Eli & Edythe Broad Foundation at number four with $37 million, and the Brown Foundation at number fifty with grants totaling $5.5 million (The Foundation Center, 2009a). For the three foundations on which we will focus much of our attention, Ford awarded $10 million, and the Carnegie Corporation of New York awarded $17 million. Rockefeller did not make the top fifty list because elementary and secondary education is no longer a priority at the foundation. A search of the Foundation Center's *Foundation Directory Online* in March 2009 with the search words "science AND education" yielded a list of 439 foundations that fund both of these areas, across an array of individual, family and corporate foundations, although there is no descriptor for "science education" in the database. There can be little doubt, however, that a considerable number of foundations do give grant awards in STEM education.

What Drives the Activities of Foundations?

Every foundation seeks to manage its portfolio of activities with some reference, explicit or implicit, to the six factors we describe below.

1. **Donor Intent**: The most important factor that foundations continuously manage is fidelity to donor intent. Except for community foundations, which fund a wide range of programs for the local community, all foundations have their origins in the minds of specific donors who brought the foundation into being, whether they are individuals, families, or corporations. Donor intent is a phrase that refers to the specific aims and goals that the founder(s) of a

foundation established at its inception and/or modified during their lifetime that guide the activities, decisions, and programs of the foundation in question (Bork & Nielsen, 1993). Over the years, there have been significant and sometimes protracted and expensive legal struggles over donor intent and the degree to which a foundation's leadership, including its board of trustees, is maintaining fidelity to the intent of the donor(s), and courts generally have been very cautious in overruling the intent of a donor. Fortunately, many donors have wisely decreed that future trustees whom they will not live to see can make adjustments in the foundation's core aims and activities as the situation seems to demand. Obviously, there are occasions when the original need on which the foundation was focused (e.g., apprenticeships in the case of Benjamin's Franklin's first foundation in America) is no longer relevant within society. Donor intents are frequently spelled out by foundations in their various grant application documents or orally by program officers as they work with potential grant recipients to craft an acceptable program. An inordinate number of proposals for funding received by foundations fall outside of donor intent and are "dead on arrival." The submitters in these cases simply did not do their homework to identify the funding interests of that particular foundation.

2. **Reputation**: A foundation chooses what to fund and what not to fund in part to maximize its reputation among other foundations and the wider public because reputational capital can be exchanged for leverage in policy deliberations, access to high ranking officials and prestigious organizations, and the accumulation of accolades for achievements, which further enhances one's reputation. Because reputation is based on the perceptions of others, foundations attempt to build their reputations through a variety of means including inviting the media to cover their grantmaking activities, involving high-ranking officials in the launch of their programs, profiling particularly important and successful grants on their Web site or in public meetings, giving away free books that highlight their work, and commissioning professional writers to produce articles or books about their efforts.
3. **Networks**: A foundation works within a larger philanthropic structure. Many projects are co-funded across multiple foundations, and others are passed on to more appropriate foundations when a proposed project does not fit within the mission of the foundation. For that reason, a good deal of energy goes into cultivating a network of interconnected organizations and individuals with whom foundation officials interact in the course of conducting their business.

These networks provide leverage points for specific foundation activities, dissemination nodes for information or policies the foundation wishes to highlight, and the means by which the ideas of others who appear at the foundation's doors might be advanced, with or without the direct financial support of the foundation in question.

4. **Ideas**: A great deal of a foundation's time and energy is spent in the analysis, synthesis, and dissemination of ideas. Large and prestigious foundations often have grand and noble ambitions to solve some of the most intractable problems of human society at a global scale such as disease, illiteracy, and international conflict resolution. But even on a more modest scale, foundations do what they can to make the world, locally and internationally, a better place to live. This requires a careful and thoughtful weighing of alternative ideas, making best estimates of the likelihood of success of proposed projects, and combining ideas into new syntheses. The work of foundations is largely to trade in ideas that are garnered from many sources, to analyze and modify them, and to repackage them for dissemination or application. Some ideas are promoted without modification because they are believed to be immediately useful to others and the individual or organization in question does not possess adequate means to convey them to much wider and more disparate audiences who also would find them useful. All this means that foundations must have a constant source of new ideas in order to function. Much of their efforts go into soliciting ideas from others and continuously scanning various policy and research environments for the best new ideas.
5. **People**: Ideas, networks, and reputations do not exist without people. Human capital is a stock in trade of foundations. Human capital resides in the vast numbers of individuals that a foundation interacts with through proposals that come to them from those individuals, and through conversations with people at the many public events foundation officials attend. A foundation's human capital also includes the people who work for the foundation itself. Foundations hire and cultivate talent to enhance their own reputation, help the foundation build essential networks, and bring a high level of insight and intelligence to the assessment and promotion of new ideas. To maximize the quality of their own staffs, foundations engage talent scouts and seek referrals from others whom the foundation trusts, including their peers in other foundations. Finally, human capital includes the people who benefit from a foundation initiative. Most of what a foundation funds are programs that

increase the talents, health, and overall well being of members of society.

6. **Money**: The sixth and final factor that foundations manage is their money, a fact usually foremost in the eyes of those who approach foundations. Obviously, foundations cannot exist without the financial resources needed to fund important projects and their own operations. There is, of necessity, an interaction between monetary resources and the previous five factors, and although many leaders within the foundation world would say that a focus on money is not the most important factor in advancing a foundation's objectives, a foundation's chief legal reason for being is to distribute money for charitable purposes. This means that a foundation must be a careful steward of the resources that created the foundation and needs to invest those funds wisely, both in the choice of programs it funds but also in terms of how its assets are protected and allowed to grow so that the foundation can continue its work well into the future. Although the manner in which it acquires, increases, and disperses its funds varies by foundation type and leadership, in all cases the acquisition, safe keeping, and distribution of money is at the heart of what a foundation does. Because people are seeking financial support of their idea, project, or program, the distribution of money is foremost in the minds of many who interact with foundations, yet foundations themselves consider all of these six factors to be important as they deliberate and engage in their grant making activities.

Modes of Interaction Between Foundations and the Public

The public often has expectations concerning the activities of foundations because many of those activities involve the distribution of private funds for public purposes. Public officials frequently express their frustrations, publicly or privately, with particular foundations when they perceive that foundations are withholding funds from worthy causes or failing to support the actions of government in domains where the foundation in question has signaled its grant making interest. This has led some to call for government actions to remake autonomous private foundations into pools of public monies where there is an express obligation to contribute substantially to social causes when there is an economic downturn or government support is stressed beyond capacity. Although there is no legitimacy to these sweeping claims that foundations are not acting in the public interest (Brody & Tyler, 2009, 2010), it is true that foundations frequently are misaligned with government or nonprofit program directions. Why are foundations not more frequent partners with other agencies and organizations within the public sphere? We offer, here, our thoughts about these matters as a guide for how one might think about the specific actions that foundations take in particular circumstances.

We can think of an array of actions or *modes of being* that present themselves to a foundation as it considers how to negotiate the public space in pursuit of its particular goals and desires. We have created a seven-P alliterative scheme to enable us to better delineate and articulate the distinct types of activities that a foundation supports and the manner in which it chooses to do so. We believe these seven modes of being for a foundation in relation to both the public sphere and the nonprofit sector comprehensively cover the array of activities in which foundations engage. Government officials and many nonprofit leaders generally view the first four modes positively, but the last three modes frequently cause those same officials to cringe. Yet, is it likely that all seven modes are essential to a healthy relationship between the work of foundations and the public and nonprofit sectors, in addition to making for a healthier civil society.

Our seven posited modes of being for foundations are:

1. **Pioneer**: Perhaps the majority of both public officials and the general public accept that one of the roles of foundations in civil society is to serve as a venture capital fund for the development of new programs, ideas, and strategies that seek to ameliorate problems both large and small. In the remainder of the chapter, we will provide examples about the ways in which the Ford, Rockefeller, and Carnegie foundations have exemplified this investment role in science education over the decades.
2. **Planner**: Foundations frequently serve as a convener of organizations and persons with mutual interests in solving societal problems and then use their intellectual and financial means to coordinate and plan for collective action. The Carnegie Corporation of New York played this role magnificently over several decades in regards to middle grades education; and the Lumina Foundation, Gates Foundation, and Goldman Sachs Foundation have partnered with many other foundations to play a similar role around issues related to college access and high ability disadvantaged youth.
3. **Partner**: Sometimes foundations partner with the public or nonprofit sectors to improve the likelihood that an endeavor is going to succeed. But, because there is no single way in which the foundation partner should act in this arrangement, misunderstandings can develop with their partners over the nature of the partnership. The Bill & Melinda Gates Foundation and the Annenberg Foundation

are foundations that have partnered with large school districts as part of their urban school reform efforts, and these partnerships have had admittedly mixed success.

4. **Preserver**: Foundations, which derive their existence independent of public funds, often serve as the chronicler and keeper of institutional, community, or problem-specific memory. They catalogue powerful stories that illustrate change, hope, challenge, or myriad other important concepts and principles for others. They frequently compile, analyze, synthesize, and maintain the history of society's efforts over time and communicate these histories and stories to others in hopes of informing future actions. The Carnegie Foundation for the Advancement of Teaching (Lagemann, 1983; Lagemann & de Forest, 2007) and the Carnegie Corporation of New York, through well-researched, well-written, and widely disseminated reports dealing with many issues in both K–12 and higher education, are outstanding examples of fulfilling this vital role. Community foundations frequently play this role for their local communities across the United States decade after decade.
5. **Provocateur**: As government officials can sometimes attest, foundations may find it necessary in pursuit of their objectives to act as a provocateur in the public sphere. Actions might include explicitly challenging reports or assertions of governmental or nongovernmental officials through targeted reports; compilation and release of contradictory evidence challenging the claims being advanced by those officials; interviews that publicly question the motivation or actions of agencies, organizations, or officials; and withholding funds for actions deemed unacceptable but not necessarily illegal. For example, the Heinz Endowments, Pittsburgh Foundation, and Grable Foundation called a highly visible public press conference in July 2002 to announce they were immediately suspending all grants to the Pittsburgh Public Schools because of their dissatisfaction with the leadership of both the Pittsburgh Board of Public Education and the Superintendent. After a series of widely-watched changes, the foundations restored funding 2 years later (Oliphant, 2004).
6. **Protector**: As public administrations and nonprofit leaders change in local and national settings, foundations sometimes choose to act as the protector of past initiatives that showed promise but then fell out of favor. Many programs have been sustained by foundation support that were decidedly out of favor in the public sphere, to ultimately see renewed public support at a later time. This role has been

fulfilled many times over by corporate, private, and community foundations as various federal or state government funded programs have been eliminated due to changes in public administrations, whose hallmark agenda often is "out with the old, in with the new" just to have something the new administration can point to as its own.

7. **Promoter**: Foundations frequently take kernels of ideas and develop them into widely accepted knowledge, promote individuals from obscurity to rock star status in a particular domain or problem area, and promote programs from fledgling endeavors into widespread adoption. For example, the Wallace Foundation some years ago decided to focus on school principals, and the Broad Foundation focused on urban superintendents and urban school districts. The foundations built up a large body of knowledge around their area of interest and made it publicly available on their Web sites in addition to funding national programs to improve the skills and talent pool in these two critical classes of school leaders. In this regard, foundations have been absolutely critical to the development and spread of U.S. FIRST Robotics and Project Lead the Way, to give two examples related to STEM education. Sometimes the projects that foundations promote do not fit well with the government's own reform agenda and, therefore, cause government officials concern.

All of the above factors and modes of being regarding foundations can be applied to thinking about the role of foundations in science education policy both past and present. Leaders in the K–12 science education community may find this seven-Ps alliterative scheme useful in considering their own interactions with foundations, including what specific role they are seeking for the foundation to play in advancing a particular policy or program.

CONTRIBUTIONS OF FOUNDATIONS TO SCIENCE EDUCATION POLICY DEVELOPMENT

The remainder of this chapter will highlight the roles three large foundations (Ford, Carnegie, and Rockefeller) and their subsidiaries have played in the past several decades in impacting science education policies and practices in K–12 schools, while also noting contributions of contemporary foundations that do not have this long history but also are active in similar efforts today. Our focus on these three foundations has been dictated largely by the fact that during the time that they were most intensively engaged in K–12 science education, there were very few other national foundations working in this area. These foundations additionally have a supply of historical analyses associated with their activities. Consequently, their specific roles and impacts can be more readily described (e.g., Buss, 1980; Fosdick, 1952; Lagemann, 1989, 1999; Richardson, 2005; Woodring, 1970).

As will be seen, these three foundations also exercised influence at the national level well beyond that of any set of contemporary foundations, due in no small measure to the fact that there was a time when federal dollars played a much less significant role in promoting innovation in the country. Now, with the tremendous growth in government at national, state, and local levels since the end of World War II, the government has become a major funder of educational development in the country. Real per capita federal expenditures increased 3.5 times from just below $2,000 per capita in 1947 to $7,100 per capita by 2004, and 7 of 16 federal cabinet departments were added since 1952: Health and Human Services (1953), Housing and Urban Development (1965), Transportation (1966), Energy (1967), Education (1979), Veterans Affairs (1987), Environmental Protection Agency (1990), and Homeland Security (2002). These newer cabinet departments, along with the departments established earlier, put substantial resources into K–12 science education, which together dwarf the investments of all current foundations in this arena. In 2004 alone, for example, 13 federal agencies spent about $2.8 billion to fund over 200 programs related to STEM education, covering the spectrum from elementary through graduate levels (Ashby, 2006; Garrett & Rhine, 2006). Due to the size of government and its spending, it is unlikely that any set of foundations today can exercise the amount of influence in society that Carnegie, Ford, and Rockefeller were able to exert in the 1940s through the early 1960s when federal funding made up a much smaller percentage of total funding for education.

In addition to supporting innovative programs in science education, foundation officials also testify before relevant decision-making bodies such as Congress, state legislatures, and local school boards. Foundations are uniquely positioned to obtain information highly relevant to decision makers through their various networks, information these decision makers often do not have ready access to themselves. In the sections that follow, we illustrate some of the areas in which foundations have been involved in the support of science education. Some of this support operates directly, but much of it operates indirectly through institutions foundations create to benefit science education along with other fields of interest to society (e.g., public libraries). Our lists and selections of examples are illustrative rather than exhaustive, and they point to a number of areas where researchers might fruitfully probe to expand our knowledge of these intersections.

Support for Public Libraries

Historically, much of the support of foundations for science education has been through contributions made to public libraries, museums, and science centers. Benjamin Franklin, for example, established a fund of 4,000 pounds sterling to support apprentices in technical trades (Langley, 2008). As the need for apprentices decreased over time due to changes in occupational development patterns, a portion of the vastly increased sum ultimately went to establish public libraries in Philadelphia as well as the Franklin Institute, the first professional organization of mechanical engineers and professional draftsmen. The Franklin Institute eventually became both an active science research organization and a science learning center, active to this day.

In 1886, Andrew Carnegie gave a gift of nearly $2 million for library buildings in cities where Carnegie industries had facilities. Believing that libraries could survive only if they had the support of their local communities, he insisted that the communities themselves had to stock the libraries with books and staff them. He also commissioned a study of libraries in the United States, which was published as *The American Public Library and the Diffusion of Knowledge* by William S. Learned in 1924. The report articulated the important roles that libraries play in communities and called for the professionalization of library staff to better serve their patrons (Cremin, 1988). His largesse provided books and other materials on the sciences to untold millions of readers over the years, including children, youth, and their teachers.

Carnegie's attitude toward and support of libraries was, in turn, an inspiration to the W. F. Kellogg Foundation, which launched a highly popular program among rural communities throughout Michigan in the 1950s. Under this program, for every five outdated books that were turned in by local libraries, one free new book was supplied by foundation funds, a boon to subject areas such as the sciences where knowledge increased rapidly (W. F. Kellogg Foundation, 1955). The Bill & Melinda Gates Foundation has been a worthy national successor to Carnegie's support of libraries, with its' digital libraries initiative that has equipped libraries across the country for the digital age, greatly expanding its reach to a truly global audience of users. More recently, the Gates Foundation has concentrated on supporting libraries in developing countries, where libraries continue to play a vital role in local education, economic development, and access to resources. Clearly, libraries through their print and multimedia resources, specialized experts, and public programs have been and remain a vital means by which individuals learn about science.

SUPPORT FOR SCIENCE CENTERS AND SCIENCE MUSEUMS

One of the most significant and visible contributions of foundations to science education in the United States has been their support of informal science learning. By establishing and maintaining science centers and science museums, foundations have had a tremendous impact on enhancing the public's understanding and engagement in science and technology. For example, in 1895, Andrew Carnegie deployed part of his vast fortune to establish the Carnegie Institution in Pittsburgh, which has grown to encompass four museums including the Carnegie Science Center and the Carnegie Museum of Natural History, the latter being one of the six largest museums in the nation with over five million specimens. James Smithson, a British citizen, provided the initial endowment that gave birth to the Smithsonian Institution complex of museum and research facilities in Washington, DC. The Field Museum in Chicago benefited enormously in its earliest days from the contributions of railroad supplies magnate and board chair Edward E. Ayer, who then convinced Marshall Field, the owner of Chicago area department stores, to provide an endowment of $1 million in 1894 with a subsequent $8 million to be provided to the museum upon Field's death. Similarly, the struggling Chicago Museum of Science and Industry was the recipient in 1926 of a $3 million pledge from Julius Rosenwald, chairman of Sears, Roebuck, and Company. Countless other science museums were started by philanthropists who sought to engage the public in science education. A driving motivation to do so was a by-product of living in an era when scientific advances and technological innovations fed into a general belief that science and technology could forever eliminate pressing social problems and alleviate all human suffering. The widow of Russell Sage, a nineteenth century New York financier who made a fortune investing in railroads, inherited $75 million in 1906 (about $1.42 billion in 2000 dollars) that helped launch and/or advance the American Museum of Natural History ($1.6 million), the New York Botanical Garden (over $800,000), the New York Zoological Society (over $800,000), as well as the National Audubon Society and its Junior Audubon Leagues, unleashing untold benefits in science learning to thousands of young people, families, and adults (Crocker, 2003).

The Carnegie Corporation, in a similar fashion to what they did for libraries, commissioned a national survey of almost 2,500 museums. The report appeared in 1939 and helped solidify the museum movement, and it advanced thinking about how museums could be used to advance the education of the public (Cremin, 1988). The post World War II era saw a proliferation of museums across the nation with philanthropic support being either the major driver or used to leverage local, state, or national sources of funding. Many of these museums included science in their exhibits and educational missions. In 1989, the Carnegie Corporation commissioned another national study of museums. One report from the study, by J. Myron Atkin and Ann Atkin of Stanford University in 1989, focused on "science-rich" institutions that had received support to engage with local schools (Atkin & Black, 2003). Their report presented 30 specific case studies. The report argued for additional federal, state, and foundation support for these institutions and the local alliances they represented in order to improve science learning on the part of students, teachers, and communities (Atkin & Atkin, 1989).

Science and technology museums in the United States continue to play a vital role in science education for children, youth, and the general public with many like the Exploratorium in San Francisco, the Boston Museum of Science, the Chicago Museum of Science and Industry, and the Lawrence Hall of Science justly world famous for their seminal contributions to science education over decades both in terms of programs they offer at the local and national level and the contributions they make to science education policy discussions at all levels of the educational system. These four museums, and many like them across the nation, keep their doors open, their exhibits fresh, and their annual visitors engaged because of the contributions of foundations, individual benefactors, and the general public. The Association of Science and Technology Centers (2009) founded in 1973 comprises 540 members in 40 nations. The association's 2009 annual sourcebook of statistics for science and technology center reports that science and technology centers on average receive 43% of their support from fees, rentals, and earned income, 25% from public funds, 28% from private donors (including foundations), and 4% from endowments. It estimates that science and technology centers in the United States provided learning experiences for 13.4 million students on school-sponsored field trips in 2008.

Support for Scientists and Scientific Research Organizations

A by-product of foundation and government funding of scientists engaged in basic and applied research in the physical and life sciences has been that these scientists frequently also have ventured into the public policy arena to advocate for various approaches, policies, and programs in science education, although with decidedly mixed effects (DeBoer, 1991; Kelves, 1992; Montgomery, 1994; Rudolph, 2002). Foundations also have contributed indirectly to this participation in science education by scientists through establishing and supporting scientific research organizations, which then influence science education policy and practice. The Carnegie

Corporation of New York, for example, created a series of Carnegie Institutions that focus on basic scientific research, but many of their chief scientists over the years also have been invited to chair or participate in important national discussions concerning science education policy.

Carnegie also gave $5 million in 1919 to help support the nascent National Research Council (NRC), which had been founded in 1916, with funding from the Engineering Foundation, to become the operating arm of the National Academy of Sciences and to direct the Academy's various research activities. The National Academy of Sciences (NAS) had been established by a congressional charter and signed by President Abraham Lincoln in 1863 to address critical national issues and give advice to the federal government and the public on scientific and technical matters. The National Academy of Sciences was composed of a select group of scientists who were chosen on the basis of their scientific accomplishments. During World War I, there was an increased need in the country for advice from the scientific and technical community, so, in response, the NRC was established by the Academy at the request of President Wilson to recruit specialists from the larger scientific and technological communities to participate in that work. Today, the National Research Council carries out most of the studies initiated by the NAS as well as those of the National Academy of Engineering (NAE) and the Institute of Medicine (IOM). For more on the history of the NRC, go to http://www.nationalacademies.org/about/history.html. Carnegie played a significant role in the development of the NRC both through its financial support and through senior staff of the foundation promoting the idea that the federal government needed to make more effective and coordinated use of the insights of the nation's scientific and technical community. The National Academy of Engineering was added in 1964 and the Institute of Medicine in 1970. Together the four organizations are collectively referred to as the National Academies, and the NRC today is administered jointly by the NAS, NAE, and IOM through the NRC Governing Board. (See also Earle, Chapter 5, this volume.)

The Rockefeller Foundation followed the lead of Carnegie in supporting the NRC, and eventually the federal government began to support the NRC directly as the major independent research unit advising the federal government on matters related to science, technology, engineering, and medicine, including matters related specifically to science education policy (Lagemann, 1989). In addition to issuing hundreds of reports in areas of critical national concern in science, medicine, and engineering, the NRC also has published many notable works that have affected science education policy, including: *The National Science Education Standards* (1996), *Inquiry and the National Science Education Standards* (2000), *Technically Speaking: Why All Americans Need to Know More about Technology* (2002), *How Students Learn Science in the Classroom* (2005), *Systems for State*

Science Assessment (2006), *Tech Tally: Approaches to Assessing Technological Literacy* (2006), *Taking Science to School* (2007), *Rising Above the Gathering Storm: Energizing and Employing America for a Brighter Economic Future* (2007), and *How People Learn Science in Informal Settings* (2009). Foundations, including Carnegie, but now joined by many additional foundations who have interests in these domains such as Gates, Hewlett, Packard, Kauffman, MacArthur, Wallace, Sloan, and Mellon, fund these peer-reviewed studies organized by the NRC. The study reports often are highly influential in shaping and informing public policy decisions, stimulating public discourse, and providing arguments needed to justify federal expenditures for the recommended activities.

Carnegie also continues to use its historic role of forming commissions itself to develop high profile reports on science education, which in many respects rival those of the NRC. The foundation's most recent foray into mathematics and science education is the Carnegie Corporation of New York Institute for Advanced Study's Commission on Mathematics and Science Education which released its full report in July 2009, *The Opportunity Equation: Transforming Mathematics and Science Education for Citizenship and the Global Economy.* Based on its assembling and analyzing data from a variety of sources, the commission report advocates for "fewer, clearer, and higher" common standards across states, improving teaching in these disciplines, and redesigning schools and systems.

In a similar manner to the NRC, the National Bureau of Economic Research owes its origins to the Commonwealth Fund, which provided an initial grant in 1920, and the Carnegie Corporation, which provided even more generous support a few years later (Lagemann, 1989, pp. 59ff.). Carnegie also was instrumental in launching the Institute of Economics, which became the Brookings Institution after a series of mergers (Lagemann, 1989). These two economics-oriented research organizations have, from time to time, issued substantial and influential reports that relate to science education policy concerns. A recent report by the Brown Center on Education Policy at the Brookings Institution, *How Well are American Students Learning* (2009), for example, provides policy guidance on a number of critical issues facing science and mathematics educators.

Support for the Public Accountability Movement Through Student Assessment

The Origins of Large-Scale Testing

The twentieth century can be thought of as the century of measurement in education reform. The Carnegie Corporation of New York and its subsidiaries were early and enthusiastic funders of research and development on student assessment. Their earliest funding provided support to the College Entrance Examinations Board (CEEB) in 1917. The CEEB had been formed in 1900 by the Association of Colleges and Secondary Schools of the Middle States and Maryland by a committee of 12 collegiate members of the association and representatives from secondary schools. It formed as a response to the varied admission requirements that colleges and universities had at the beginning of the twentieth century, which made it difficult for secondary schools to adequately prepare their students for the different requirements. (See http://ia341316.us.archive.org/3/items/cu31924031758109/cu31924031758109.pdf for a list of committee members and proposed areas for testing.)

The following year, the CEEB provided its first tests, a weekend battery of essays in various subjects (Bennett, 2005). Some years later, CEEB also funded Carl Brigham of Princeton University to develop a psychological examination that would accompany the essays, which he produced in 1926. Brigham, a strident eugenicist at an earlier point in his career, maintained in the first testing manual of this new Scholastic Aptitude Test (SAT), that "there is a tendency for … scores in these tests to be associated positively with … subsequent academic attainment … [but] … this additional test … should be regarded merely as a supplementary record" (Bennett, 2005, p. 5). Nearly 9,000 high school students took the SAT in 1931 (Kridel & Bullough, 2007, p. 237). Providing support for work of this nature sensitized Carnegie officials to the importance of assessment and the manner in which tests could influence curricular and instructional decisions of teachers. This also was a period where the ascendancy of the sciences, including the young but rapidly developing area of psychology, provided hope that many formerly intractable problems would yield to advances in scientific understanding of human behavior.

As was true of the initial CEEB tests, formal statewide testing programs in secondary schools up through the 1920s consisted largely of essays, brief written responses, and drawings provided by students upon demand with appropriate labeling. (See, e.g., Cheek & Quiriconi, Chapter 7, this volume, for a discussion of early testing in New York.) As part of a larger movement within American culture toward efficiency and standardization, a debate ensued about the utility of the new forms of testing that relied on true/false, multiple choice, and fill-in-the-blank modes of assessment. Carnegie, already supporting work on this form of assessment in higher education, sponsored a series of conferences to take up questions of assessment in secondary schools as well. These conferences included an international component because there was considerable concern during this period about whether American students were being as well educated as their counterparts in Europe. Under the direction of Paul Monroe of Columbia University Teachers College, these meetings were intended to inform Americans about European approaches to student assessment, although the meetings also involved exchanges about curriculum and instruction in addition to assessment (Monroe, 1931).

The Carnegie Foundation for the Advancement of Teaching then commissioned a lengthy report from Professor Isaac Leon Kandel at Teachers College (1936) to provide a fuller exploration of these matters. They also funded a series of practical studies in conjunction with the State Department of Public Instruction in Pennsylvania between 1925 and 1938 that involved 45,000 students from high school through college and a smaller cohort of students who were beyond college. Altogether, 55,000 high school achievement tests were administered, which included about 70% of all seniors in secondary schools in Pennsylvania (Kridel & Bullough, 2007). The so-called Pennsylvania Study, reported on in *The Student and His Knowledge* by William Learned and Ben D. Wood in 1938, led to the conclusion that large test batteries could be employed to measure student knowledge in place of essays, grades, or Carnegie units, all of which were found to be less reliable predictors of student future achievement in comparison to the large-scale test batteries (Norton, 1980; Kridel & Bullough, 2007). Learned was a full-time employee of the Carnegie Foundation for the Advancement of Teaching's Division of Educational Enquiry, and Wood was a professor at Columbia University Teachers College. Wood had been working with Edward Thorndike at Columbia on multiple choice achievement tests since the late 20s in a series of investigations sponsored by the CEEB (Bennett, 2005)

Establishing Large-Scale Testing Companies

Along with the debates and experiments that took place in educational testing during the first few decades of the twentieth century, we also see the beginnings of the development of the large-scale testing companies. Foundations were very instrumental in the development, organization, and early support of these testing companies.

Cooperative Testing Service. Carnegie, and the Rockefeller Foundation through its General Education Board, helped to create a number of testing services during the late 1920s and 1930s to address new ways to identify students who were most suitable for college admission and to experiment with various new approaches to student assessment in colleges and secondary schools. Sometimes collaborators and sometimes competitors, these testing organizations laid the groundwork for developments in testing throughout the twentieth century, and many of their innovations are still with us today.

The American Council on Education (ACE) was established in 1918 as a federation of fourteen national educational associations to coordinate higher education's resources so as to meet national wartime needs and to ensure that wartime conditions would not too sharply curtail the number of students who would attend colleges and universities. Their Advisory Committee on College Testing, chaired by Max McConn, Dean of Education at Lehigh University, arranged for ACE in 1924 to commission Louis Thurstone of the Carnegie Institute of Technology (later merged with the Mellon Institute of Industrial Research to become Carnegie-Mellon University) to develop a new college admissions test that focused not on measuring discipline-specific knowledge acquired in high school but on intelligence. The test was modeled after the Yerkes' 1918 Army Alpha Test, which had been used by the military to screen and sort recruits. A few years later, the ACE Advisory Committee on College Testing recommended that ACE create a Cooperative Testing Service (CTS). With Rockefeller funding, the testing service was established in 1930 under the direction of Ben Wood from Teachers College. The CTS, building upon Wood's work on achievement tests for the CEEB, produced standardized objective achievement tests, including the tests that were administered as part of the Pennsylvania Study (the program directly run by the Carnegie Foundation for the Advancement of Teaching) and the College Sophomore Testing Program sponsored by Carnegie involving 140 institutions.

Issues of Standarization. Both the College Entrance Examination Board, with its use of the SAT and essay tests, and the Cooperative Testing Service, with its use of standardized achievement tests and Thurstone's Psychological Examination, argued that they were freeing high schools from slavish attention to the Carnegie unit as the chief measure of high school academic attainment as well as opening up new opportunities for students who had ability but lacked opportunity to learn (Kridel & Bullough, 2007). It should be noted that the term "Carnegie unit" is a little bit of a misnomer because the standard actually was created by the National Education Association (NEA) Committee on College Entrance Requirements and the North Central Association of Colleges and Secondary Schools to represent 120 teacher contact hours (now redefined as 130 contact hours). It is called the "Carnegie unit" because the Carnegie Foundation for the Advancement of Teaching strongly encouraged its use once it was created since it allowed colleges and universities to interpret more readily high schools students' exposure to subject matter content, and also provided standardization across and within high schools. This move toward standardization was actively resisted by many of the more innovative secondary schools, including the most experimental of the schools, well known because of their participation in the "Eight-Year Study."

The majority of high schools that took part in the Eight-Year Study did not accept the argument that standardized tests would relieve them of the pressures of standardization. With the help of Ralph Tyler, they resisted the movement on the part of the Carnegie Foundation and the testing services to subject their students to these batteries of external high school examinations, taking a final and definitive stand against the Carnegie Foundation and its agents in 1934. The result of the schools' unwillingness to have their students tested was that Carnegie elected not to supply further funds for the Eight-Year Study. With the consent of Carnegie, the Rockefeller Foundation's General Education Board took over funding of the Study (Kridel & Bullough, 2007). The fear of principals and teachers was that these standardized tests would narrow the curriculum and place it beyond the control of those persons who best knew their students and local contexts. Tyler provided the ammunition the schools needed, arguing that tests should be used principally to inform teaching and learning, not for standardization purposes. He used his central role as an expert advisor to the Eight-Year Study to establish the field of educational evaluation, using his study of the participating schools as his initial laboratory for what would be a long and distinguished career in evaluation and education reform (Kridel & Bullough, 2007).

Educational Records Bureau. The Educational Records Bureau (ERB) was established with Carnegie funding in 1927, originally to organize conferences where educators could discuss the various issues related to the use of intelligence tests, achievement tests, and aptitude tests. Acting also as a brokerage service for schools, ERB identified and sold tests to schools and helped develop and promote the Cumulative Record Card, which reached its zenith through the work of Eugene Randolph Smith (Kridel & Bullough, 2007). The Cumulative Record Card provided a way to systematically record important information about students that could then be passed on the future teachers to aid them in guiding each student's educational development. Ben Wood of Teachers College was directly involved in ERB as well, and the Cumulative Record Card was extensively employed in the Carnegie-funded Pennsylvania Study. Carnegie also supported the development by ERB of machine-aided marking of achievement tests with the IBM 805 first being used in 1936 to score the enormous number of achievement tests administered in Pennsylvania and elsewhere under the auspices of the Cooperative Testing Service of the American Council on Education.

Educational Testing Service. James Conant, then President of Harvard University, suggested at a 1937 Educational Records Bureau conference that the various national testing services should be combined into a single "nationwide cooperative testing service." This call, echoed by Learned of Carnegie and Wood who headed the ACE's Cooperative Testing Service, was strongly resisted by Princeton professor Carl Brigham (creator of the SAT), who believed that such a move was premature given the many problems with testing and measurement from a scientific standpoint. He also was concerned that lack of competition would stifle further innovations; that marketing would take precedent over careful investigation and requisite and impartial scrutiny; and that the move would jeopardize the primary purpose of assessment, which was to strengthen classroom teaching and learning, not to administer external examinations merely to make decisions about college admissions. Brigham's concerns, clearly articulated and marshaled with many examples of "psychologists [who] have sinned greatly in sliding easily from the name of the test to the functions or trait measured" led to the postponement of any merger of the testing services until after his death (Bennett, 2005, pp. 6-7).

Then, after protracted negotiations among CEEB, ACE, and the Carnegie Foundation's Division of Educational Enquiry orchestrated by Henry Chauncey, the three organizations agreed to turn over various testing programs and assets to a new entity, the Educational Testing Service (ETS). The new testing service was granted a charter in December 1947 by the New York State Board of Regents. The CEEB provided ETS with the SAT and the Law School Admission Test along with 139 employees and its Princeton office. Carnegie provided 36 employees, its Pre-Engineering Inventory, and the GRE. ACE donated its entire Cooperative Testing Service, the National Teacher Examinations, Thurstone's Psychological Examination, and 37 employees. Henry Chauncey became the first President, and James Conant was made the Chairman of the Board of Trustees (Bennett, 2005). Carnegie also provided start-up funds for the ongoing work of ETS. The Educational Testing Service has now grown to a global company with 9,000 locations in 180 nations, 5,000 employees (including its wholly-owned subsidiary Prometric), and an annual administration of 50 million achievement and admissions tests, as well as ongoing work to create new forms of measurement, new teaching tools, and new applications of technology in assessment. The organization's IRS reporting form 990 for 2007, the latest publicly available, shows revenues in 2006 of $880 million, expenses of $786 million, and net assets of $324 million.

National Assessment of Educational Progress (NAEP)

The story of NAEP begins with the appointment of Francis Keppel, Dean of the Harvard Graduate School of Education, to be U.S. Commissioner of Education in 1962 in the Department of Health, Education, and Welfare. Keppel's father, Frederick, had been the longtime President of the Carnegie Corporation of New York (1923–1941). Keppel had received numerous grants from Carnegie, Ford, and Rockefeller during his years at Harvard. Shortly after taking up his new responsibilities, he discovered in the Congressional charter of 1867, which governed the operations of his office, that the Commissioner was to "determine the progress of education." In the spring of 1963, he asked Ralph Tyler, then Director of the

Center for Advanced Study in the Behavioral Sciences at Stanford University, to propose a way this could be accomplished. Tyler responded in July of that year with a detailed memorandum, and in October Keppel secured from John Gardner, then President of the Carnegie Corporation of New York, money for two conferences to explore Tyler's proposal.

By June of 1964 discussions and preparatory work for a national test had proceeded to the point where Carnegie authorized a grant of $100,000 for the establishment of an Exploratory Committee for the Assessment of Progress in Education (ECAPE), and Ralph Tyler was named the chair for its first meeting in August. A Technical Advisory Committee under the Princeton statistician John Tukey was established in February of 1965, and contractors were chosen to develop national assessments in Citizenship (American Institutes for Research), Reading (Science Research Associates), and Science (ETS). ECAPE itself was incorporated by the New York Board of Regents (who had earlier granted a charter to ETS) in June of 1965 for 5 years, and in December the Ford Fund for the Advancement of Education (FAE) provided a grant of $496,000 alongside a new grant from the Carnegie Corporation of $2.3 million. Technical and operational work on these national assessments of education progress continued into 1966.

The first opposition to the testing program surfaced when the Executive Committee of the American Association of School Administrators (AASA) urged its membership to refuse to participate in any work regarding NAEP, a position they later softened by February of 1967 when they established a joint committee with ECAPE to explore participation. Then, in July of 1967 the National Education Association passed a resolution to "withhold cooperation" from NAEP (Lehmann, 2004). Opposition to NAEP was at least in part opposition to the expanding role of the federal government in education.

Education Commission of the States (ECS). As these events were unfolding, there was a parallel movement afoot to counterbalance the expanding role of the federal government in education, which had grown significantly under the Johnson Administration. The formation of an organization of states was first broached in 1964 in a Carnegie-funded study by James B. Conant called *Shaping Educational Policy.* Conant argued that such an organization: (1) would help states with diverse interests, needs, and traditions speak with a common voice in Washington, (2) facilitate cooperation and collaboration in education, and (3) focus national attention more fully and precisely on the needs of education. Beginning in 1965, John Gardner, President of the Carnegie Corporation, worked with Terry Sanford, former Governor of North Carolina, to take the necessary steps to realize Conant's dream of such an organization of states. They began working with governors, state education department officials, and legislators to draft a Compact for Education. By 1967, the Compact had been endorsed by representatives from all 50 states, approved by Congress, and was able to open offices in Denver with support from the Carnegie Corporation. By 1969, the Education Commission of the States (ECS) had secured substantial grants from both the Ford and the Carnegie foundations and from various state legislatures.

Although private foundations are forbidden by the IRS tax code to lobby legislators regarding specific pieces of legislation, they can provide information to policymakers and the public that is influential in shaping opinions and informing actions. The creation of the Education Commission of the States is a good example of how foundations sometimes perform this advocacy role by giving birth to new institutions, like ECS, that subsequently directly engage in the creation of public policy. The chief purpose of ECS in its earliest days was to overcome opposition on the part of administrators and teachers to the administration of NAEP. This opposition was often part of a resistance to the massively enlarged federal role in education (Buss, 1980; Weischadle, 1980), which certain leaders hoped to mitigate. In April, 1969, ECS ultimately agreed to govern NAEP. Because of that decision, the concerns of states and local school districts were effectively assuaged, and the first administration of NAEP for 9 and 13-year-olds took place in December 1969 in citizenship, science, and writing.

Foundation interest in student and teacher assessment, along with more recent developments in public accountability brought on by federal education legislation (No Child Left Behind Act of 2001), continues unabated as foundations, both regional and national in scope, fund small-scale and large-scale studies, pilot efforts, and conferences that take up issues about the worth, saliency, and impact of assessment policies and practices, including those targeted to science education. The Carnegie Corporation of New York, for example, funded the recent *Meaningful Measurement: The Role of Assessments in Improving High School Education in the Twenty-First Century* from the Alliance for Excellent Education (Pinkus, 2009) returning to a theme it has revisited several times since its earlier work in the 1920s and 1930s.

Support for Science Teachers

Summer Institutes for Teachers

The recognition that science teachers needed advanced training in science and ways to maintain currency in their science knowledge throughout their teaching careers was realized early on by heads of major American foundations, in part because many of them had scientific or technical backgrounds, or they realized through their involvement in

World War II the vital importance of these fields and the difficulty of staying abreast of new developments in fields which advance so rapidly.

The General Electric Corporation (GE) launched one of the first summer schools for high school science teachers in 1945 at Union College in its backyard of Schenectady, New York. The 6 weeks were devoted to modern physics content, unfamiliar to virtually all high school science teachers at that time because much of it had developed rapidly during the war due to the Manhattan Project, radar, and other top secret research and development efforts, and hundreds of teachers applied for the forty slots. This interest led GE to expand the program to three other campuses the following year, and its name became a by-word among science teachers, especially given that a tour of the nearby GE facility was included as part of the package. Westinghouse launched its own program with the Massachusetts Institute of Technology in 1949, also focused on physics. The Ford Foundation followed with a program for high school physics teachers at the University of Minnesota in 1952. Then, only 3 years after its founding, the fledgling National Science Foundation picked up the banner and began funding summer institutes for science teachers in 1953 (Rudolph, 2002). It should be noted that there is no evidence during this period that any of the funders believed that science teachers needed help with instructional or assessment strategies. The focus was solely on keeping teachers up on the content knowledge of the disciplines. (See also Chapter 3 in this book, Kahle and Woodruff.)

New Certification Programs

The Ford Foundation also was influential in advancing the Master of Arts in Teaching (MAT) degree for able liberal arts students as a means of getting better-qualified teachers into school teaching, especially in technical areas. The program started in Arkansas as a pilot program in the early 1950s with funding from Ford, but then Francis Keppel, Dean at the Harvard School of Education, recommended to the Ford Foundation that it support a national program. Ford responded by investing $29 million, beginning in 1953, to spread the program to 52 college and university campuses, including Harvard, Reed, Duke, Wesleyan, Middlebury, Emory, and Johns Hopkins (Buss, 1980; Woodring, 1970). Today MAT programs remain a fixture in many colleges and universities. Another legacy of these efforts to attract liberal arts students into teaching is Teach for America (TFA). The TFA program was initially funded by various foundations (including the Carnegie Corporation of New York, the Goldman Sachs Foundation, the Walton Foundation, the Robertson Foundation, and the Joyce Foundation) and now draws upon state and federal funds in addition to those from foundations for its support. Although not limited to attracting science teachers, TFA attracts many able STEM bachelor's degree graduates from highly selective universities and colleges into its applicant pool and annually places them in the nation's high need urban and rural science classrooms following a condensed and accelerated program to prepare them to teach.

Setting Standards for Teaching and Teacher Education

In 1986, a Carnegie Commission on Education and the Economy published *A Nation Prepared*, which reaffirmed the teacher as the "best hope" for ensuring educational excellence in elementary and secondary education. An outgrowth of this work was Carnegie's support for the establishment of the National Board for Professional Teaching Standards (NBPTS) the following year. The NBPTS sets standards for attaining teaching certification that reflects a high degree of excellence. It is a performance-based system of attainment. Candidates must submit a portfolio of their teaching, demonstrate their content knowledge, and be evaluated by a group of their peers. The intent is to attract and retain superior teachers by establishing high standards of excellence. The idea was first broached by American Federation of Teachers (AFT) President Albert Shanker in 1985. The effort resulted in a set of criteria for what a high quality teacher should know and be able to do at various levels within elementary and secondary education. Over 74,000 teachers have successfully demonstrated their proficiencies since the first certificates were offered in 1987. This work also has influenced states in making changes to their regular science teacher certification systems and teacher preparation institutions to develop more outcomes-based science education programs. Many regional and national accreditation programs have also been influenced by the work of NBPTS.

Founded in 1954, the National Council for the Accreditation of Teacher Education (NCATE) is one of two recognized professional accrediting bodies for colleges and universities that prepare teachers and other professional personnel for work in elementary and secondary schools. The other accrediting organization is the Teacher Education Accreditation Council (TEAC), which originated in 1997 due to dissatisfaction with NCATE. NCATE is a coalition of 33 professional organizations of teachers, teacher educators, content specialists, and local and state policymakers. NCATE received its first foundation funding in 1991 and has received over $10 million from a variety of foundations including the Carnegie Corporation of New York, Ford, and Hewlett. TEAC received substantial initial support from the Pew Charitable Trusts and additional support from the John M. Olin Foundation, Atlantic Philanthropies, and the Carnegie Corporation of New York. Recently, the foundations that fund the two organizations proposed a merger and agreed to support merger discussions and to move it forward. In 2009, the two accrediting organizations jointly announced that they would work on a merger over the coming year as their goals and approaches are now more closely aligned. We see in this development once again the unique role that foundations play in fostering innovation and creating partnerships. For the Carnegie Corporation, supporting the improvement of teacher preparation programs through NCATE and TEAC was a continuation of its earlier investment in James B. Conant's 1963 study, *The Education of American Teachers*, which offered policy suggestions for increasing standards, strengthening and rationalizing control of teacher certification, improving continuing and in-service education, developing true reciprocity among states, and a wealth of other changes that were based on a 2-year study of certification policies in 16 states and personal visits by Conant to 77 teacher preparation institutions around the country. It strongly endorsed "the use of summer institutes for bringing teachers up to date in a subject-matter field" as "perhaps the single more important improvement in recent years in the training of secondary school teachers" (Conant, 1963, p. 207f.). The National Science Foundation as well as many other federal agencies in more recent years has been especially prominent in the creation and expansion of these content-focused programs for teachers.

Professional Development for Inservice Teachers

The Noyce Foundation, the Knowles Science Teaching Foundation, and the Merck Institute for Science Education are examples of foundations decidedly focused on fellowships and/or professional development for science teachers. While the Noyce Foundation's principal investment is in student achievement in mathematics, science, and literacy in the Silicon Valley, it also has supported the work of New Leaders for New Schools and Teach for America. In addition, the Robert Noyce Teacher Scholarship Program, which was first authorized under the National Science Foundation Authorization Act of 2002 and reauthorized in 2007 under the America COMPETES Act, provides federal scholarships and stipends for teachers of STEM subjects who in turn complete two years of teaching in a high-need school district for each year of support The Knowles Science Teaching Foundation supports individuals with strong mathematics and science backgrounds to become teachers. Support is provided over 5 years and covers the costs of obtaining the necessary preparation, the costs of credentialing, and support until the teacher gains tenure. Recipients also are provided access to relevant research and further learning opportunities. The Merck Institute for Science Education has a Merck Fellows program and a Teacher Leader Institute, both of which help teachers or prospective teachers increase their knowledge and skills at employing inquiry-based teaching strategies in their classrooms.

Teacher Retention

Teachers also need support of a different kind for teaching to be an attractive profession. With that in mind, Carnegie established a pension fund for college professors in 1918 that ultimately became the Teachers Insurance and Annuity Association—College Retirement Equities Fund (TIAA—CREF) to increase the prestige and security of the profession. This early fund became a model upon which many state retirement systems for K–12 public school teachers were created to help stabilize the teacher profession. Yet, teacher retention remains a significant issue in K–12 schools. Foundations continue to be significant funders of studies and efforts to understand and improve teacher retention in the sciences as well as other subject areas that are in critical demand or where shortages of supply exist. For example, the Carnegie Corporation of New York and the Rockefeller Foundation jointly financed the National Commission on Teaching and America's Future whose 1996 report, *What Matters Most*, provided a national framework and agenda for teacher education reform across the nation.

Support of Science Curriculum Reform Efforts

In the first 2 decades of the twentieth century, the Rockefeller Foundation was active in reforming the secondary science curriculum, seeking both increased standardization and specialization, including in laboratory science. Concerned about urbanization, the influx of large number of immigrants, the rapid growth of scientific endeavors, and the need for more highly able people in science and medicine, Rockefeller used a national commission and a series of funded projects to introduce more rigor into the school science curriculum, more theoretical work rather than simply science as applied to everyday life situations (a common theme in school curricula in the first half of the twentieth century), argued for standardization of courses of study in the nation's schools with a particular concern for rural America and the changes happening in agriculture, and advocated for laboratory work for students who demonstrated aptitude for science (Heffron, 1988). These activities would set a tone and benchmark for efforts by numerous foundations throughout subsequent decades.

Differentiated Curriculum and the Gifted and Talented

The issue of how to educate students who had special aptitude for science and mathematics has been a concern throughout our educational history, and foundations have lent their support to informing and shaping that debate. Ford, for example, funded a major study of the secondary school curriculum in Portland, Oregon, working with the public schools and nine Oregon colleges and universities. Based on the advice of the 50 participating college teachers, including those in mathematics and science, the goal was to devise an entire curriculum that would fit students with the requisite aptitude for success in college. Their full body of work appeared in 1959 as *The Gifted Child in Portland* (Portland Public Schools, 1959). This focus on special programming for the gifted and talented was consistent with the recommendations of James B. Conant (1959) that came out of his study of American high schools, a study that had been funded by the Carnegie Corporation. Based on visits to 55 high schools in 18 states, the Conant report advocated a much more rigorous set of courses in mathematics and sciences for able students, as well as vocational and career education for less able students. It strongly advanced the notion, first raised in the 1918 *Cardinal Principles of Secondary Education* from the NEA-appointed Committee on the Reorganization of Secondary Education (Spring, 2008), that only large comprehensive high schools could effectively offer this type of differentiated curriculum, leading to further high school consolidation efforts across the country (Hampel, 1986; Montgomery, 1994). This orientation toward differentiated curricula in high schools, with different tracks for different students, was unquestionably embraced by presidents of Ford, Rockefeller, and Carnegie as well as people like Francis Keppel, James Conant, and others who received foundation grants.

Establishing the Need for Curricular Reform

In 1951, the Ford Foundation funded the establishment of the National Manpower Council to help guide the nation in improving its human resources, especially in technical fields. Its 1953 report, *A Policy for Scientific and Professional Manpower,* was widely discussed at the national level. The Council realized from their studies that data on the nation's human resources in scientific, technical, and professional areas was sorely lacking in quality, and they advanced the notion that the federal government needed to work with professional societies to create a much more reliable data gathering capability. This urgent need led to the National Science Foundation being tasked the following year with collecting this information for the first time, an exercise that would ultimately become today's biannual *Science and Engineering Indicators* report. The Council also called for a strengthening of the institutions that provided education and training for scientists, including K–12 schools. They called for fuller and wider support for youth to gain college diplomas in critical fields of study and for better use of the existing talent pool of scientists, engineers, and other science professionals. A Central Intelligence Agency and National Security Council briefing to the Eisenhower administration that same year finally galvanized the necessary support for science education efforts, given that reports of Soviet mastery of advanced scientific and technical subjects in high schools had led to the belief that the U.S. could be overwhelmed by a technically superior foe. After much haggling among federal agencies (including the Department of Defense), the National Science Foundation (founded in 1950) was chosen as the institution that should take up the challenge of improving American science education (Rudolph, 2002).

While these various reports were being generated for federal policymakers, senior officials at Ford, Carnegie, and Rockefeller were convinced that more rigorous and scientifically up-to-date curriculum materials were needed if the Soviet threat was going to be effectively answered. The Carnegie Corporation of New York provided funding for the University of Illinois Committee on School Mathematics (UICSM), organized by Max Beberman late in 1951, which led to a curriculum focused more on mathematics as understood and practiced by mathematicians (including set theory and formal logic) rather than the applied mathematics that is more likely to be used and appreciated by the general public. The deliberations of the Carnegie-supported Commission on Mathematics of the College Entrance Examination Board (CEEB), which began in 1955 and were crystallized in their influential 1959 report, *Program for College Preparatory Mathematics*, recommended a curriculum for college-bound high school students that had a focus on the mathematics and the terminology used by contemporary mathematicians. This report led to even greater investments on the part of Carnegie for the University of Illinois mathematics effort.

By 1959, Carnegie's investments in various mathematics and science curriculum projects were amounting to over $1 million a year (Atkin & Black, 2003; Weischadle, 1980). In 1958, Carnegie started investing in the fledgling Physical Science Study Committee (PSSC) curriculum project that was being developed at the Massachusetts Institute of Technology under the eminent physicist Jerrold Zacharias. With foundation support, Educational Services Incorporated (ESI) was established to administer the PSSC curriculum, and within 7 years there were various additional ESI science and mathematics curriculum development projects in the works. Then, in 1968, ESI merged with the Institute for Educational Innovation to form the Education Development Center (EDC). The EDC has been a major source of national and international influence on science education policy and practice for over 40 years (Weischadle, 1980).

A curriculum development project in the biological sciences was supported by the Rockefeller Foundation in 1954 because of the foundation's strong interest and previous investments in medicine and allied health disciplines. The committee on biology began to discuss what the modern biology curriculum should be, mimicking to a large degree the work of their mathematics and physics colleagues but unable to attain the rapid consensus witnessed in these two other disciplines.

Stimulating Additional Federal Support for Curriculum Reform

In 1958, the Rockefeller Brothers Fund funded the report, *The Pursuit of Excellence: Education and the Future of America*, which questioned whether public education could adequately respond to challenges posed by communism and a changing world. The report, coupled with skillful testimony and orchestrated advocacy on the part of the Rockefeller Foundation, resulted in Congressional passage of the National Defense Education Act, which provided historic levels of support for federal activities in science education. The innovative science and mathematics curriculum projects that already were underway with support from Rockefeller, Carnegie, and Ford provided ideal funding opportunities for the new National Science Foundation to establish its reputation and form connections with leading scientists who were engaged in developing curriculum materials for the schools. The PSSC physics and the Illinois mathematics groups were some of the first recipients of the new funds. Several attempts by Rockefeller officials, as well as Rockefeller grantees, however, to interest NSF in supporting biology curriculum development fell on deaf ears. But the 1957 launch of Sputnik had galvanized a working group of biologists, and with the involvement of the American Institute of Biological Sciences, they finally convinced NSF officials in 1959 to fund a group of eminent biologists to begin serious work on the biology curriculum. The Biological Sciences Curriculum Study (BSSC) was formed, modeled on the influential work of the PSSC physics team.

With the development of these curriculum projects there were serious questions arising about the extent to which students could be expected to learn the scientific information and the technical skills that this new approach to science education demanded. In response, the Carnegie Foundation funded the 1959 Woods Hole conference, chaired by noted psychologist Jerome Bruner, whose conclusions were brilliantly summarized by him in the 1960 classic, *The Process of Education*. It argued that any subject or concept could be taught in an *appropriate* manner to *any* student. This conference profoundly influenced the support that the various new science and mathematics curriculum projects received, and many of Bruner's core ideas remain in circulation in the science education community and are enshrined in curricular and instructional approaches.

Following the strong foundation support of science curriculum during the 1950s and 60s, there was a period of time when foundation support for science education became more limited, but in more recent decades there has been renewed investment on the part of corporate foundations in precollege science education. This change can be traced to the work of the Business Roundtable (BR) that partnered with the President of the United States in the Goals 2000 effort in 1989, coupled with the BR's 1992 publication to its members, *Agents of Change: Exemplary Corporate Policies and Practices to Improve Education*. Members took many of its provisions to heart, and corporate giving to precollege education virtually doubled in the next decade (Himmelstein, 1997). Corporate involvement continues today in light of the No Child Left Behind legislation, and corporations are actively involved in STEM policymaking efforts at national levels through the Business Roundtable, Business-Higher Education Forum, and various industry alliance groups in addition to efforts at state levels in concert with the offices of the respective governors, legislative leadership, state departments of education, and with local school districts where their companies have a corporate presence. The GE foundation alone, for example, invested $19 million in 2007 in elementary and secondary education in its local communities, much of it focused on STEM education; while its corporate parent General Electric Company, gave grants in 2008 totaling $133.7 million to K–12 schools, colleges, and other nonprofit organizations (Barton, Jones, & Combs, 2009; The Foundation Center, 2009b).

Although most of the investment in science education curriculum reform has been through federal agencies since the curriculum reform efforts of the 1960s, foundations continue to be a source of funds for the development of curriculum materials at state and local levels. Many of these curricula are tailored to site-specific situations and environments and customized in a manner that national curricula cannot provide. For example, the Kauffman Foundation provided assistance for several districts in the Kansas City metropolitan area to develop science curriculum materials involving local business and industry partners and tailored to local research and development activities and expertise. The Abbot Fund and the MacArthur Foundation have supported similar work in Chicago. Environmental education materials are another example where foundations across the country support local development efforts due to the specific flora, fauna, geology, and other conditions that are site specific.

Foundation Support for the Development of Science Content Standards

Foundations also have been at the nexus of public debates about science content standards. The modern standards movement in K–12 education, including the sciences, was stimulated and sustained in part by philanthropic support emanating from private foundations. Project 2061 of the American Association for the Advancement of Science (AAAS), for example, has received millions of dollars from national foundations including Carnegie, Mellon, Pew, MacArthur, Noyce, and IBM to develop content standards in science and tools to help teachers implement them. Project 2061, receiving its name from the fact that students in schools in 1989 could be expected to see the predicted perihelion of Halley's Comet in 2061, has produced a wide variety of products over the years focused on science education and STEM more broadly, including *Science for All Americans* (AAAS, 1989); *Benchmarks for Science Literacy* (AAAS, 1993), which heavily influenced the subsequent National Science Education Standards effort of the NRC (1996); the two-volume *Atlas for Science Literacy* (AAAS, 2001, 2007), which map the relationships among standards; highly influential reports on the state of mathematics and science textbooks and curriculum materials in use in the nation's schools; and fundamental research and development projects focused on curriculum, teaching, and assessment.

In addition, the Pew Charitable Trusts and the MacArthur Foundation provided a grant in 1996 for the National Center on Education and the Economy to hold the First Annual Standards-Based Reform Conference. Pew funded the development of national standards in civics and government and in early childhood. In 1996 they provided 4-year grants to seven urban school districts to implement standards-based systemic reform. IBM's chief executive Louis Gerstner (CEO from 1993–2002) was instrumental in pushing for academic content standards and was a critical player in the Goals 2000 conference, which ultimately led to the Goals 2000 Educate America Act in 1994. Xerox CEO, David Kearns (CEO from 1982–1990) was another early corporate champion of school reform, including standards and state testing; Kearns was subsequently named Deputy Secretary of the U.S. Department of Education and served from 1991–1993. Both CEOs as well as others in their companies over the years, have been heavily involved in leading corporate support for standards and assessments at state and national levels.

Finally, the 2009 Carnegie Foundation's *The Opportunity Equation* report discussed earlier continues this long tradition of foundations being centrally involved in debates and efforts focused on standards and assessments. Many foundations will undoubtedly continue to fund studies and supply reports of various kinds to members of Congress, state policymakers, advocates, and lobbyists regarding the state of American education as the reauthorization of federal education legislation moves forward.

Support for New Technologies in Science Teaching and Learning

Educational television was a major interest of both the Ford and Carnegie foundations in the 1950s. Ford alone devoted over $100 million to this effort, a considerable proportion of its total expenditures on education (Buss, 1980; Murphy & Gross, 1966). Groups it helped fund started lobbying as early as 1948 for the Federal Communications Commission to set aside channels for noncommercial use. By 1952, a total of 242 channels were reserved for that purpose, and the foundation's funds were used in a matching program to launch the first set of stations, which grew to 99 by 1964. Carnegie threw its weight behind this effort by creating a commission focused on this topic, which recommended to Congress in 1967 that it establish a Corporation for Public Television. President Johnson proposed its creation with initial federal funds of $9 million, and the Corporation for Public Broadcasting (CPB) came into existence that same year. The Public Broadcasting Service (PBS) stations, especially three of its largest and most influential ones (WGBH in Boston, WNET in New York City, and WETA in Washington, DC) have produced countless science programs to the delight of viewers around the globe—almost all of which were supported by CPB as well as by funds from individual viewers private and/or corporate foundations, and NSF.

More recent years have witnessed the expansion of foundation support into new technologies in addition to television and public radio. Ford's Fund for the Advancement of Education funded the pioneering 1967 study by Don D. Bushnell and Dwight W. Allen, *The Computer in American Education*. Computers in schools, computer software, laboratory probes, microchemistry, Web sites about specific subjects, three dimensional (3-D) immersive worlds, mobile science education applications, GPS units, and a plethora of technology-enabled science learning now occurs in schools, museums, and parks, in part due to the funding by foundations. The MacArthur Foundation alone has pledged $50 million to these types of learning technologies (including gaming and simulations), with many other private foundations also engaged at substantial levels. Corporations also have been heavily involved in this arena, as the 2008 giving patterns to K–12 schools, colleges, and other nonprofit organizations of four corporations in the Information Technology sector illustrate (Barton, Jones, & Combs, 2009) (see Table 4.2).

SUMMARY

This brief overview of the role of foundations in science education policy establishes the many ways in which foundations have conducted and

Table 4.2. 2008 Giving by Corporations in the Information Technology Sector

Corporation	*Cash Giving in 2008*	*Cash and Product Giving in 2008*
CISCO Systems	$54,101,248	$91,959,094
Dell	$20,533,931	$24,540,489
Hewlett-Packard Company	$24,780,685	$52,325,096
IBM	$50,145,954	$186,545,954

continue to conduct their operations in relation to science education policy and practice. It should be obvious from this overview that foundations have played a very important role in American science education and that they were often at the forefront of important policy debates and initiatives. Many American institutions that are today active in science education efforts owe their origins to philanthropic giving. Many major reform efforts in science education were initiated by the prompting of foundations or supported by foundation dollars long before any governmental entity was willing to extend support. Some of these programs were controversial, experimental, or risky. Foundations often choose to invent the new rather than spending their limited resources on the status quo.

Foundations differ substantially in how they conduct their affairs. Some solicit ideas widely from others about what type of work they might undertake in science education. Others make all deliberations and decisions behind closed doors with their boards of trustees and go public only when they have determined their course of action. Some set policies in light of contemporary, widely accepted ideas about what changes are needed in science education. Others are purposefully contrarian or adopt a contrarian position because they believe the widely accepted view is misplaced or ill-informed.

There has been considerable criticism about foundations regarding their lack of transparency, and there are fundamental disagreements even among foundation officials themselves over how transparent foundations can or should be. Some regard their relative autonomy as absolutely critical to their role in birthing the new. Others view this attitude as mere posturing or fundamentally unjustifiable given that a tax benefit is conferred in America for philanthropic actions.

Many foundations view their funds as investments, albeit sometimes risky ones. These foundations tend as a group to believe that they are focusing on root causes and that their role is to eliminate these problems from systems. Some foundations exercise no more risk than their very public government counterparts. For those who do seek a measurable return, their expectations can manifest themselves in written outcomes, matching dollars, required signed commitments of changes that will occur, the feasibility of sustainability beyond the period of the grant, or leverage of subsequent action by others.

As varied as they may be, one thing is certain, and that is that foundations in the United States have a major impact on the development of policy that affects science education, often at the highest level of our society. This is true today and has been true from the earliest days of our country. It also is clear that foundations need a steady supply of new ideas to function well. This is where researchers and educational entrepreneurs can contribute to policy development through the generation and dissemination of new ideas through their own work. In addition, researchers can contribute to the body of new ideas by studying the role of foundations as influencers of policy. More information about the specific roles that foundations can and do play in science education policy is needed. This is a topic that so far has been of little interest to the science education research community. Only a few documented examples exist in comparison to the size and scope of activity within this sector, many of which are cited in this chapter. It is hoped that we have provided some orientation for those who seek to work with foundations, added a modicum of insight for those who already work within foundations, and those who wish to study them.

REFERENCES

Abell, S. K., & Lederman, N. G. (Eds.). (2007). *Handbook of research on science education*. New York, NY: Routledge.

Ashby, C. M. (2006). *Testimony before the Committee on Education and the Workforce, House of Representatives*. Washington, DC: United States Government Accountability Office, GAO-06-702T. Retrieved May 3, 2006, from www.gao.gov/new.items/d06702t.pdf

American Association for the Advancement of Science. (1989). *Science for all Americans*. New York, NY: Oxford University Press.

American Association for the Advancement of Science (1993). *Benchmarks for science literacy*. New York, NY: Oxford University Press.

American Association for the Advancement of Science (2001). *Atlas of science literacy. Volume 1*. New York, NY: Oxford University Press.

American Association for the Advancement of Science (2007). *Atlas of science literacy. Volume 2*. Washington, DC: American Association for the Advancement of Science and National Science Teachers Association.

Association of Science and Technology Centers (2009). *2008 sourcebook of statistics & analysis*. Washington, DC: Author.

Association of Colleges and Secondary Schools of the Middle States and Maryland (1900). *Plan of organization for the College Entrance Examination Board of the Middle States and Maryland and a statement of subjects in which examinations are proposed*. Retrieved from http://ia341316.us.archive.org/3/items/cu31924031758109/cu31924031758109.pdf

Atkin, J. M., & Atkin, A. (1989). *Improving science education through local alliances. A report to the Carnegie Corporation of New York*. Santa Cruz, CA: Network Publications. (ERIC document ED 314 244)

Atkin, J. M., & Black, P. (2003). *Inside science education reform: A history of curricular and policy change*. New York, NY: Teachers College Press.

Barton, N., Jones, C., & Combs, A. (2009, July 2). Charitable giving at 105 big companies. *The Chronicle of Philanthropy*, *8*(9), 11.

Bennett, R. E. (2005). *What does it mean to be a nonprofit educational measurement organization in the 21st century?* Princeton, NJ: Educational Testing Service.

Bork, R., & Nielsen, W (1993). *Donor intent*. Washington, DC: The Philanthropy Roundtable.

Brody, E., & Tyler, J. (2009). *How public is private philanthropy? Separating reality from myth*. Washington, DC: The Philanthropy Roundtable.

Brody, E., & Tyler, J. (2010). Respecting foundation and charity antonomy: How public is private philanthropy? *Chicago-Kent Law Review*, *85*(2), 571–617.

Bruner, J. (1960). *The condition of education*. Cambridge, MA: Harvard University Press.

Bushnell, D. D., Allen, D. W. (with Sara S. Mitter). (Eds.). (1967). *The computer in American education*. New York, NY: Wiley.

Business Roundtable. (1992). *Agents of change: Exemplary corporate policies and practices to improve education*. Washington, DC: Author.

Buss, D. C. (1980). The Ford foundation in public education: Emergent patterns. In R. F. Arnove (Ed.), *Philanthropy and cultural imperialism: The foundations at home and abroad* (pp. 331–362). Boston, MA: G. K. Hall.

Commission on Mathematics of the College Entrance Examination Board. (1959). *Program for college preparatory mathematics*. New York, NY: College Entrance Examination Board.

Conant, J. B. (1959). *The American high school today: A first report to interested citizens*. New York, NY: McGraw-Hill.

Conant, J. B. (1963). *The education of American teachers*. New York, NY: McGraw-Hill.

Cremin, L. A. (1988). *American education: The metropolitan experience, 1876-1980*. New York, NY: Harper & Row.

Crocker, R. (2003). From gift to foundation: The philanthropic lives of Mrs. Russell Sage. In L. J. Friedman & M. D. McGarvie (Eds.), *Charity, philanthropy, and civility in American history* (pp. 199–215). New York, NY: Cambridge University Press.

DeBoer, G. E. (1991). *A history of ideas in science education: Implications for practice*. New York, NY: Teachers College Press.

Fosdick, R. B. (1952). *The story of the Rockefeller foundation: Nineteen thirteen to nineteen fifty*. New York, NY: Harper & Brothers.

Friends of the Society of Servants of God, 75. T.C. 209, 213, 1980

Garrett, T. A., & Rhine, R. M. (2006). On the size and growth of government. *Federal Reserve Bank of St. Louis Review*, *88*(1), 13–20.

Giving USA Foundation. (2009). *Giving USA 2009: The annual report on philanthropy for the year 2008*. Glenview, IL: Author.

Heffron, J. (1988). *Science, southerners, and vocationalisms: Rockefeller's 'comprehensive system' and the reorganization of secondary school science education, 1900–1920.* PhD dissertation, University of Rochester, New York.

Hampel, R. L. (1986). *The last little citadel: American high schools since 1940*. Boston, MA: Houghton Mifflin.

Himmelstein, J. L. (1997). *Looking good and doing good: Corporate philanthropy and corporate power.* Bloomington, IN: Indiana University Press.

Internal Revenue Service. (2009). Life cycle of a public charity/private foundation. Retrieved from http://www.irs.gov/charites/charitable/article /0,,id=136459,00.html

Kandel, I. L. (1936). *Examinations and their substitutes in the United States. Bulletin No. 28.* New York, NY: The Carnegie Foundation for the Advancement of Teaching.

Kelves, D. J. (1992). Foundations, universities and trends in support for the physical and biological sciences, 1900–1992. *Daedalus, 121*(4), 195–235.

Kridel, C., & Bullough, R. V., Jr. (2007). *Stories of the eight-year study: Reexamining secondary education in America.* Albany, NY: State University of New York Press.

Lagemann, E. C. (1983). *Private power for public good. A history of the Carnegie Foundation for the Advancement of Teaching*. Middletown, CT: Wesleyan University Press.

Lagemann, E. C. (1989). *The politics of knowledge: The Carnegie Corporation, philanthropy, and public policy.* Middletown, CT: Wesleyan University Press.

Lagemann, E. C. (Ed.). (1999). *Philanthropic foundations: New scholarship, new possibilities*. Bloomington, IN: Indiana University Press.

Lagemann, E. C., & de Forest, J. (2007). What might Andrew Carnegie want to tell Bill Gates? Reflections on the hundredth anniversary of The Carnegie Foundation for the Advancement of Teaching. In R. Bacchetti & T. Ehrlich (Eds.), *Reconnecting education & foundations: Turning good intentions into educational capital*. (pp. 49–67). San Francisco, CA: Wiley.

Langley, J. M. (2008). The moral of the Ben Franklin story. *The Langley Angle on Philanthropy.* July 22, 2008. Retrieved December 7, 2008, from http://thelangleyangleonphilanthropy.blogspot.com/2008/07 /moral-of-ben-franklin-story.html

Lawrence, S., & Mukai, R. (2009). *Foundation growth and giving estimates: Current outlook, 2009 edition*. Washington, DC: The Foundation Center.

Learned, W. S., & Wood, B. D. (1938). *The student and his knowledge*. New York, NY: Carnegie Foundation for the Advancement of Teaching.

Lehman, I. J. (2004). The genesis of NAEP. In L. V. Jones, I. Olkin (Eds.), *The nation's report card: Evolution and perspectives*. (pp. 25–92). Bloomington, IN: Phil Delta Kappa Educational Foundation.

Monroe, P. (1931). *Conference on examinations under the auspices of the Carnegie Corporation, the Carnegie Foundation, and the International Institute*. New York, NY: Bureau of Publications, Teachers College, Columbia University.

Montgomery, S. L. (1994). *Minds for the making: The role of science in American education, 1750–1990*. New York, NY: The Guilford Press.

Murphy, J., & Gross, R. (1966). *Learning by television*. New York, NY: Fund for the Advancement of Education.

National Center for Charitable Statistics. (2008). *501(c)(3) private foundations in 2008*. Washington, DC: Urban Institute. Retrieved June 26, 2009, from Nccsdataweb.urban.org/pubapps/nonprofit_overview_segment.php?t=pf

National Center for Education Statistics. (2007). *Digest of education statistics 2007*. Washington, DC: U.S. Department of Education.

National Commission on Teaching and America's Future. (1996). *What matters most: teaching for America's future*. New York, NY: Teachers College, Columbia University.

National Manpower Council. (1953). *A policy for scientific and professional manpower.* New York, NY: Columbia University Press.

Norton, R. J. (1980). *Private foundations and the development of standardized tests, 1900-1935*. EdD. dissertation, University of Massachusetts at Amherst.

Oliphant, G. (2004). Making the grade. *h magazine*, 33–37 (the magazine of The Heinz Endowments).

Pinkus, L. M. (2009). *Meaningful measurement: The role of assessments in improving high school education in the twenty-first century.* Washington, DC: Alliance for Excellent Education.

Portland Public Schools. (1959). *The gifted child in Portland*. Portland, OR: Author.

Richardson, T. (2005). Reconfiguring schools as child welfare agencies: Rockefeller boards and the new program in general education at the secondary school. *American Educational History Journal*, *32*(2), 122–130.

Rockefeller Brothers Fund. (1958). *The pursuit of excellence: Education and the future of America. Panel Report V of the Special Studies Project*. New York, NY: Author.

Rudolph, J. L. (2002). *Scientists in the classroom: The Cold War reconstruction of American science education*. New York, NY: Palgrave.

Spring, J. (2008). *The American school: From the Puritans to No Child Left Behind* (7th ed.). Boston, MA: McGraw-Hill.

Stannard-Stockton, S. (2007). The evolution of the tactical philanthropist. In S. U. Raymond & M. B. Martin (Eds.), *Mapping the new world of American philanthropy: Causes and consequences of the transfer of wealth* (pp. 39–47). Hoboken, NJ: Wiley.

The Foundation Center. (2009a). *Top 50 foundations awarding grants for elementary and secondary education, circa 2007*. Washington, DC: Author.

The Foundation Center. (2009b). *Highlights of foundation giving trends*. Washington, DC: Author.

Weisbuch, R. (2007). Robin Hood with a difference: Operating foundations and universities and schools. In R. Bacchetti & T. Ehrlich (Eds.), *Reconnecting education & foundations: Turning good intentions into educational capital*. (pp. 381–407). San Francisco, CA: Wiley.

Weischadle, D. E. (1980). The Carnegie corporation and the shaping of American educational policy. In Arnove, R. F. (Ed.), *Philanthropy and cultural imperialism: The foundations at home and abroad* (pp. 363–384). Boston: G. K. Hall.

W. F. Kellogg Foundation. (1955). *The first twenty-five years*. Battle Creek, MI: Author.

Woodring, P. (1970). *Investment in innovation: An historical appraisal of the Fund for the Advancement of Education* Boston: Little, Brown.

CHAPTER 5

HOW DO FUNDING AGENCIES AT THE FEDERAL LEVEL INFORM THE SCIENCE EDUCATION POLICY AGENDA?

The Case of the National Science Foundation

Janice Earle

INTRODUCTION

Attention to education policy has increased over the past few decades as policymakers at all levels—district, state, and federal—have become interested in the connection between education and economic competitiveness. In the early years of the country, education policy was usually left to local districts, but over the years there has been greater and greater state and federal involvement and control over various aspects of the educational system in the United States (see Cheek & Quiriconi, Chapter 7, this volume). Since the 1980s, the creation of state and national standards, the use of high stakes state assessments, the increased use and reporting of results from the National Assessment of Educational Progress

The Role of Public Policy in K–12 Science Education, pp. 117–146

(NAEP), the establishment of national education goals, and, most recently, the passage of the No Child Left Behind Act of 2001 (NCLB, 2001), all provide examples of how the activity level has heightened. Although the National Science Foundation (NSF) does not propose education legislation, it can have an impact on other issues of policy interest such as curriculum, pedagogy, and assessment across the education system.

With the increased prominence of international assessments such as the Third International Math and Science Assessment in the 1990s, followed more recently by the Trends in International Mathematics and Science Study (TIMSS) and the Program for International Student Assessment (PISA) in the past decade, it is clear that the education policy landscape has become global. Now, international assessments can be used by countries to compare features of their educational system with those of more successful countries, which often influence the direction of education policy in those countries. As in the case of the United States, where these results have been disappointing compared to those of other countries, this often means a greater focus on science and mathematics education and discussions at the national level of policies that will improve the performance of U.S. students (National Science Board [NSB], 2008).

While there has been an abiding concern about the state of reading in the United States, it is science and mathematics education that is currently at the center of many education policy debates. Part of this heightened focus is derived from the current policy emphasis on advancing and strengthening the U.S. science and technology enterprise. A recent example of this is President Obama's address to the National Academy of Sciences, April 29, 2009, where he identified science and technology as top priorities of his administration. Among other things, he announced an intention to finish the 10-year doubling of three key science agencies—the National Science Foundation (NSF), the Department of Energy (DOE) Office of Science, and the National Institutes of Standards and Technology (NIST), along with a joint initiative by the Department of Energy and the NSF to "inspire tens of thousands of American students to pursue careers in science, engineering and entrepreneurship related to clean energy" (Office of the Press Secretary, 2009) According to the *New York Times* (Revkin, 2009), the President "presented a vision of a new era in research financing comparable to the Sputnik-period space race, in which intensified scientific inquiry, and the development of the intellectual capacity to pursue it, are a top national priority." What is interesting about this reference to the Sputnik-era space race is that it was the Sputnik era, beginning in 1957, that propelled the newly created NSF to initial prominence.

This chapter analyzes the various ways that a federal agency such as the NSF impacts STEM education policy and practice. It includes background on the creation and mission of NSF and a discussion of the mechanisms NSF has available to influence the direction of research and development in science education. It covers: (1) how the Directorate for Education and Human Resources (EHR) budget process affects STEM education priorities, (2) the role that Congress can play in identifying priorities and directions for NSF and science education, (3) how NSF staffing policies, specifically the use of the Intergovernmental Personnel Act (IPA) influences the direction that programs take, (4) the kinds of interactions NSF has with various STEM education communities, and (5) how program solicitations created by the Foundation determine the kinds of projects that are called for and supported. Although the NSF's primary mission is to support research, not influence policy, the research that it supports often provides policy makers with the information they need to make informed policy decisions, including policy decisions involving STEM education.

A BRIEF HISTORY OF THE NATIONAL SCIENCE FOUNDATION THROUGH AN EDUCATION LENS

The creation of the National Science Foundation in 1950 was fueled by a concern that the federal government needed to strengthen the scientific research potential in the United States, in part by strengthening the education that students received in science. Much like the arguments heard today, there was a growing national concern following World War II about the ability of the country to maintain technological superiority and so ensure national security. In 1945, at the request of President Truman, Vannevar Bush, who then served as the Director of the Office of Scientific Research and Development, was asked to prepare a report on how to accommodate these concerns. His report, *Science: the Endless Frontier* (Bush, 1945), called for federal support of science and technology, and it called for support of higher education as well. One of the recommendations in the report was to organize a federal agency to fund and lead these research and educational efforts. That recommendation ultimately led to the establishment of the NSF. In that report, Bush argued that attention to science is within the proper sphere of government. "The Federal Government should accept new responsibilities for promoting the creation of new scientific knowledge and the development of scientific talent in youth" (p. 31). The argument was that these responsibilities are the proper concern of government as they affect health, jobs, and national security.

> Science offers a largely unexplored hinterland for the pioneer who has the tools for his task. The rewards of such exploration both for the Nation and the individual are great. Scientific progress is one essential key to our security as a nation and to our better health, to more jobs, to a better standard of living, and to our cultural progress. (p. 2)

In part in response to that report, the NSF was founded in 1950. The 1950 NSF Act said:

> The Foundation is authorized and directed to initiate and support basic scientific research and programs to strengthen the scientific research potential and science education programs at all levels. (National Science Foundation Act of 1950)

The unique opportunities afforded to NSF to affect STEM education today result from these founding documents. They include:

1. The active engagement of experts in the science, technology, engineering, and mathematics (STEM) disciplines to improve STEM learning.
2. A focus both on basic research and research to improve education practice.
3. An emphasis on innovative research (Ferrini-Mundy, 2008).

Although the establishment of the NSF opened the door for significant involvement by the federal government in science education that is tightly coupled with advances in science, early education efforts at NSF were generally confined to promoting science fairs and clubs, and funding summer teacher institutes (Dow, 1991). However, following the launch of Sputnik in 1957 and the signing of the National Defense Education Act by President Eisenhower in 1958, NSF took a more activist approach to education. What followed first was a very significant foray into curriculum development. The School Mathematics Study Group, the Biological Sciences Curriculum Study sponsored by the American Institute of Biological Sciences, the Chemical Bond Approach Project, and the Chemical Education Materials Study were all curriculum development projects aimed at bringing scientists together to improve high school curricula According to Dow, "Never before had university research scholars taken such a deep interest in the improvement of instruction in the public schools" (p. 28).

The early justification for the federal government's involvement in science and science education by way of the NSF continues to influence the Foundation today. The most recent *NSF Citizen's Report* (NSF, 2009b), the Foundation's performance and accountability report to the public, has the subtitle, "Advancing Discovery, Innovation, and Education." The text under NSF's strategic goal on learning states:

> Leadership in today's knowledge economy requires world-class scientists and engineers, and a national workforce that is scientifically, technically, and mathematically strong.Investments in Learning aim to improve the quality and reach of science, engineering and mathematics education and enhance student achievement. (NSF, 2003, p. 10)

The NSF is now the premier federal agency supporting basic research at the frontiers of discovery in the STEM fields (NSF, 2009a). The Foundation is organized along disciplinary lines into six Directorates—Biological Sciences, Computer and Information Science, Education and Human Resources, Engineering, Geosciences, Mathematics and Physical Science as well as several Offices—the Office of Polar Programs, the Office of International Science and Engineering, and the Office of Cyberinfrastructure. The primary directorate concerned with education research and development is Education and Human Resources (EHR), although education programs are also funded across the Foundation.

The mission of the Education and Human Resources Directorate is "to achieve excellence in U.S. science, technology, engineering, and mathematics (STEM) education at all levels and in all settings (both formal and informal) in order to support the development of a diverse and well-prepared workforce of scientists, technicians, engineers, mathematicians and educators and a well-informed citizenry that have access to the ideas and tools of science and engineering. The purpose of these activities is to enhance the quality of life of all citizens and the health, prosperity, welfare and security of the nation" (NSF, 2009a). To that end, EHR developed a brochure, *Promoting Excellence, Transforming Education* that lays out four themes for the Directorate. They are: broadening participation, cyber-enabled learning, teacher enrichment, and public understanding of science. Under girding all the themes is a focus on promoting learning through research and evaluation (NSF, 2009a).

The Division of Research on Learning in Formal and Informal Settings (DRL) is the primary division within EHR for conducting research on K–12 education. Its mission is:

> promoting innovative research, development, and evaluation of learning and teaching across all STEM disciplines by advancing cutting-edge knowledge and practices in both formal and informal learning settings. DRL also promotes the broadening and deepening of capacity and impact in the educational sciences by encouraging the participation of scientists, engineers, and educators from the range of disciplines represented at NSF. Therefore, DRL's role in the larger context of Federal support for education research

> and evaluation is to be a catalyst for change, advancing theory, method, measurement, development, and application in STEM education. The Division seeks to advance both early, promising innovations as well as larger-scale adoptions of proven educational innovations. In doing so, it challenges the field to create the ideas, resources, and human capacity to bring about the needed transformation of STEM education for the 21st century. (NSF, 2009a)

COMPARISONS TO OTHER FEDERAL AGENCIES

It is interesting to compare and contrast the origins and focus on science education of the NSF with that of organizations such as the National Academy of Sciences and other federal agencies such as the National Aeronautics and Space Administration (NASA), the National Oceanographic and Atmospheric Administration (NOAA), and the U.S. Department of Education, all of which also influence policies that affect how science is taught. NSF differs from agencies such as NASA and NOAA (commonly called "mission" agencies) primarily because of its much broader mandate that includes all of science and engineering (with the exception of the health-related sciences, which are primarily the domain of the National Institutes of Health). Because of its broader scope, NSF also has a significantly larger budget for research and development in science education than these other agencies, primarily because education research is less of a focus in the other agencies.At the other end of the spectrum, the U.S. Department of Education has the broadest mandate of all—to cover all parts of the education system. The Department of Education has a significant research component in the Institute for Education Sciences (IES).

The Department of Education

Although historically science education has not received as much attention from the Department of Education as other areas have, there has been an increasing emphasis on mathematics and science education over the past few years. Examples include the Department's Mathematics Science Partnership Program, which makes awards to states to improve mathematics and science teaching, and the Institute of Education Science's recent focus on mathematics and science education research. Indications of the broader educational mission of the Department of Education can be seen in the areas of educational reform that are included in the current Education Secretary's priorities. These programs, which are part of the Department's $4.35 billion *Race to the Top* initiative, will be funded through the Secretary's discretionary funds provided through the American Recovery and Reinvestment Act of 2009 (U.S. Department of Education, 2009). States that apply for the funds are asked to address the following four broad areas of reform:

- putting the best teachers in schools where they are most needed
- closing chronically underperforming schools and creating better ones
- creating data systems that track students from the cradle to college and that link student results to teacher performance
- the development of world-class standards to help states build their reforms

Although the mandate covers broad categories of funding, and those categories of funding can relate to any part of the curriculum, it is significant that at the time of this writing, the draft solicitation that is undergoing review gives "competitive preference" only to applicants that place added emphasis on STEM through programs that will offer rigorous courses, partner with industry experts, museums and research centers, and those that prepare students for advanced study in STEM fields including addressing the needs of underrepresented groups and of women and girls (Federal Register, 2009).

National Academy of Sciences

Unlike the NSF and the U.S. Department of Education, the National Academy of Sciences (NAS) is not a federal agency, and it does not fund basic research in science or science education. It was chartered by President Lincoln in 1863 to provide independent, balanced, and objective advice to the government. The NAS Incorporation Act states that the National Academy of Sciences "shall, whenever called upon by any department of the Government, investigate, examine, experiment, and report upon any subject of science or art" (NAS Incorporation Act, 1863). At present, the National Academy of Sciences, the National Academy of Engineering (created in 1964), the Institute of Medicine (created in 1970), and the National Research Council (created in 1916) all fall under the title of The National Academies.

The National Research Council (NRC), which is the operational arm of the Academies, operates through its disciplinary units—the Division of Behavioral and Social Sciences and Education, the Division of Earth and Life Studies, the Division of Engineering and Physical Sciences, the

Institute of Medicine, Policy and Global Affairs, and the Transportation Research Board. NRC activities mostly include workshops and consensus studies. Workshops involve the convening of experts to discuss national issues or problems, which results in workshop reports. Consensus studies involve a more intensive effort and can take years to complete. All members of consensus study committees serve as volunteers, and there is a rigorous report review process. There are also a small number of operational programs at the NRC that include education and training, research surveys, and fellowship and associateship programs.

Because of the special relationship that the Academies have with the federal government, as outlined in the NAS charter, it is possible for federal agencies to request independent advice or the gathering of information from the Academies on any issue of particular interest. The NRC does not receive direct federal appropriations for its work. Instead, individual projects are funded by federal agencies, foundations (see Cheek & Quiriconi, Chapter 4, this volume), other governmental and private sources, and the institution's endowment. When submitting requests for funding to the NSF, the National Research Council submits a proposal to NSF, which is then sent out for external review.As noted above, these proposals are typically to conduct consensus studies or to organize workshops and workshop reports.

A consensus study typically involves identifying national experts and academy members to study available research about a particular issue over a specified time frame, and then to write a report on their findings. The report involves a synthesis of what is known about the issue from the information gathering activities of the study group, along with findings and recommendations. These studies typically take two or more years and include not only the findings of the committees undertaking them, but an elaborate and thorough review process by those external to the work. Some recent examples of consensus studies funded by NSF are: *Knowing What Students Know; Learning and Understanding* (co-funded by NSF and the U.S. Department of Education; NRC, 2001)*; Systems for State Science Assessment* (NRC, 2006)*; Taking Science to School* (co-funded by NSF and the National Institute of Child Health and Human Development (NRC, 2007); and *Learning Science in Informal Environments* (NRC, 2009).

In addition to consensus studies and workshops, the NRC also organizes roundtables, workshop series with participation by federal funders, and letter reports, which are quick-turnaround products that can inform federal thinking and decision making. An example of a roundtable conducted during 2009–2011 was a Roundtable on Climate Change Education. The roundtable explored critical issues in teaching and learning, workforce development, and public understanding of the science involved in climate change. For more information see the National Academies and the list of NRC reports published by the National Academy Press on their websites.

Through its initiatives, the NRC can bring issues of national importance before the science education community, and they can identify priorities for research in science education. Similarly, the NSF can inform policy and practice in science education through the particular NRC consensus studies, workshops, and roundtables that it supports.

MECHANISMS FOR INFORMING STEM EDUCATION POLICY

Using the Budget to Set Priorities

One of the key ways that agencies such as the NSF communicate priorities to the broader field is through the budget process. At NSF, these priorities are largely set by the Office of the Director and are consistent with the Foundation's strategic plan. Individual directorates then fill in details for their programs through a negotiation process with NSF leadership and the Office of Management and Budget (OMB), which is the executive office that advises the President on the federal budget. OMB comments on the budget with regard to its fit with the current administration's priorities and eventually sends a budget request for NSF to Congress.

The NSF Budget Building Process

The budget process at NSF calls for staff members to think about three fiscal years simultaneously. For example, in September 2009, NSF staff were making final end-of-year decisions about expenditures from the fiscal year (FY) 09 budget, which ended September 30, 2009. They also were planning for expenditures covered by the FY 2010 budget that was developed throughout the previous year to be sent to Congress October 1, 2009. Finally, they were having conversations about the 2011 budget. Drafts of the 2011 budget are then be sent to OMB throughout the year, until OMB makes a final decision on the President's funding request for the directorates and programs within NSF.

Current Expenditures—The Fiscal Year 2010 Budget Request

The total budget request for the NSF for FY 2010 was just over $7 billion, an increase of $555 million or 8.5% over the 2009 year. NSF also received a one-time appropriation of $3 billion from the American Recovery and Reinvestment Act of 2009 (ARRA) because of the Administration's acknowledgement that investments in science and technology foster economic growth and create high-tech, high-wage jobs. The Education and Human Resources directorate received $100 million of the $3 billion.

These funds were apportioned to the Robert Noyce Scholarship Program ($60 million), the Math and Science Partnership Program ($25 million), and for the creation of a Science Masters' Program ($15 million). Specifics of the budget for Education and Human Resources can be seen in Table 5.1:

Table 5.1. Education and Human Resources Funding

	FY 2010 Request (in Millions)	*Change in Amount Over 2009 (in Millions)*	*Change in Percent Over 2009*
Research on Learning in Formal and Informal Settings (DRL)	$229.50	$3.00	1.3%
Undergraduate Education (DUE)	$289.91	$.68	2.4%
Graduate Education (DGE)	$181.44	$0.06	–0.0%
Human Resource Development (HRD)	$156.91	$2.88	1.9%
Total EHR	$857.76	$12.50	1.5%

Adapted from NSF (2009c)

The 2010 budget request document also lists directorate-wide EHR priorities for the coming year. The priorities are:

1. Advanced Technical Education (ATE), which focuses on educating technicians with the knowledge and abilities to support science and engineering
2. Graduate Research Fellowships to encourage more highly talented students to pursue graduate education in science and engineering
3. Climate Change Education (in collaboration with the Directorates of Geosciences, Biological Sciences, and the Office of Polar Programs) to support a multidisciplinary, multifaceted climate change education program enabling a variety of partnerships in multiple settings
4. Integrative Graduate Education and Research Traineeships Program (IGERT), an NSF-wide program that will provide awards in critical areas such as clean renewable energy and water and climate change education
5. Louis Stokes Alliances for Minority Participation (LSAMP) to increase the number and quality of underrepresented minorities completing undergraduate degrees in STEM

6. STEM Talent Expansion Program (STEP) to increase the number of students receiving degrees in established or emerging fields within science, technology, engineering, and mathematics
7. Project and Program Evaluation (PPE) that emphasizes planning and oversight of third-party evaluations of EHR programs and thematic STEM evaluation studies
8. Research and Evaluation on Education in Science and Engineering (REESE), a program that supports basic and applied research to enhance understanding of STEM learning and teaching (NSF 2009c)

By requesting funding for these programs, NSF is making a statement that these programs address some of the most important and critical issues in STEM education that the nation faces. This is particularly significant when the funding leads to new program development, research and evaluation studies, and findings that can be used by policymakers.

Priorities Set by Congress

Sometimes priorities or changes in funding levels are directed by Congress through legislation. An example of how federal legislation determines priorities and funding levels is the America Competes Act (Public Law 110-69, August 9, 2007). While there is much in this Act that affects NSF as a whole, there are specific calls for increases in programs that affect K–12 science education in particular. Two of the programs affecting K–12 education that received increases under this act were the Robert Noyce Scholarship Program (section 7030), a scholarship program that provides "support to colleges and universities to recruit and train mathematics and science teachers and to provide scholarships and stipends to individuals participating in the program" (U.S. Code, Title 42, Chapter 16, Section 1862n-1), and the Mathematics and Science Partnership Program (MSP), which funds partnerships between universities and schools/districts to improve student achievement in STEM.

In other instances, Congress directs new funds to NSF to support specific kinds of activities. NSF then creates programs that meet the intent of the congressional initiative. An example is the H-1B Nonimmigrant Petitioner Fees Funding that began in 1999 (Public Law 105-277). The H-1B visa program is used to temporarily employ foreign workers in specialty occupations. Some of the fees collected from applicants to that visa program have been made available to NSF by Congress over the past decade. One of the programs that NSF created from these funds is the Innovative Technology Experiences for Students and Teachers (ITEST) program. "ITEST invests in K-12 activities that address the current concerns about shortages of STEM professionals and information technology workers in the U.S. and seeks solutions to help ensure the breadth and depth of the STEM workforce, including education programs for students and teachers that emphasize IT-intensive careers" (NSF, 2009c, p. 19).

ITEST as an Example of a Program Changing Over Time

Following the broad, general direction identified in the congressional H-1B initiative, the Elementary, Secondary, and Informal Education Division (ESIE) at NSF developed a program solicitation to address the issues called for in the legislation. Once approved at the Division level, the solicitation went through a clearance process that included the Office of the Assistant Director for EHR, Budget and Financial Affairs, and the Office of the Director. An early, 2002, ITEST solicitation (NSF, 2002) identified three kinds of potential projects: (1) Youth-Based Projects that included both school year and after-school or summer programs/interventions and that focused on information technology over two-or more years. (2) Comprehensive Projects for Students and Teachers, a program that supports projects to adopt, adapt, or infuse informational technologies into STEM courses commonly available in schools and provide resources for teacher materials as well. (3) Resource Center(s) to support the projects and engage in technical support and research that can improve information technology experiences in both formal and informal venues.

The early solicitation had a heavy focus on developing IT interventions for students in specific content areas and helping teachers use IT-related approaches in their STEM teaching. But programs that are relatively long term, such as the H-1B fee-supported ITEST program, typically evolve over the life of the program. In this case, the program has evolved into a more comprehensive approach to information technology education and one that has a stronger research component than there was in the early version of the program. The 2009 solicitation (NSF, 2008) identifies important questions of interest that the projects should address:

> What does it take to effectively interest and prepare students to participate in the science, technology, engineering and mathematics (STEM) workforce of the future? What are the knowledge, skills, and dispositions that students need in order to participate productively in the changing STEM workforce and be innovators, particularly in STEM-related networked computing and information and communication technology (ICT) areas? How do they acquire them? How can the Nation's burgeoning cyberinfrastructure be harnessed as a tool for STEM learning in classrooms and informal learning environments? What will ensure that the nation has the capacity it needs to participate in transformative, innovative STEM advances? How can we assess and predict the inclination to participate in the STEM fields and how can we measure and study the impact of various models to encourage that participation? (p. 2)

To address these questions, the 2009 solicitation outlines three kinds of projects that can be pursued: (1) *Research Projects* that produce empirical findings and research tools that contribute to the knowledge base about approaches that are most likely to increase STEM capacity of the future workforce; (2) *Strategies Projects* to design, develop and study strategies that encourage K–12 students to be prepared for STEM and information technology fields; (3) *Scale-up Projects* that examine effective steps in expanding the adoption of successful interventions. Over time, these projects will yield insights about the factors that motivate K–12 students to develop an interest in and persist in STEM-related information technology education past the high school years, into postsecondary education or the workforce (NSF, 2008, p. 2). This greater emphasis on research, the identification of factors that lead to successful outcomes, and understanding what is required to bring a program to scale are consistent with broader efforts along these lines in the Education and Human Resources directorate and provide an example of how programs change to fit into the larger framework of evolving policy priorities at NSF.

How Maintaining Contact With the Field Affects Research Priorities

Another way that NSF affects the direction of STEM education is through the relationships it has with scientists and educators throughout the country, and through the mix of scientists and educators that are employed at the Foundation. As stated earlier, from its beginning, NSF has worked closely with scientists in universities on the education projects funded through the Foundation, and these close connections between education and disciplinary scientists characterize NSF today as well. This means that NSF-funded projects have a strong focus on the content of the science disciplines, but it also means that the disciplinary content is viewed through the eyes of professional educators who have both theoretical and practical knowledge about what is appropriate to teach students at various ages and the ways to teach it.

Staffing Policies at NSF

One of the ways that NSF maintains its connection with scientists in the field and ensures that scientists and professional educators can work together in developing education programs is through its staffing practices. For many years, a significant number of NSF's professional staff has been funded through the Intergovernmental Personnel Act (IPA) (U.S.

Office of Personnel Management, 2009). The act provides for the temporary assignment of personnel between the federal government and state and local governments, colleges and universities, Indian tribal governments, federally funded research and development centers, and other eligible organizations. Professional staff hired under this act are called "rotators" because of the limited time they are at NSF. A rotation, which is paid for by NSF, typically lasts for one, two, or three years, after which the rotators return to their universities or other institutions. Rotators have the same responsibilities as other Assistant Directors, Division Directors, or Program Officers.

Currently, approximately 40% of the professional staff at NSF is hired through the Intergovernmental Personnel Act (H. Sullivan, personal communication, September 1, 2009). Because of this program, there is a constant cycle of scientists, mathematicians, engineers, computer scientists, social scientists, and mathematics and science educators constantly coming in and out of the Foundation. For the education directorate, this gives program officers access on a daily basis to disciplinary scientists and educators of all kinds, both within the education directorate and the research science directorates.

Although rotators do not have the institutional memory and longstanding connection to the cyclical work of the Foundation in terms of budgets, program solicitations, and the proposal review process, having a large percentage of the staff hired on a temporary basis has some significant advantages. It connects active researchers with the permanent NSF staff members in the NSF decision-making processes in setting direction and priorities, selecting peer reviewers, identifying projects to be funded, negotiating and managing awards, and re-thinking program solicitations. The program is important because it ensures a constant flow of new ideas into the Foundation.

Workshops

Related to the staffing policies as a way of bringing new ideas into the Foundation, another important strategy for engaging the community in assessing approaches, themes, and the current state of thinking around issues in STEM education research and development is through workshops. Sometimes workshops are proposed by NSF staff, but often they are proposed by researchers in the field and then reviewed by NSF staff. If awarded, they provide opportunities for experts to gather to discuss key ideas (NSB, 2009, p. 6). At other times, policymaking bodies like the NSB, the governing board of NSF, support workshops to explore similar issues. One example is the August 24–25, 2009 Expert Panel on Preparing the Next Generation of STEM Innovators. The workshop heard from cognitive scientists, experts on gifted education, psychologists, and educators about the current state of knowledge and possible future directions for research. Some of the topics addressed at the workshop included a summary of the state of the field in gifted and talented education, innovation, creative thinking, and the learning sciences; the characterization and development of future STEM innovators; developing STEM innovators through both formal education and in informal settings; identifying under-developed pools of STEM talent; and innovation ecology and entrepreneurship.

A number of workshops have been proposed by NSF staff. For example, between 2006 and 2008, two NSF organizations (EHR and the Office of International Science and Engineering (OISE), in collaboration with the U.K.-based Wellcome Trust, hosted a series of three meetings all aimed at exploring key issues in science education. The first (in 2006) included representatives from the United States, the United Kingdom, and Canada to discuss current and promising future directions for attracting and retaining students and teachers of STEM. The impetus for the meeting were two reports: *Rising Above the Gathering Storm* (National Academies, 2006) and, in the United Kingdom, *Set for Success* (Roberts, 2002) both of which raised concerns about the supply of people with the STEM skills needed for the twenty-first century (Marcus, 2007). The second workshop, held in 2007 in the United Kingdom, was concerned with the *National Value of Science Education* and explored issues and national policies related to science literacy, career trajectories for those interested in science careers, and issues of curriculum, assessment and professional development of teachers (Wellcome Trust, 2007). Finally, a third meeting was held in the United States in 2008 where the focus was on *Assessing Youths' Interests in STEM: What Do We Need to Know?* This workshop continued the themes of evidence of students' low interest in and engagement in science and approaches to adapting school curriculum and pedagogy to increase the science career supply. The workshop included sessions on measuring students' interest and engagement in STEM and the need to improve measures around motivation and interest (Duschl & Earle, in preparation).

In addition to these broad-based policy-focused workshops, researchers can propose workshops that address important issues related to the specific programs in the Directorate.

The Role of Education Research and Development in EHR

One of the most important ways that NSF informs STEM education policy is through the research projects it funds. Setting priorities for research affects what knowledge will be made available to the field, which then affects the quality of policy decisions that can be made in those areas. At all times, NSF seeks to fund education research in areas that are thought to be of the greatest national significance. This is particularly important when the results of this research can be used by policymakers to make informed policy decisions. It is often not specifically one funded project that will inform policy as it is the creation of programs of work or a strand of education research that matures and advances knowledge over time. As stated in NSF's founding documents and continued activities, NSF is primarily a research and development organization. Although there have been forays into the world of STEM education interventions in the education directorate—for example, Statewide Systemic Initiatives, Local Systemic Change Projects—NSF does not have either the budget or the mechanisms to take such education programs broadly to scale. And even when they do occasionally get involved in intervention projects, those projects too have a research and development component. The culture of the agency is to conduct basic research and program development and leave broader implementation to other agencies, states, school districts, or schools.

The Place of STEM Education Research

NSF maintains a strong presence in education research compared to other federal agencies. In a 2006 report examining the STEM education research investment of federal agencies, only three agencies were identified with "a significant and targeted focus on funding research in STEM education": the U.S. Department of Education, the National Institutes of Health (NIH) and the National Science Foundation (NSF). For the year 2003, the study identified a total of 29 programs across those three agencies—14 in Education's Institute of Education Sciences, 2 in the National Institute of Child Health and Human Development at NIH, and 13 at the NSF. For that baseline year, total expenditures were approximately $190 million, with NSF supporting 82% of STEM education research funding (National Science and Technology Council [NSTC], 2006). Since the data for this report was collected several years ago, and since the U.S. Department of Education now has a STEM focus in some of the programs operated by the Institute of Education Sciences (IES), the percentage of education research supported by NSF has most likely decreased from the 82% number.

Programs at NSF undergo fairly regular revision, but the fundamental activities of curriculum development, basic research on teaching and learning, innovative technology use, and approaches to motivating students to engage in science, engineering, mathematics, and technology continue. Findings from these R & D projects accumulate in various ways. For example, they often make their way into the large consensus studies conducted by the NRC. Recent examples include *Taking Science to School* (NRC, 2007), which referenced significant bodies of research conducted with NSF funding on learning progressions, uses of technology in learning, and argumentation and discourse.

Currently, the education directorate's research emphasis can be seen in the following categories of programs:

- ***Programs with a specific research focus.*** The broadest of these programs is Research and Evaluation on Education in Science and Engineering (REESE), which supports research on teaching and learning at all ages and in all settings; Gender Studies in Education (GSE), which supports research on the participation of women and girls in STEM; Research on Disabilities Education (RDE), which investigates how to support students with disabilities in STEM; and the Research, Evaluation and Technical Assistance (RETA) program, which supports research on the Math Science Partnerships Program.
- ***Programs that require a research component.*** Examples of programs that require a research component include the Discovery Research K–12 Program, which supports the development of models, resources, and tools but also requires a research component to understand the nature of the impact of those tools and resources on learning; the Mathematics Science Partnerships Program (MSP), which implements partnerships between disciplinary departments in higher education institutions and school districts, and funds specific studies on issues of interest to the MSPs through the Research, Evaluation and Technical Assistance (RETA) component of the program; and the Science, Technology, Engineering Talent Expansion Program (STEP) II, which has a research track to study undergraduates' STEM baccalaureate attainment.
- ***Programs that have an education research strand.*** Examples of programs that have a research strand include the previously discussed Innovative Technology Experiences for Students and Teachers (ITEST) program, which encourages interventions to increase K–12 students' interest and engagement in STEM through information technology. This program has an entire strand devoted to research on the factors that attract and retain K–12 students to STEM courses and careers.

INFORMING POLICY: EXAMPLES FROM RESEARCH AND DEVELOPMENT

The education directorate's research and development programs represent one of the most important ways that research can inform policy. The influence is not always direct, but often grows from the accumulated portfolio of research and interventions funded by various programs over time. As knowledge accumulates around particular topics, it often begins to affect national, state, or district policies.

An Example From National Research Council Consensus Studies

Over the past several years, the education directorate has supported several National Research Council consensus studies. These studies draw upon expertise in the field to develop a report that synthesizes the available research and provides recommendations for the direction that programs and policy should take both for federal agencies and the field in general.

One example is the study on assessment of student learning, *Knowing What Students Know* (NRC, 2001). This study examined what is known about assessing student learning from the fields of STEM education, the cognitive and learning sciences, and psychology. According to the report, advances in both the cognitive and measurement sciences have created opportunities to fundamentally rethink student assessment. Therefore, it is now possible to move beyond "a focus on component skills and discrete bits of knowledge to encompass more complex aspects of student achievement" (NRC, 2001, p. 3). Studies on long-term memory, novice-expert differences, metacognition, and students' prior knowledge all provide a foundation for creating more nuanced approaches to creating assessments. Changes in the knowledge base of measurement and statistical modeling now make it possible to characterize student achievement in terms of multiple aspects of proficiency rather than a single score. Assessment can also be improved by bringing together research on cognition and learning with these newer measurement techniques, particularly when connected to technologies that permit individualization and interactivity through approaches such as modeling and simulating complex reasoning tasks.

Two of the key recommendations from the study include: (1) Assessments should "sample the broad range of competencies." (2) Policymakers are encouraged "to recognize the limitations of current assessments and support new systems of multiple assessments" when making educational decisions (NRC, 2001, pp. 13–14). Thus, the study provides tools for how assessment can be reconceived and recalibrated so that they can be used to better reflect what students understand about science. This focus on assessment was carried further by the study *Systems for State Science Assessment* (NRC, 2006), which includes specific guidance to states on how to place state assessment into a larger *system* of assessing student learning.

The ideas contained in these studies have not yet resulted in an overhaul of state assessment systems, but there are some examples of how the ideas in these reports are being used in ways that could impact the policy environment. For example, Evidence Centered Design (ECD) is an approach to assessment that is prominently featured in the NRC's (2006) *Systems for State Assessment*. The approach is based on evidentiary argument, a theoretical approach previously described by Messick (1994). According to the NRC report, Evidence Centered Design means that "assessment must be designed from the start around the intended inferences, the observations and performances that are needed to support those inferences, the situations that will elicit those performances, and a chain of reasoning that will connect them" (NRC, 2006, p. 88). The approach has subsequently been adopted by the College Board Advanced Placement Program as part of its rethinking of AP science courses and exams (Mislevy & Haertel, 2006; Pellegrino, 2008). Although the ECD approach may have been implemented by the College Board without the NRC report, it is also likely that the report gave the approach additional credibility and visibility. This is a good example of how NSF funding of an NRC study brought an idea to the attention of educators who then adopted it to significantly impact a major educational program, the Advanced Placement Program of the College Board (College Board, 2008). The policy implications of changes in the AP Program could be considerable and affect both the high school teaching of science as well as introductory undergraduate courses in science. Based on the same model, the College Board has also developed standards for college readiness that lay out what it would take for students from middle school and high school to be ready to participate in the redesigned AP courses (College Board, 2009)

Affecting Policy Through Program Solicitations

The next two examples show how NSF can influence the direction of educational research and policy development through its own program solicitations. The first is a solicitation to develop high quality middle school science curriculum materials. The solicitation was developed by the staff members of what was then the Division of Elementary, Secondary, and Informal Education (ESIE). The intent of the program was to challenge the field to rethink the process of curriculum development. There were no mandates to do this from Congress or from the leadership of NSF or the education directorate. The solicitation was the result of discussions among program officers, leadership in the Division of Elementary, Secondary, and Informal Education, and leadership in the education directorate, as well as various interactions between NSF staff and the field. The second, to create "learning progressions" in science and mathematics, grew out of a desire to determine how student learning of science and mathematics can be most effectively organized and sequenced over time. The program was first offered as part of a 2005 Instructional Materials Development (IMD) program solicitation, which introduced the concept on a trial basis. This program, too, came from within the division, not because of a congressional mandate or budgetary priorities set by the Foundation. These examples represent the kinds of approaches and strategies that NSF can use that can help frame debates and discussion by anticipating where the field might (or should) be going and providing opportunities for research and development through directions provided in program solicitations and the funding of individual projects.

Example 1: Creating a New Solicitation for Middle School Science

A new Program Solicitation was issued in 2000 (NSF, 2000) with the title, *Middle School Science Instructional Materials Initiative*. The solicitation was unusual in that it was considerably more directive than previous solicitations. There were several reasons for creating the solicitation. One was that the American Association for the Advancement of Science (AAAS) Project 2061 had just completed a study of high school science materials and concluded that very few of the materials that were widely available to the field passed muster in terms of rigor, quality, and degree to which they were tied to important and widely accepted learning goals (Stern & Roseman, 2001, 2004). In addition, U.S. student performance on both national and international tests indicated that there were shortcomings in students' understanding of basic scientific concepts (National Center for Education Statistics, 1997; TIMSS, 1996). This, plus the fact that there had been no new directions in comprehensive K–12 science curriculum development for some time, led NSF program officers to create a solicitation that would address these issues.

Under the new solicitation, materials development was to follow what was referred to as "assessment-led design." This meant that the design was to be an iterative process in which the results of student testing would inform revisions to the curriculum materials. In addition, the design was to begin by stating the learning goals that students were expected to achieve. The curriculum was also to address issues of student motivation and, in particular, the "centrality of technology in the student learning experience." The development process should develop sequentially and include: (1) identification of learning outcomes, (2) development of appropriate assessment tasks, (3) and the inclusion of content, instructional strategies, and instructional materials based on the first two (NSF, 2000).

There were to be two phases of development. In phase I, which would last three and a half years, one semester's worth of material from a single middle school grade level would be designed. In phase II, which would last five years, a full Grades 6–8 curriculum would be developed. This second phase required a new proposal, which had to include the results of pilot and field tests that had been gathered during Phase I.

In summarizing the new program, the solicitation focused on this new model of curriculum development:

> The importance of a standards-based design, the guiding purpose of assessment in the design process, and the centrality of technology in student learning experiences need to be seriously addressed. It is expected that the prototype materials will be evaluated in pilot- and field-tests and that the efficacy of the curriculum materials and their design in enhancing student learning will be evident before Phase II proposals are received. (NSF, 2000)

This solicitation was conceived as being "one-shot" rather than one that would recur over several years. It was not designed to become a "core" program for the Directorate, and only a handful of projects were to be funded. However, these projects were required to follow a model of development that was strongly tied to learning outcomes for students and that followed an iterative design approach in which modifications were made following the collection of data on the outcomes of the various parts of the program.

One of the projects to come out of this solicitation was *Investigating and Questioning Our World through Science and Technology (IQWST)* developed by a team of researchers headed by Principal Investigators Joseph Krajcik and Brian Reiser (Krajcik, McNeill, & Reiser, 2008). This project was based on years of prior work from an established research group that held to a theoretical social constructivist perspective and stressed a situation-bound theory of cognition, i.e., how learning takes place in the "real" world (Blumenfeld, Marx, Patrick, Krajcik, & Soloway, 1997; Lave & Wenger, 1991).

The IQWST model has three primary components:

1. SPECIFIED LEARNING GOALS based on (1) an unpacking of national standards, and (2) the development of learning performances to describe learning expectations for students.
2. MATERIALS DEVELOPMENT in which (1) the units are contextualized through a driving question (2) learning tasks are identified,

(3) an instructional sequence is produced, and (4) assessments and rubrics related to the learning goals are developed.

3. FEEDBACK on the development process through the collection of (1) student test data and (2) reactions from teachers (Krajcik, McNeill, & Reiser, 2008).

These materials focus on a few "big ideas" in science that were framed within "driving questions" such as *How can I make new stuff from old stuff?* Materials were also deliberately developed to be "educative" for teachers, i.e., the materials could be used to help teachers deepen their content knowledge around the important science ideas in the materials. Other design principles included: (1) taking into account prior student learning and making connections between the new ideas and their prior ideas, (2) using a set of complementary approaches rather than a single one (e.g., face-to-face workshops with online follow-up) and (3) balancing curricular prescription and teacher autonomy. The learning technologies include data collection, communication, visualization and modeling (Krajcik, McNeill, & Reiser, 2008). It is too soon to state that these materials and others like them that focus on "big ideas" and learning technologies will have a significant influence with schools and hence on education policy. However, again, as educators and the education policy community once again consider new standards in mathematics, reading, and science it appears that a focus on "big ideas" is gaining traction, for example, Secretary of Education's Arne Duncan's call for fewer, higher, clearer standards. It is too early to tell how this will play out in the development of new science education standards. However, it is instructive that the NSF funded the National Research Council to hold a meeting in August 2009 called *Expert Meeting on Core Ideas in* Science. This meeting served as a prelude to what may become a larger science standards initiative. Curriculum materials such as IQWST are organized around big ideas, and if the science standards movement gains momentum, these kinds of materials could be well positioned to provide guidance to the field.

Example 2: Encouraging a New Strand of Work Within an Existing Program Solicitation—Learning Progressions

In the 2005 solicitation of what was then the Instructional Materials Development (IMD) program in EHR (NSF, 2005), a section was inserted that called for projects to develop "learning progressions." Learning progressions were defined as descriptions of the successively more sophisticated ways of thinking and reasoning in a content domain (Smith, Wiser, Anderson, & Krajcik, 2005) and as a strategy for organizing science instruction and curricula to focus content on the most important science ideas, for connecting teaching and learning, and for defining additional research that supports students' understanding of core ideas (NSF, 2005).

The solicitation drew heavily on two papers developed for the *Systems for State Science Assessment* consensus study (NRC, 2006). The papers by Smith, Wiser, Anderson, and Krajcik (2005) and Catley, Lehrer, and Reiser (2005) were particularly helpful because they included examples of possible learning progressions in science. Another precursor to the learning progressions program being developed at NSF was the *Atlas for Science Literacy* (AAAS, 2001, 2007), which identified strands of science ideas that could be linked together over time to provide students with a coherent understanding of science.

The idea of doing research in the area of learning progressions was intriguing for several reasons:

- Research on learning progression addresses concerns that existing STEM curricula include too many discrete pieces of information that are not linked together in a coherent way and do not allow for the alignment of curricular content with instruction and assessment.
- Research on learning progressions requires that STEM concept selection must be based on a strong empirical base about how learning takes place.
- Research on learning progressions draws on research that has been done in the cognitive and the learning sciences that is relevant to science education.

Some of the language from this 2005 solicitation describing learning progressions appears below:

> To build the foundation for more coherent curricula in science and technology, a pilot initiative will support the development of learning progressions (Catley, Lehrer, & Reiser, 2005) of STEM content and processes that build student understanding over time. These would build models of instruction and demonstrate how instructional materials and professional development would look over multiple grade bands (K–5, 6–8, 9–12) when intended to support students building deep understanding of a key concept over time.
>
> The ultimate goal is to significantly enhance student learning through classroom instruction that is driven by learning progressions.
>
> For this initiative, the learning progressions are limited to three process strands—modeling, engineering design, and inquiry—developed in the context of important content as specified in national standards (AAAS,

1993; NRC, 1996, International Technology Education Association [ITEA], 2000–2002).

> The development of instructional frameworks centered on learning progressions must be based on the fundamental ideas of STEM and understandings from the cognitive and educational sciences literature. Such exemplars must support learning and constructing knowledge that are consistent with the developmental level of the students, build incrementally over grade bands, and culminate in deep understanding of core ideas.

Following the solicitation, the Division arranged a conference in January of 2006 for potential applicants interested in the ideas presented in the learning progressions solicitation. The workshop focused on how the development of learning progressions was being done in the field of history because learning progression research had not yet been done in science.

The new solicitation received a positive response from researchers, and since its inception in 2005, interest in this area of research has grown significantly and has already gained traction in the policy arena. For example, the Consortium for Policy Research in Education (CPRE) held two workshops to assess the state of the art in learning progressions research, and it sees the potential for learning progressions to support a more empirical approach to deriving standards. Typically, standards are created deductively out of the major ideas that define a science discipline, whereas a learning progressions approach would be based on empirical evidence on how students progress or achieve mastery around key science ideas (e.g., Smith, Wiser, Anderson, & Krajcik, 2005). A learning progressions approach to standards writing also differs from current approaches to standards writing in that while it may trace a trajectory, it does not specify when any particular idea will be learned. Further, progressions can try to make the "interactions between content and practices explicit in a ways that current standards and assessments often do not, and this in turn provides direction for more effective instructional responses" (Corcoran, Mosher, & Rogat, 2009, p. 21).

Again, the research on learning progressions is quite new and there have been few studies that have had time to fully play out their hypotheses. While some are quite enthusiastic about the potential of learning progressions to inform curriculum and instruction, others are more cautious and argue that "it also places a great deal of weight on empirical research to establish these progressions." They make the argument for more research while pushing ahead with practice and revision of progressions in the meantime (Shavelson, 2009).

I selected these examples for this chapter because I think they have the potential to influence the field of science education, and will have an impact on science education policy. Specifically, both of these efforts are related to the current policy debate on revising the national science standards. Secretary of Education, Duncan, has repeatedly called for the creation of state standards that are "fewer, clearer, and higher." This call has been echoed by others; notably the Council of Chief State School Officers (CCSSO), the National Governors Association (NGA) and Achieve, three organizations that have already been working collaboratively to develop a process and a product around new common standards for mathematics and English language arts. There is also the possibility that common standards will be developed in science. Along these lines, the National Science Teachers Association (NSTA) has argued for fewer standards in science and has conducted initial studies on the feasibility of developing those standards through what they are calling the "science anchors" project (NSTA, 2009) A common core has also been specifically called for in the joint report commissioned by the Carnegie Corporation of New York and the Institute for Advanced Study (IAS), *The Opportunity Equation* (Commission on Mathematics and Science Education, 2009). One of the key recommendations is to: "Establish common mathematics and science standards that are fewer, clearer, and higher and that stimulate and guide instructional improvement in math and science and lead the way toward preparing all American students for a global economy" (p. 21). The report also calls for the development of more sophisticated assessment and accountability systems that stimulate improved instruction and innovation.

If there is going to be an effective re-working of the existing national standards documents, an improvement in student assessment systems, and improved ways of developing curriculum materials, those changes need to be informed by the results of research. The mission of the NSF is to fund that research so that policy makers can make the most informed decisions possible.

CONCLUSIONS

An agency like the NSF cannot directly affect science education policy through legislation like the U.S. Department of Education did with its support of the No Child Left Behind Act. Nor does NSF provide direct funds to states or local school districts except as they might be awarded through the grants submission and peer review processes. Neither does NSF have direct links to the state departments of education that oversee graduation requirements, teacher credentialing, or state standards and assessments. NSF's role is more properly described as one that can inform policy. The priorities that NSF establishes through the budget development process, the development of program solicitations that call for particular kinds of work addressing national needs, and through the proposal submission and award process are all mechanisms that, over time, can result in incremental shifts in policy direction. NSF's structure that permits large numbers of IPAs helps to create a culture of currency and innovation between what's happening in the agency and what's going on in the fields of research and development that NSF supports, including education.

This chapter has focused more on current and future areas of influence than past ones. If one goes back 30 years or so, one can see NSF's influence on science curricula through the support of materials that have engaged the participation of scientists in their creation and encouraged instructional resources that help students think like scientists and engage in activities that call for observation and analysis of phenomena.

The recent examples described here include some promising candidates for what may affect science education policy going forward, including NRC publications such as *Knowing What Students Know* (2001) and *Taking Science to School* (2009); the funding of curricula such as IQWST that focuses on "big ideas" and the work of those investigating Learning Progressions where researchers are identifying learning sequences at increasing levels of sophistication across grades. Potential areas of influence include new ways to organize instructional resources, and building an evidence base for curricula, standards, and possibly assessment for years to come.

AUTHOR'S NOTE

Any opinion, finding, conclusion, or recommendation expressed in this material are those of the author and do not necessarily reflect the views of the National Science Foundation.

REFERENCES

American Association for the Advancement of Science (1993). *Benchmarks for Science Literacy.* New York, NY: Oxford University Press.

American Association for the Advancement of Science. (2001). *Atlas of Science Literacy* (Vol. 1). Retrieved from http://www.project2061.org/publications/atlas/default.htm

American Association for the Advancement of Science. (2007). *Atlas of Science Literacy* (Vol. 2). Retrieved from http://www.project2061.org/publications/atlas/default.htm

America Competes Act, 20 U.S.C. § 9801 (2007).

Blumenfeld, P. C., Marx, R. W., Patrick, H., Krajcik, J., & Soloway, E. (1997). Teaching for understanding. In B. J. Biddle, T. L. Good, & I. Goodson (Eds.), *International Handbook of Teachers and Teaching* (pp. 819–878). Dordrecht, The Netherlands: Kluwer.

Bush, V. (1945). *Science: The endless frontier.* Washington, DC: United States Government Printing Office. Retrieved from http://www.nsf.gov/od/lpa/nsf50/vbush1945.htm

Catley, K., Lehrer, R., & Reiser, B. J. (2005). *Tracing a prospective learning progression for developing understanding of evolution.* Paper commissioned by the National Academies Committee on Test Design for K–12 Science Achievement.

College Board. (2008). *AP Science Redesign Update*. Unpublished report prepared for the National Science Foundation. New York, NY: Author.

College Board (2009). *Science College Board standards for college success.* New York, NY: Author.

Commission on Mathematics and Science Education. (2009). *The opportunity equation: Transforming mathematics and science education for citizenship and the global economy.* Report commissioned by the Carnegie Corporation of New York and the Institute for Advanced Study. New York, NY: Carnegie Corporation.

Corcoran, T., Mosher, F. A., & Rogat, A. (.2009). *Learning progressions in science: An evidence-based approach to reform*. Report prepared for the Center for Continuous Instructional Improvement. New York, NY: Consortium for Policy Research in Education.

Dow, P. B. (1991). *Schoolhouse politics: Lessons from the Sputnik era*. Cambridge, MA: Harvard University Press.

Duschl, R. A., & Earle, J. M. (in preparation). *Assessing youths' interests in STEM: What do we need to know?* (Report of an NSF-funded workshop conducted Oct. 30–Nov. 1, 2008).

Federal Register. (2009, July 29). Document No. FR Doc E9-17909. *Federal Register, Volume 74,* Number 144, pp. 37803-37837. Retrieved from http://www.ed.gov/legislation/FedRegister/proprule/2009-3/072909d.html

Ferrini-Mundy, J. (2008, April). *NSF's involvement in educational research: A retrospective look*. Paper presented at the annual meeting of the American Educational Research Association. New York, NY.

International Technology Education Association. (2000–2002). *Standards for technological literacy: Content for the study of technology.* Reston, VA: Author. Retrieved from www.iteawww.org/TAA/PDFs/xstnd.pdf

Krajcik, J., McNeill, K. L., & Reiser, B. J. (2008). Learning-goals-driven design model: Developing curriculum materials that align with national standards and incorporate project-based pedagogy. *Science Education*, *92*(1), 1–32.

Lave, J., & Wenger, E. (1991). *Situated learning: Legitimate peripheral participation*. Cambridge, England: Cambridge University Press.

Marcus, S. (2007, May 4). *The challenge for K–12 science and math education: Attracting and retaining students and teachers*. Unpublished workshop summary report.

Messick, S. (1994). The interplay of evidence and consequences in the validation of performance assessments. *Education Researcher, 23*(2), 13–23.

Mislevy, R. J., & Haertel, G. D. (2006). Implications of evidence-centered design for educational testing. *Educational Measurement: Issues and Practice, 25*(4), 6–20.

National Academies. (2006). *Rising above the gathering storm: Energizing and employing America for a brighter economic future*. Washington, DC: National Academy Press.

NAS Incorporation Act. (1863). 37th Congress, Session 3, Chapter 111, 12 Stat. 806. Retrieved September 17, 2009, from http://www7.nationalacademies.org/archives/nasincorporation.html

National Center for Educational Statistics. (1997). *Pursuing Excellence: A study of U.S. fourth-grade mathematics and science achievement in international context, NCES 97-255*. Washington, DC: U.S. Govenment Printing Office.

National Research Council. (1996). *National Science Education Standards*. Washington, DC: National Academy Press.

National Research Council. (2001). *Knowing what students know: The science and design of educational assessment*. Committee on the Foundations of Assessment. J. Pelligrino, N. Chudowsky, & R. Glaser (Eds.). Washington, DC: National Academy Press.

National Research Council. (2006). *Systems for state science assessment*. Committee on Test Design for K–12 Science Achievement. M. R. Wilson & M. W. Bertenthal (Eds.). Washington, DC: The National Academies Press.

National Research Council. (2007). *Taking science to school: Learning and teaching science in grades K–8*. Committee on Science Learning, Kindergarten Through Eighth Grade. R.A. Duschl, H. A. Schweingruber, & A. W. Shouse (Eds.). Washington, DC: The National Academies Press.

National Research Council. (2009). *Learning science in informal environments: People, places, and pursuits*. Committee on Learning Science in Informal Environments. P. Bell, B. Lewenstein, A. W. Shouse, & M. A. Feder (Eds.). Washington, DC: The National Academies Press.

National Science and Technology Council. (2006). *Review and appraisal of the Federal investment in STEM education research*. Report of the NSTC Committee on Science, Subcomittee on Education and Workforce. Washington, DC: Author.

National Science Board. (2008). *Science and Engineering Indicators 2008* (Vol.1; Document No. NSB 08-01). Arlington, VA: National Science Foundation.

National Science Board. (2009). *Memorandum to members and consultants of the National Science Board: Summary Report of the September 24, 2009*. (Document No. NSB 09-87). Retrieved from http://www.nsf.gov/nsb/meetings/2009/0923/summary_report.pdf

National Science Foundation Act, 42 U.S.C. § 1861 *et. seq*. (1950).

National Science Foundation. (2000). *Middle Grades Science Instructional Materials Initiative: Program Solicitation* (Document No. NSF 00-80). Arlington, VA: Author. Retrieved from http://www.nsf.gov/pubs/2000/nsf0080/nsf0080.htm

National Science Foundation. (2002). *Information Technology Experiences for Students and Teachers* (Document No. NSF 02-147). Retrieved from http://www.nsf.gov/pubs/2002/nsf02147/nsf02147.pdf

National Science Foundation. (2003). *National Science Foundation Strategic Plan: FY 2003-2008*. Arlington, VA: Author. Retrieved from www.nsf.gov/pubs/2004/nsf04201/FY2003-2008.doc

National Science Foundation. (2005). *Instructional Materials Development* (Document No. NSF 05-612). Retrieved from http://www.nsf.gov/pubs/2005/nsf05612/nsf05612.htm

National Science Foundation. (2008). *Innovative Technology Experiences for Students and Teachers* (Document No. NSF 09-506). Retrieved from http://www.nsf.gov/pubs/2009/nsf09506/nsf09506.pdf

National Science Foundation. (2009a). [website]. Retrieved from http://www.nsf.gov

National Science Foundation. (2009b). *Citizens' report: FY 2008 summary of performance and financial results* (Report No. NSF 09-03). Arlington, VA: Author.

National Science Foundation. (2009c). *Fiscal Year 2010 budget request to Congress, Education and Human Resources*. Retrieved from http://www.nsf.gov/about/budget/fy2010/pdf/29_fy2010.pdf

National Science Teachers Association. (2009). *Science Anchors*. Retrieved from http://scienceanchors.nsta.org/

No Child Left Behind Act of 2001, Public Law 107–110 (2001).

Office of the Press Secretary. (2009, April 27). *Fact sheet: A historic commitment to research and education* [Press release]. Retrieved from http://www.energy.gov/news2009/7347.htm

Pellegrino, J. W. (2008). *Using construct-centered design to align curriculum, instruction, and assessment development in emerging science*. Paper presented at the 2008 International Conference for the Learning Sciences, Utrecht, The Netherlands.

Revkin, A. C. (2009, April 27). Invoking the sputnik era. Obama vows record outlays for research. *The New York Times*. Retrieved from http://www.nytimes.com

Roberts, G. (2002). *SET for Success: The supply of people with science, technology, engineering and mathematics skills*. Report to the Treasury, UK. Retrieved from http://www.hm-treasury.gov.uk/ent_res_roberts.htm

Shavelson, R. (2009, June). *Reflection on learning progressions*. Paper presented at the Learning Progressions in Science (LeaPS) Conference, Iowa City, IA.

Smith, C. L., Wiser, M., Anderson, C. W., & Krajcik, J. (2005). Implications of research on children's learning for standards and assessment: A proposed learning progression for matter and the atomic-molecular theory. *Measurement, 14*(1&2), 1–98.

Stern, L., & Roseman, J.E. (2001). Textbook alignment. *The Science Teacher, 68*(3), 52–56.

Stern, L., & Roseman, J. E. (2004). Can middle-school science textbooks help students learn important ideas? Findings from Project 2061's curriculum evaluation study: Life science. *Journal of Research in Science Teaching, 41*, 538–568.

U.S. Department of Education. (2009, July 24). *President Obama, U.S. Secretary of Education Duncan announce national competition to advance school reform* [Press release]. Retrieved from http://www.ed.gov/news/pressreleases/2009/07/07242009.html

U.S Department of Education (2009). *Mathematics and science partnerships.* Retrieved from http://www2.ed.gov/programs/mathsci/index.html

U.S. Office of Personnel Management. (2009). *Intergovernmental Personnel Act Mobility Program*. Retrieved from http://www.opm.gov/PROGRAMS/IPA/

Wellcome Trust. (2007). *The national value of science education conference*. More information at http://www.wellcome.ac.uk/Education-resources /Teaching-and-education/International-Science-Education-Conference/ index.htm

CHAPTER 6

LA MAIN À LA PÂTE

Implementing a Plan for Science Education Reform in France

Jean-Pierre Sarmant, Edith Saltiel, and Pierre Léna

INTRODUCTION

Learning by Doing (La main à la pâte [Lamap]) is an inquiry-based science teaching program launched in 1996 by Georges Charpak, winner of the Nobel Prize for Physics in 1992. The program was supported by the French Academy of Sciences (*Académie des sciences*), which has managed it since then with the help of the National Institute for Pedagogical Research (*Institut National de Recherche Pédagogique*), and the *Ecole Normale Supérieure*, an elite institution of higher education originally founded to train high school teachers, but now an institution that also trains researchers, professors, high-level civil servants, as well as business and political leaders. In 2000, the French Ministry of Education decided to implement an ambitious "national plan of renewal of science teaching" at the primary level, inspired by *La main à la pâte*. Since then, *Lamap* also has been engaged in the elaboration of new national standards and best practices

The Role of Public Policy in K–12 Science Education, pp. 147–171

for science education in France. This chapter describes how *La main à la pâte* moved from vision to national program and considers the prospects of that vision for the future. The *Lamap* story is an example of how a single program was able to contribute to the transformation of science education in a highly centralized educational system. We describe both the *Lamap* program and the impact it has had on French science education, and then we discuss the reasons it was able to achieve national recognition and so significantly affect national policy and practice. But, first, we provide some background information on the French educational system, particularly at the elementary school level, which may not be familiar to most readers.

STRUCTURE OF THE FRENCH EDUCATIONAL SYSTEM

The French education system is a unified one, whose present three-part structure was gradually put in place during the 1960s and 1970s. The levels of schooling correspond to the U.S. grade levels in the following way: primary schools serve students in U.S. Grades 1–5; *colleges,* Grades 6–9; and *lycées,* Grades 10–13. Schooling is compulsory until age 16, with around 13 million pupils from age 6 to 16 attending school in France. Since the 1970s, France also has had an outstanding record with respect to the development of preschool education for all 3 to 5-year-olds. Preschool education is not compulsory but is attended by more than 95% of the children.

France has 60,000 primary schools that cater to more than 6.5 millions pupils during their first 5 years of formal education, with the first three grades providing grounding in the basic skills. The next stage takes the children up to the end of primary school at about age 10 or 11. The private sector educates approximately 15% of primary school pupils (about a million students), a percentage that has remained stable over the past decade. Most of the private schools are Catholic and operate under contracts with the state (which, among other responsibilities, pays staff salaries). There are about 50,000 pupils in private schools that do not operate under those contracts, and families of those students pay higher fees. Both public and private schools are managed under the aegis of the National Ministry of Education (*Ministère de l'Education Nationale).*

National tests assess students' progress in French and mathematics for all children in the third year of primary school (*cours élémentaire 2* [CE2]) when students are about 8 years old, and just before the end of primary school in the fifth year of primary school (*cours moyen 2* [CM2]), when students are about 10 years old. There is no testing in science.

State Control and Policies

Since passage of the 1982 and 1983 Decentralization Acts, there has been greater diversity and a more flexible organization to what was once a fully monolithic educational system. Despite this increasing decentralization, the central government still retains fundamental powers when it comes to defining and implementing education policy and national education curricula. With the exception of the relatively small private sector mentioned above, French primary education is run at the national level. The curriculum is the same for all schools and is developed by committees appointed by the Ministry of Education and approved by the minister himself. Beyond controlling the national curriculum, the ministry also often sends various recommendations to the local levels.

Since the 1990s, as part of the decentralization movement, greater power has been given to 30 regional administrators (*recteurs d'académie*) who are directly under the authority of the National Education Minister. Those 30 regional administrators are responsible for the schools in their region *(académies)*. Each year, the administrators receive a sum of money from the national ministry, which they then allocate to the various educational establishments in their region. Since 1999, decentralization of the management of teachers has also given these regional administrators new and important responsibilities for assigning teaching posts and promoting and moving teachers between schools within their region. Each region typically encompasses three to six smaller geographical areas (*departments)*. As far as primary school is concerned, the practical aspects of education within these areas are run by a local inspector (*Inspecteur d'Académie*, or IA).

The Decentralization Acts of 1982 and 1983 also significantly increased the role of the elected local authorities, that is, regional, departmental, and municipal assemblies, which have substantial budgets of their own. Today, about 20% of the total cost of education is funded locally, and the national government provides about two thirds of the total funding for the education system, principally as pay for teachers but also to support various educational programs. Each tier of local authority is responsible for one level of education. Municipalities, for example, are responsible for primary- and nursery-school buildings, equipment, and maintenance, and for paying the non-teaching staff.

The Recruitment, Training, and Supervision of Teachers

There are approximately 364,000 primary school teachers in public and private schools, with each teacher responsible for providing instruction in all subjects. There are no teachers who specialize in science or any other content area. The size of a classroom is currently about 26 students in nursery schools and 23 students in primary schools. The central government is responsible for the recruitment, training, and salaries of teachers, most of whom are civil servants trained at university-level schools of education, the *Instituts universitaires de formation des maîtres* (IUFM). Established in 1991, these university-level schools of education provide training to both primary and secondary school teachers. Since 2009, these schools of education have been integrated into the French universities, and the entire scheme of teacher education is being modified to encompass a total of five years of study and training at the university.

Before 2009, teachers were selected through a competitive examination, with about seven students who wanted to become teachers for every one who was admitted into a teacher training program. The competitive examination was available only to students who already had studied for three years or more at the university after receiving their high school diploma, the baccalauréat.

Once hired, teachers must attend in-service training sessions every year for three sessions of four hours each. The topics they study depend on choices made by the local school inspector among priorities defined at the national level. There are additional in-service sessions offered to the teachers as well, but they are not compulsory. This situation of in-service training is a major problem for the development of science education, which must rely on the conviction of the teachers themselves to choose the various optional offerings.

School inspectors (*Inspecteurs de l'Education Nationale* [IEN]) work under the authority of their local inspector (*Inspecteur d'Académie,* or IA) for organizational matters and under the authority of the Inspector General for National Education (*Inspection Générale de l'Education Nationale* [IGEN]) for matters having to do with pedagogy and curriculum. Each school inspector supervises about 300 teachers. Each teacher is visited by his/her inspector every 3 to 5 years. The inspector gives advice and verifies if the teaching is in accordance with the curriculum and the recommendations from the ministry. The teachers who are judged of outstanding quality get a modest pay rise.

THE NEED FOR CHANGE

By the mid-1990s, officials in the French education system acknowledged that the teaching of science in primary schools was in a state of crisis. Difficulties were highlighted during a detailed investigation carried out in 1997 with children in the first year of high school: 15% were poor readers and 4% were nearly illiterate (Ministry of Education, December 1998). The national tests introduced at that time were designed specifically to identify pupils struggling in school. Because of these difficulties with basic literacy skills and efforts to correct them, science teaching was often neglected in order to focus on the so-called basic skills of speaking, reading, and writing. When science was not neglected—that is, when the number of teaching hours for science specified by the national standards was met (and that was done by no more than 3% of the teachers in 1995)—the teaching of science was often bookish. Scientific experiments in the classroom were rare and almost never carried out by the children themselves. Even in science classes, teaching was focused on the basic skills of speaking, reading, writing, and counting. For science, as for other subjects, the teaching technique was "frontal" and "vertical," meaning that scientific truths were explained by the teacher. There were no investigations into natural phenomena, and scientific knowledge was to be accepted, recorded, and memorized. Moreover, the teacher acted in isolation; there was essentially no link between the school teachers and the scientific community.

Various inquiries had shown that interest in science was plummeting at all levels of schooling in France, and in the general population as well. A 1992 study reported in *The Child and Science* (*L'Enfant et la science*) (Charpak, Léna, & Quéré, 2005) found that a majority of students described themselves as "fed up" with science by the end of middle school, and in 1995 the Third International Mathematics and Science Study (TIMSS) showed that French 13-year-olds were weak in their knowledge and skills in natural sciences compared to mathematics (Charpak, Léna, & Quéré, 2005). This decline of science literacy and the anticipated decrease in the number of future scientists and engineers raised concerns about the long term economic prosperity of the country. (See Osborne, Chapter 2, this volume, for a discussion of similar concerns throughout the European Union.) In France, there was a general feeling that science education needed improvement and there had to be a willingness to do something dramatically different. Nothing truly significant at the national level had been successful thus far. So the time was right for the introduction of a new approach to science teaching, and *La main à la pâte* was that approach.

ABOUT THE *LA MAIN À LA PÂTE* PROGRAM

The international development of inquiry-based science teaching in primary schools owes much to the efforts of the scientific community. It was Leon Lederman, Nobel Laureate for Physics in 1988, who introduced hands-on science into schools in poor neighborhoods of Chicago. French physicists Georges Charpak, the Nobel Laureate for Physics in 1992, along with Pierre Léna and Yves Quéré visited these schools and saw for themselves how the hands-on approach had fired the children's enthusiasm for science. Upon their return to France, the three French physicists, all members of the French Academy of Sciences decided to launch *La main à la pâte,* their own version of inquiry-based science teaching.

Previous to this action, several research groups in the United States had carried out extensive research, development, and testing of inquiry-based science curriculum materials; the concept of inquiry was also introduced into the National Science Education Standards in the United States by the National Research Council of the National Academy of Sciences in Washington, DC in the mid-1990s (National Research Council [NRC], 1996). French teachers visited a number of inquiry-based programs in the United States and studied the work done by Karen Worth at the Educational Development Center (EDC) in Boston and by Jerry Pine at the California Institute of Technology. The first experimental science curriculum in France was translated from the *Insights* curriculum developed by EDC (1997). Intended to revitalize the teaching of science in French primary schools, *La main à la pâte* received the unanimous support of the French Academy of Sciences in July 1996, and that support has been sustained ever since. Also in 1996, with the strong support of a diverse range of public and private scientific and educational organizations in addition to the Academy of Sciences, the Ministry of Education launched *La main à la pâte* as an experimental program involving 344 primary school teachers in five of the ninety French administrative subdivisions (*départements*). The program, which included the new curriculum, an emphasis on science in the training of teachers, additional funds to purchase pedagogical and scientific equipment, and tools to help teachers, was encouraged by the Department of School Education (*Direction de l'enseignement scolaire*) of the Ministry of Education, which organized a special training for the relevant school inspectors.

La Main à la Pâte as a Teaching Strategy

The general idea of *La main à la pâte* is to encourage children to participate in the discovery of natural objects and phenomena directly through observation and experimentation, to stimulate their imagination, to broaden their mind, and to improve their command of language. In 1998, "Ten principles of *La main à la pâte*" were established by the Academy of Sciences and approved by the Ministry of Education:

1. Students observe an object or a phenomenon in the real, perceptible world around them and experiment with it.
2. During their investigations, students argue and reason, pooling and discussing their ideas and results, and building on their knowledge. Manual activity alone is insufficient.
3. The activities suggested by the teacher are organized in sequence for learning in stages. The activities are covered by the program but leave a large measure of pupil autonomy.
4. A minimum of two hours per week for several weeks is devoted to the same science theme. Continuity of activities and teaching methods is ensured throughout the entire period of schooling.
5. Students keep a science notebook in which they make notes in their own words.
6. The prime objective is the gradual acquisition by pupils of scientific concepts and skills through written and oral expression.
7. The family and community are solicited for their ideas about activities that can be done by students in class.
8. At the local level, scientific partners (including scientists in universities) support class work by making their science skills and knowledge available to teachers and students.
9. Teaching colleges in the vicinity give teachers the benefit of their experience.
10. Teachers are able to obtain teaching modules, ideas for activities, and replies to queries via the Internet.

They also can take part in a dialogue with colleagues, training officers, and scientists via the Internet. Available in print and supported by a website (www.lamap.fr), these principles guided the program's development and implementation at the national level. By 1999, about 3% of French schools had adopted *La main à la pâte*, but its' success went far beyond this figure. A study carried out by the Ministry of Education brought to light the very positive effects of *La main à la pâte* on children's construction of scientific knowledge and on the development of their language, cognitive, and social skills as well (Sarmant, 1999). The *La main à la pâte* principles also informed teachers' practices and the design of classroom lessons that emphasize the engagement of children through their senses and their intellect, thereby encouraging their enjoyment of science learning.

In a typical *La main à la pâte* lesson, teachers set the ball rolling by quizzing the children about inert objects (e.g., rocks, water, and the sky), living beings (e.g., insects, the human body, and plants), and natural phenomena (e.g., winds, tides, and climate), asking "What do you think?" thus inviting them to advance their own "hypotheses." Even the most naïve of these hypotheses is accepted at face value. The next stage is investigation. This usually involves four or five children working in a small group. The children decide which is the best hypothesis and, ideally, answer the question that they had posed. To round off the lesson, the children are encouraged to express themselves by noting their practical and intellectual adventure in their science notebook. It is not hard to imagine how this exercise develops children's imagination and observation skills. As they come to understand the concept of evidence-based reasoning, children learn to question preconceived ideas expressed by others or rooted in their own prejudices. Thus, children acquire a new awareness of the utility and explanatory powers of scientific principles and the logic of science. In addition, children's mastery of language improves—science has no time for imprecise words or unfounded assertions—they discover the virtues of teamwork, and they acquire the skills needed to prepare and carry out an experiment (see Figure 6.1).

FROM EXPERIMENTATION TO NATIONAL PROGRAM

In June 2000, because of positive information about *La main à la pâte*, and in particular on the basis of the 1999 report mentioned above (Sarmant, 1999), the Minister of Education determined that all children in France should benefit from *La main à la pâte*. He launched a national program to revitalize the teaching of science and technology in primary schools using the *La main à la pâte* strategy. The program was initiated in the final three years of primary school with the further goal of extending it to the whole of primary education, including preschools. The program included:

Example of the Teaching Strategy

The following lesson was designed to take advantage of children's curiosity about the concept of "rhythm." The children made rudimentary pendulums out of weights attached to a piece of string. Swinging the pendulums back and forth, they then measured the periodic regularity of the motion (the oscillation) and tried to work out why the pendulums moved at different speeds. Each child (age 9–11) was convinced, and could not hide a certain complacency, that their idea was the "right" one. It took some time and a lot of patient reflection on the separation of the different parameters before the truth was revealed: the time it took for the pendulum to swing back and forth in a full cycle did not depend on the weight, nor on the thickness of the string, nor on the initial impetus, nor on how the knot was tied (all hypotheses suggested by the children) but solely on the length of the pendulum. This was a superb lesson in that it taught the children much more than a simple relationship in the physical world. It also taught them the importance of careful observation, measurement, and empirical validation before claims are made.

Figure 6.1. Example of the teaching strategy.

- Developing a new curriculum for the primary grades modeled after *La main à la pâte*
- Increasing the emphasis on science in the training of teachers
- Providing schools with funds to purchase pedagogical and scientific equipment (about 10 millions dollars in the first year)
- Developing and supplying tools to help teachers implement the new curriculum

At the national and regional levels, the program was organized to coordinate the activities of directors of education, inspectors, institutes for teacher training, and scientific institutions (Ministry of Education, 2000), and national workshops were organized to train teachers to use the new materials.

A Plan for Cooperation

In August 2000, the Minister of Education appointed a National Committee for the Program of Renewed Teaching of Science and Technology. The task of this committee was to coordinate the program at the national level. Its 20 members represented the various constituents involved with the program including the Director of the Department of School Education, members of the Academy of Sciences, inspectors at the national and local levels, directors of teacher training centers, teacher trainers, and teachers themselves. In September 2000, an official document signed by the Department of School Education of the Ministry of Education, the Academy of Sciences, and J. -P. Sarmant, then President of the National Committee for the Program of Renewed Teaching of Science and Technology, stated:

> *La main à la pâte* is continued in the new context in the hands of the Academy of Sciences and the National Institute for Pedagogical Research. It keeps its own organization; it keeps also its special programmatic features such as scientific partnerships. By the way, it must be stressed that, in the context of the Plan, scientific partnership is desired but not compulsory: science and technology must be taught everywhere and according to the new strategy including parts of the country where scientific partnership is not yet available.

As an element of the whole program for renewed teaching, *La main à la pâte* is a hub for innovation and a center for diffusion of skills. The main aim now is to support the development of the program in order to reach 100% of the French schools, starting from the 3% we have today. (Ministry of Education, 2000)

Cooperation Between *La Main à la Pâte* and the Ministry of Education

One example of the cooperation between *La main à la pâte* and the Ministry of Education's program to renew science and technology education was the publication of *Teaching Science in Primary School (Enseigner les sciences à l'école)* (Ministry of Education, 2002), a book designed to help teachers adopt the strategies that had been used successfully in *La main à la pâte* classrooms. Co-authored by a *Lamap*-Ministry team, the book was produced with public funds, and more than 500,000 copies were distributed free to teachers. From 2003 on, this book played a major role in the implementation of the national program to renew the teaching of science. This enterprise was important and unusual in the French educational context. Typically, the ministry does not publish textbooks itself. (See Figure 6.2 for excerpts from the introduction to the textbook.)

Besides publishing this book on the teaching of science in primary schools, there were other cooperative efforts that the *Lamap* developers engaged in as well. For example, working with the National Institute for Pedagogical Research, the Academy of Sciences also provided French schools with a website that enables the teachers involved in *La main à la pâte* to link up with one another and with the world of research. The website,

Excerpts From the Introduction to the Textbook Teaching Science in Primary School

As a tool for the implementation of the renewal plan and the new programs, this book is designed to assist teachers in their development of teaching that is based upon questioning and upon experimentation by the pupils themselves.

This book is designed to help the teacher to teach science and technology in a reformed manner, from the point of view of both the educational method and the required elements of scientific knowledge.

It is in no way a manual for the teaching of the sciences in primary schools. The modules described, whose themes are taken from the very heart of the programs, are intended to provide the committed schoolteacher with a tool with which to embark upon the path of renewing the teaching of the sciences. It is assumed that the teacher will consolidate his/her approach over the course of these modules and will be progressively able to proceed with the aid of the tools already available and those that will continue to be provided in the future.

Consideration of the development of the pupils' ability to express themselves, both orally and in writing, lies at the center of the teaching encouraged by the science and technology program. The section "Science and language in the classroom" … presents various recommendations to that effect. As far as the French language is concerned, this aspect is developed throughout the modules described in this document.

Figure 6.2. Excerpts from the introduction to the textbook *Teaching Science in Primary School.*

expanded and still very active in 2010 (www.lamap.inrp.fr), provides access to:

- a *La main à la pâte* network of national and regional sites where teachers can find locally produced pedagogical and scientific resources and information about the project
- a network of *scientific consultants* who answer science questions raised by teachers
- a network of *training officers/teaching specialists* who respond to questions on teaching and education

Another tool to support teachers in their self-training is the DVD, *Teaching School Science and Technology* (*Apprendre la science et la technologie à l'école*), jointly published by the Academy of Sciences and the Ministry of Education (2008) to be distributed to all schools in France (National Center for Educational Documentation [CNDP], 2008).

In addition, since 2000, the Academy of Sciences has run 15 "pilot centers" in cooperation with the Ministry. These centers are intended to develop inquiry-based science education locally through the creation of various resources. The centers function as sites for innovation and testing where teacher training, assessment, and resource development can be investigated and implemented. *Lamap* is then able to disseminate these high-quality products and validated practices at the national and local levels.

A NEW CURRICULUM FOR TEACHING SCIENCE AND TECHNOLOGY

The national plan to improve science education also led to a national curriculum developed under the authority of the Ministry of Education (2002). After some debates, it was decided at the Ministry level that the investigative approach being introduced in science should be adapted for teaching technology as well, and a unified curriculum was designed for science and technology. The curriculum that was developed is highly consistent with the approach to teaching science and the principles behind *La main à la pâte*. In contrast to the traditional French curriculum for teaching science and technology, which had been focused on a list of topics to be taught, the new document emphasizes both the science and technology content to be learned and the strategies for teaching it. The excerpts from the national curriculum that appear in Figure 6.3 illustrate the ways in which it is consistent with the teaching strategies and science skills introduced by *La main à la pâte*. The document is quoted in considerable detail to demonstrate how thoroughly and consistently the national curriculum follows the principles of *Lamap*.

Excerpts From the National Science Curriculum 2002

How should sciences and technology be taught at school?

The teaching of sciences and technology at school should allow pupils to build up their own knowledge.

The teaching of sciences and technology at school should allow pupils to take part in building their own knowledge, and should satisfy their curiosity as well as exploiting it. Teaching methods should involve pupils in research and active participation based on the pupils' questioning, prompted and guided by the teacher for whom extensive scientific training does not need to be considered essential.

The method proposed for teaching sciences is incorporated in the learning of fundamental skills (speaking, reading, writing, and counting) and can be carried out with very simple equipment. It is consistent with the "Plan for renewing the teaching of sciences and technology at school," adopted in June 2000 (Ministry of Education, 2000).

Experimental approaches to investigation

The approach that should be adopted obeys the principles of unity and diversity.

Unity:

The method is structured around pupils' questioning of a real world phenomenon or object, living or non-living, natural or man-made. The questioning leads to the acquisition of knowledge and know-how as a result of investigations led by pupils who are guided by their teacher.

Diversity:

Various methods can be used for investigation by pupils, including the following during the same session:

- direct experimentation (should be given preference wherever possible)
- making things (to seek a technical solution)
- observation, direct or instrument-assisted, with or without measurement
- document research
- conducting interviews and using questionnaires

These methods of gaining knowledge should be used in balanced proportions depending on the subject of study.

Science and language

Mastering language is an essential aspect of the method employed, from the questioning and exchanges to which it gives rise, to the comparison of the results obtained with established knowledge, and writing them up.

Science is first of all a way of discussing the world; it names objects and phenomena, and allows us to talk about the world in both its present and future state.

Matter, whether living or non-living, is described by precise laws; discussion is incorrect if it is not itself precise and if it is not based on the logic and rigor of these laws. Speaking in scientific terms therefore requires both exact lexical choice (scientific terminology) and well-constructed, clear syntax.

Many teachers have noted progress in mastering language by pupils studying experimental sciences who have to talk and write (in their science notebook) about their scientific experiences. The need for them to match what they say and what they write to a logical sequence of precise actions (experimental protocol) and circumscribed thought (reasoning constrained by facts) forces them to use restrained, accurate language. Thought is, in fact, structured throughout the approach by the act of writing. The role of writing varies between two extremes (not necessarily mutually exclusive):

- writing "for oneself," to guide one's own approach and one's own reasoning;
- writing "for others," which is often reconstructed, the role of which is communication. Communication by e-mail allows pooling of information with other classes.

These various forms of writing have a different status in relation to knowledge. Personal writing represents the pupil's knowledge at a given time. Writing produced after comparison is shared with others. Writing produced after comparison with the class group (including the teacher) has the status of knowledge. Each form can help pupils to gain access to better-founded aspects of knowledgePupils use various methods of communication and representation (texts, tables, drawings, diagrams, graphs, etc.). The use of software (word processing or spreadsheets) can help to develop the work produced by pupils. Electronic software employs information and communication techniques in a logical manner in addition to (and not instead of) real techniques. Wherever possible, real objects should be preferred to substitutes (paper documents, photographs, etc.), which does not exclude initiation into the various methods of documentary research.

Figure 6.3. Excerpts from the National Science Curriculum 2002.

The excerpts in Figure 6.3 point to ways in which the *Lamap* team and the Ministry of Education incorporated this new curriculum into an educational system where basic skills—especially the skill of reading—were considered fundamental to a child's primary education. Throughout those excerpts from the national curriculum, we also see the efforts to balance hands-on experimental science with the development of linguistic skills. (For a discussion of similar efforts to incorporate linguistic skills into science education in the United States (see Halverson, Feinstein, & Meshoulam, Chapter 13, this volume).

WHAT LED TO THE USE OF *LAMAP* AS A MODEL FOR SCIENCE EDUCATION IN FRENCH PRIMARY SCHOOLS?

Although the current state of science education in French primary schools is far from satisfactory, the rise and expansion of *La main à la pâte* has been the result of an impressive effort and is an example of how a good idea can take hold and spread. In this section, we discuss some of the main factors responsible for the success of the program.

A Coherent Idea That Was Well Described

La main à la pâte is based on a set of coherent and thoroughly described educational principles. From the beginning, the developers of this program had a clear sense of what they were trying to accomplish and the rationale for it. This made it easy for them to be consistent in the messages they presented to the Academy, to the Ministry of Education, to the teachers, and to the public. They also invested the time needed to communicate their message clearly to these groups. Although not themselves specialists in pedagogy, the developers of *Lamap* were fully aware of the rich French tradition in science education, beginning with the object lessons (*leçon de choses*) from the nineteenth century, and later the influences of Célestin Freinet (1896–1966), Jean Piaget (1896–1980), and Lev Vygotsky (1896–1934.) See, for example, Freinet (1949/1993), Piaget (1928), and Vygotsky (1962, 1978). In addition, one of the developers of *Lamap* had been president of the board of the National Institute for Pedagogical Research. This greatly helped their efforts to apply sound pedagogical principles to their work and to describe them clearly.

Support and Cooperation Among Various Partners

Although the Academy of Sciences took the lead in organizing the *Lamap* program and promoting it throughout the country, to realize its objective the Academy also built networks of support among the larger scientific community, teachers, and the Ministry of Education. In particular, it established a support team of 15 full-time persons (the *Lamap* team) and a *Scientific Council* composed of representatives from partner institutions and from the Ministry of Education. The Council's role was to provide new ideas and financial support to guide the actions of the Academy.

Strong Support From the Scientific Community

A crucial factor in the program's success was the unanimous support of the Academy of Sciences, obtained from the very beginning of *La main à la pâte*. Gaining this support was due in large measure to the reputations of the founders of the program. All are notable scientists and members of the Academy, a body that enjoys much respect throughout French society. Moreover, one of the developers, Georges Charpak, added the stature of a laureate of the Nobel Prize. This support by the Academy has given the program an exceptional permanence. The Academy's present statutes include the duty "to look after the quality of science education in France." Despite many changes in leadership within the Ministry of Education (seven ministers from 1996 to 2007), *La main à la pâte* has been able to steadily pursue its development in cooperation with the Ministry, thanks to the permanence of the Academy's presence and will.

Beyond the support of the Academy, *La main à la pâte* gained the early support of large parts of the scientific community because of scientists' concerns about the poor state of science education and the resonance of the program's basic ideas with the ways in which scientists went about their work and with their beliefs about teaching and learning. To take advantage of this positive attitude among scientists, a national committee for the coordination of scientific partnership and support was set up in 2004. This committee was made up of scientists and civil servants from the ministry of education. A guide, available in English, "Supporting Teachers through the Involvement of Scientists in Primary Education," (2009) presents the different types of collaboration that are available between scientists and primary school teachers.

Support From Teachers

Another strong point of *Lamap* was the awareness of the need for special efforts to win the support of the teachers and by understanding the reason why teachers were so hesitant about teaching science: most of them (85%) were without special training in science and were afraid of their lack of knowledge of science. The founders understood that the school teachers had to be helped, and they directed the *Lamap* team to create the support tools noted above. Above all, the opening of the website convinced many teachers that this was a program worth supporting and joining.

Emphasis on Using Science to Build Basic Language Skills

One of the key reasons that the Ministry of Education decided to implement the *Lamap* program on a national basis was its emphasis on building students' language skills. As stated in the "ten principles" of *Lamap*, students use their scientific investigations as opportunities to "argue and reason, pooling and discussing their ideas and results … [and] each child keeps an science notebook, in which the children make notes in their own words." A report from the Inspector General of Education

(*Inspection Générale de l'Education Nationale*) acknowledges the importance of this feature of *Lamap*:

> Teaching science in school had declined mostly because people thought that the time spent in teaching science was taking time away from the so called "fundamental skills" (speaking, reading, writing, and counting). *La main à la pâte* has given the opportunity of bypassing this contradiction by offering a strategy of teaching science that fits into the acquisition of the fundamental skills. (Sarmant, 1999)

Effective Public Relations

From the beginning, the founders of *La main à la pâte* also have made effective use of the mass media. These efforts have included publishing books, awarding a prize each year personally presented by the Minister of Education to a dozen classrooms, organizing colloquia and symposia, and arranging visits to classrooms by government and Academy officials. Each year there are several events involving *La main à la pâte* that are recorded in the newspapers, which then give occasion for interviews on radio or television. This presence in the media has increased the reputation of the program, convinced a steadily growing circle of supporters, and helped gather new partners who themselves convinced other persons.

Good Timing

As is often the case with the successful implementation of new ideas, the timing was right. In the case of *Lamap*, there was a clear and recognized need that had to be met. Before launching *Lamap*, a report by the National Institute for Pedagogical Research (Larcher, 1995) had been very positive about the possible launching of a new program because of the concerns about the state of science education in the country. The developers of *Lamap* also were careful not to move too quickly in the wide-scale implementation of the program. Even when it became clear during the initial phases of implementation that it might be successful on a large-scale basis, and the School Director of the Ministry had proposed a national implementation of the program as early as 1997, the founders wisely refused this proposal. They were convinced that the program needed to gather strength through a longer experimental stage. In 2000, after four years of experimentation, the program was ready for a nationwide extension.

TEACHING SCIENCE IN FRANCE TODAY

In 2000, the officials who launched the plan to revitalize science teaching knew that simply publishing new regulations would not change classroom practices instantaneously. Even a program for science was designed at the national level was not seen as being sufficient to guarantee widespread implementation. Teachers have to be convinced of a new program's merits and receive support to implement it properly. It may appear paradoxical that, in such a centralized system, teachers still enjoy so much freedom in their teaching practice and have so much control over what they actually teach. This is made even more possible because national textbooks from the Ministry do not exist. Publishing textbooks is a private business and teachers can select the textbooks they wish to use. The experience gained throughout previous reforms by the Ministry of Education indicated that a real nationwide change would take several years to reach its full strength. Now, 8 years after the plan for renewed science teaching, how much progress has been made?

Common Base of Knowledge and Skills

First, the influence that *Lamap* has had on government officials' thinking about science education is obvious. For example, in 2006, the *Common Base of Knowledge and Skills* (*Socle commun de connaissances et de compétences*) was incorporated into French law and education regulations. This document contains requirements for knowledge and for social and civic skills. As far as science is concerned, the *Common Base of Knowledge and Skills* recommends all the well known characteristics of inquiry-based science teaching recommended in *Lamap*. Again, we quote extensively from this document to show how closely the *Common Base of Knowledge* reflects the principles of *Lamap* (see Figure 6.4).

Evaluation and Assessment of the Program's Success

A thorough, nationwide evaluation of the effect of *Lamap* on student learning in science does not yet exist. Only some elements of an evaluation are available:

- The Sarmant report (Sarmant, 1999), mentioned above, contains some statistical data on the success of the program.
- Better results have been achieved by students from the Lamap pilot centers on the national tests in French and mathematics (Charpak, Léna, & Quéré, 2005). Unfortunately, science is not tested on the national tests.

Much more remains to be done. Nevertheless, the new national approach to science education gives some hope for the future of education in France. For example:

- Formerly, only knowledge was tested. From 2009 on, in accordance with *the Common Base of Knowledge and Skills*, students will also be tested on practical skills and social behavior as well, and many of these skills are consistent with scientific attitudes that are the goals of *Lamap*.
- A summative report is now required on each student during the last year of primary school. In this report, teachers will provide information about their students' knowledge and skills, including their knowledge of science. This data collection at the classroom level will make it more likely to be able to conduct studies of student learning across the country.

Since 2007, *Lamap* has been carrying research of its own. In order to describe and qualify the teaching in classrooms in the pilot centers, an observation grid was designed and is being applied at a large scale with 200 classrooms already surveyed as of the end of 2009. Results are being used as a formative assessment tool to provide teachers with information they can use to modify their teaching. In the future, another aim of *Lamap* is to evaluate the work of the pupils in the pilot centers all through their primary school years.

HOW MANY TEACHERS ARE USING THE NEW APPROACH TO SCIENCE TEACHING?

Another measure of the success of the reforms in France, and of the influence of *La main à la pâte,* is the number of teachers who are using the new approach. Even though the French education system is highly centralized under the authority of the Ministry of Education, its large size makes it difficult to get a complete picture of how many teachers are actually implementing the new approach in their classrooms. In addition, the answer to the question is not a simple one. From 2004 onwards, if asked if they followed the new national curriculum, the answer given by the teachers was always "yes," even if they admitted they didn't follow it entirely. But the most important question is: "Are the teachers using the new investigative approach inspired by *La main à la pâte*?" The answer is not simply "yes" or "no" and cannot be determined by asking the teachers themselves. A form of external judgment is required. In principle, that is the task of the local inspectors who reg-

Excerpts From the Common Base of Knowledge and Skills

Scientific and technological knowledge

Experimental sciences and technologies seek to understand and describe the natural world and the man-made world, as well as changes brought about by human activity. Studying these subject areas will help students grasp the distinction between verifiable facts and hypotheses on the one hand, and opinions and beliefs on the other hand. To reach these goals, observation, questioning, manipulating, and experimenting are essential in primary school, similar to the "hands on" operation that fosters interest in science and technology at an earlier age.

Abilities

The study of experimental sciences develops inductive and deductive intelligence in various forms. The pupil should be able to:

- use a scientific approach
- know how to observe, question, formulate and validate a hypothesis, argue, and design elementary models
- understand the relationship between natural phenomena and the mathematical language that applies and helps describe them
- manipulate and experiment to test the properties of things
- help design and implement a protocol using relevant instruments, including information technology instruments
- develop manual skills and become familiarized with certain hand tools
- perceive the difference between reality and simulation
- understand that an effect can have several simultaneous causes and grasp that there may be unseen or unknown causes
- express and use results of a measurement or research, and to do so:
    - use both written and spoken scientific language
    - master the main measurement units and know how to associate them with corresponding scales
    - understand that uncertainty is attached to every measurement
    - understand the nature and validity of a statistical result
- understand the relationship between science and technology
- apply one's knowledge in practical situations, for instance, in understanding how the body works and the impact of a diet, take action by getting involved in physical activities and sports and watching out for natural, professional, or household accidents
- use techniques and technologies to overcome obstacles

Attitudes

Rational comprehension develops the following attitudes:

- observational skills
- curiosity about the causes of natural phenomena, reasoned imagination, and open-mindedness

Figure 6.4 continues on next page.

- a critical mind: making distinctions between the proven, the probable, and the uncertain; predicting and forecasting; situating an outcome or information in terms of its scientific and technical development
- awareness of the ethical implications of science and technology
- desire to follow basic safety rules in the fields of biology, chemistry, and in electricity use
- responsibility toward the environment, the living world, and health

Source: Ministry of Education (2006).

Figure 6.4. Excerpts from the *Common Base of Knowledge and Skills.*

ularly visit the classrooms, but these inspectors are responsible for the entire curriculum, and, unfortunately, their attention is less likely to be focused on assessing the teaching of science than the teaching of mathematics and French language arts.

For these reasons, we are not able to provide precise nationwide figures. Nevertheless, it is possible to obtain an estimate through the extrapolation of local inquiries. From these inquiries we estimate that about 30 to 40% of the primary school classrooms were teaching science in 2007, a percentage that is growing slowly but steadily. In those classes, there is a continuum of teaching practices, from a fully investigative pedagogy to more traditional approaches. Given these numbers, the Academy knows that its efforts are not finished. In response, in 2007, the Academy adopted and published a set of recommendations for universities regarding pre-service and in-service teacher training to support inquiry-based science education. The recommendations emphasize, in particular, the importance of and need for in-service training, which is currently poorly addressed in France. Teacher training is more than ever the most important strategic issue needed to expand inquiry-based science education in France.

Continuing Evolution of the Standards

Although much progress has been made in the area of science education since *La main à la pâte* was introduced in 1996, more work is needed, not only to continue to improve science teaching but also to maintain the gains that have been made. For example, in 2008 the Ministry introduced new standards for the primary schools. The motivation for revising the standards was to simplify their presentation, not only for science but for other subjects as well. The new version of the standards focuses again on the so called "fundamental skills" of speaking, reading, writing, and counting. As far as science is concerned, the scope of topics to be taught has been reduced, but the inquiry-based teaching technique inspired by *La main à la pâte* remains the official recommended doctrine, as required by the *Common Base of Knowledge and Skills*.

PROSPECTS FOR THE FUTURE

Extending *Lamap* to the Middle School Level?

Experimentation with the same inquiry-oriented pedagogy at the lower middle school level (Grades 6 and 7) is currently in progress, again proposed by the Academy of Sciences and supported by the Ministry of Education. Because the experimental approach could not make use of the standards and syllabi currently in use at the middle school level, new standards are being developed specifically for this experiment. Instead of having teachers teaching a single subject, which is typical in France at this level, under the experimental plan, Physics-Chemistry, Biology-Geology, and Technology will be taught together by the same teacher in close coordination with his or her colleagues. In addition to the investigative approach that characterizes *Lamap*, this experimentation emphasizes the importance of a unified presentation of science at this school level. Several evaluations of this experimentation are underway.

International Implications

Although the rules and regulations of the European Union (EU) leave primary and secondary education to its individual member states, the overall agreement on the principles of inquiry-based science education (IBSE) and the strong European cooperation in scientific research plead for joint efforts in science education as well (Osborne & Dillon, 2008; Osborne, Chapter 2, this volume). The French Academy took the initiative in this direction in 2003, and several EU projects led by *La main à la pâte* have been founded since then. These include Scienceduc 2004–2006, and Pollen 2006–2009 ("International Action," 2007), with more ambitious initiatives in preparation. There is now a steadily expanding consensus for joint action in the direction of IBSE within the EU, with increased funding being provided as well.

La main à la pâte also cooperates with 12 European countries in the project, *Pollen*, (Pollen: Seed Cities for Science, 2009). *Pollen* was launched in January 2006 and developed over a 3½ year period. With inquiry-based science education as a primary objective, the project focused on the creation of 12 Seed Cities throughout the European Union. A Seed City is an educational territory that supports primary science education through the commitment of the whole community. The major goals of *Pollen* are to: (1) provide an empirical illustration of how science teaching can be reformed on a local level within schools while involving the whole community, (2) demonstrate the sustainability and efficiency of the Seed City approach to stakeholders and national education authorities, and (3) leverage its effects more broadly. In each Seed City, *Pollen* provides material and methodological and pedagogical support compatible with the framework of the local curriculum. Beginning in January 2010, a new European project named *Fibonacci* will begin. This new project involves 25 organizations from 21 European countries, and it is under the direction of the *Ecole Normale Supérieure* and *La main à la pâte* in France. Through *Pollen* and other initiatives, the European Union is taking a greater interest in science education (Léna, 2009; Osborne & Dillon, 2008; Rocard, 2007).

Numerous other countries outside of the EU, including both some of the richest and some of the poorest, also are facing the need to revitalize their systems for teaching the sciences. The French Academy of Sciences has established a large number of collaborations on this theme, all the more powerful because the Inter Academy Panel (IAP) for international issues, which is the international federation of science academies, has also made the improvement of science education one of its priority tasks. Collaborations have been established with Brazil, China, Colombia, Egypt, Israel, Morocco, Mexico, the United States, and Vietnam. One sign of this broad interest in improving science education is that the book *La main à la pâte* (Charpak, 1996) has been translated into Arabic, Mandarin Chinese, Portuguese, and Vietnamese among others, and that the book *The Child and Science* (*L'enfant et la science*) (Charpak, Léna, & Quéré, 2005) is being translated into Spanish. The current status of *La main à la pâte* worldwide can be found on the *Lamap* website http://Lamap.inrp.fr/index.php?Page_Id=1179.

SUMMARY

In this chapter we have provided an example of how an organization outside of the national government, in this case the French Academy of Sciences, was able to influence national science education policy in a country with a highly centralized educational system. For more details about the French education system, one may consult the website http://www.discoverfrance.net/France/Education/DF_education.shtml.

Although the centralized nature of the educational system in France provided certain advantages in getting a policy implemented at a national level, there are lessons to be learned that can be applied at the state, regional, and local levels. First, we noted that there has to be a perceived need for the change that is being proposed. The time has to be ripe for change. If no one thinks change is needed, little is likely to happen. Success also depends on the involvement of a wide range of partners who can collaborate on the project. This requires that a consensus has developed around the new idea. In this case, it was widely recognized that the old-style and bookish form of science education should be replaced by something more interesting and challenging to students. As we noted, a consensus also was developing among EU countries that some form of science education that actively engaged students was needed to improve student attitudes toward science and to increase their participation in science.

The developers of *Lamap* also faced the challenge that the policy change being proposed had to fit into a school curriculum valuing the development of the fundamental skills of speaking, reading, writing, and counting in the primary grades. Anyone who hopes to have a policy initiative adopted must be aware of the context in which the new program must fit as well as of the values and beliefs of the people who are responsible for the implementation. By devoting a very significant portion of the program to the development of those fundamental skills, the *Lamap* developers and the Academy were able to convince government officials that the program would contribute to students' skill and understanding in those areas.

But, as successful as the program has been in France and throughout the world, there is another lesson to be learned, namely, that the teachers themselves, even in a highly centralized system, are ultimately responsible for the success or failure of policy initiatives such as the one presented in this chapter. In our case, we estimate that only about 40% of primary school teachers in France were teaching science at all in 2009, and far fewer—probably a quarter of them—were using the investigative methods of *Lamap*. Even though we wish these numbers were higher, this result can be viewed as a hopeful outcome, given that not too many years ago this fraction was of the order of 3% of the teachers. This significant change provides encouragement for additional programs and policy initiatives at the national level in order to better prepare teachers for more effective science teaching. Finally, the sustainability of this reform in France will depend on continuing support to primary school teachers for both inquiry pedagogy and conceptual understanding.

ACKNOWLEDGMENTS

The authors express their thanks to the whole team of *La main à la pâte*, for their detailed contributions, to the editor of the book and to the reviewers, for their helpful criticisms and comments.

REFERENCES

Charpak, G. (1996). *Learning by doing* [La main à la pâte]. Paris: Flammarion.

Charpak, G., Léna, P., & Quéré, Y. (2005). *The child and science [L'enfant et la science].* Paris: Odile Jacob.

Education Development Center. (1997.) *Insights: An elementary hands-on inquiry science curriculum.* Dubuque, IA: Kendall/Hunt.

Freinet, C. (1993). *Education through work: A model for child centered learning.* New York, NY: Edwin Mellen Press. (Original work published in 1949 and revised in 1967)

International action of *La main à la pâte*. (2007). Retrieved from http://www.Lamap.fr/bdd_image/action_int_ENGL_full.pdf

Larcher, C. (1995). *Rapport sur les expériences nord-américaines et leur compatibilité avec le contexte français* [Report on the North American experience and its compatibility with the French context]. Paris: Institut National de Recherche Pédagogique.

Léna, P. (2009). Europe rethinks education, *Science, 326,* 501.

Ministry of Education. (1998, December). *Comparison of student performance in reading-comprehension at the end of CM2* [*Comparaison des performances en lecture-compréhension des élèves en fin de CM2*]. Retrieved from http://educ-eval.education.fr/pdf/ni9839.pdf

Ministry of Education. (2000, June 15). *Plan for renewing the teaching of sciences and technology at school* [*Plan de renouvellement de l'enseignement des sciences et des technologies à l'école*]. Official Journal, No. 23.

Ministry of Education. (2002). *Teaching science in primary school* [*Enseigner les sciences à l'école outil pour la mise en œuvre des programmes 2002 cycles 1 et 2*]. Paris: CNDP.

Ministry of Education. (2006). *Common base of knowledge and skills [commun de connaissances et de compétences].* Paris: Author. Retrieved from http://media.education.gouv.fr/file/51/3/3513.pdf

Ministry of Education. (2008, June). Official Bulletin. *Schedules and curricula from primary school* [*Horaires et programmes d'enseignement de l'école primaire*]. Occasional Paper, No. 3, 2008.

National Center for Educational Documentation. (2008). *DVD: Learning science and technology in primary school* [*Apprendre la science et la technologie à l'école primaire*]. Paris: Scérén-CNDP (Centre National Documentation Pédagogique).

National Research Council. (1996). *National Science Education Standards.* Washington, DC: National Academy Press.

Osborne, J., & Dillon, J. (2008). *Science education in Europe. A report to the Nuffield Foundation*. London: Nuffield Foundation.

Piaget, J. (1928). *The child's conception of the world.* London: Routledge and Kegan Paul.

Pollen: Seed cities for science. (2009) Retrieved from www.pollen-europa.net

Rocard, M., (2007) *Science education now: A renewed pedagogy for the future of Europe* (Report EU22-845). Brussel, Belgium: European Commission.

Sarmant, J. (1999). *Report on the operation of La main à la pâte and science education in primary schools [Rapport sur l'opération La main à la pâte et l'enseignement des sciences à l'école primaire]*. Retrieved from www.cndp.fr/ecole/sciences/rap_igen0799.htm

Supporting teachers through the involvement of scientists in primary education [*L'Accompagnement en Science et Technologie à l'École Primaire*]. (2009). Retrieved from http://www.lamap.fr/astep/le_guide_de_decouverte

Vygotsky, L. S. (1962). *Thought and language.* Cambridge, MA: MIT Press.

Vygotsky, L. S. (1978). *Mind in society.* Cambridge, MA: Harvard University Press.

CHAPTER 7

THE ROLE OF STATE EDUCATION DEPARTMENTS IN SCIENCE EDUCATION POLICY DEVELOPMENT

Dennis W. Cheek and Margo Quiriconi

INTRODUCTION

In this chapter we discuss the role of U.S. state education departments in science education policy formation. The chapter is organized into three sections. We begin with a brief overview of the development of schooling in the United States and state responsibility for education. In the second section, we use New York state as a case study to describe in more detail the increasingly complex nature of educational systems, the increasing control that states have exerted over education, and the intersection of state and national concerns. In the final section, we step back and look more broadly at the role of state education departments in school reform, the effectiveness of those efforts, and reasons for this success or failure. All of this is presented in the context of science education when possible and appropriate.

The Role of Public Policy in K–12 Science Education, pp. 173–209

State education departments have become increasingly more important as the federal role in education has increased, economic demands on local schools have changed, schooling has become the norm for everyone, and American society has grown more complex. Studies of state education departments, however, are few in number (e.g., Lusi, 1997; cf. comments by Steffes, 2008), and usually cover only limited topics and time periods. Most policy studies of state education departments focus on just one single policy area such as teacher certification, educational accountability, or special education regulations. California has been the most widely studied state in recent years, followed by Kentucky, due to the substantial educational reform legislation triggered by a judgment from the Kentucky Supreme Court in 1989 that declared the entire system of public education inefficient, ineffective, and unconstitutional (e.g., Fusarelli & Cooper, 2009; LaSpina, 2009; Pankratz & Petrosko, 2000).

Although science education has been an important part of the school curriculum since the early days of the country, in comparison to other functions that state education departments perform, the science curriculum has been a relatively small part of their overall concern. Therefore, information related to the role of state education departments in science education policy development is limited.

This lack of research on the role of state education departments in science education policy development is consistent with the lack of policy research in science education in general. To see how limited science education policy research is, let alone research on the role of state education departments in science education policy development, we need only observe how little attention is paid to these topics in the science education research literature or at science education conferences. For example, the 2008 National Association for Research in Science Teaching (NARST) annual meeting in Baltimore had as its theme "The Impact of Science Education Research on Public Policy." Yet, of over 630 presented papers by our count, only 23 focused on national or state-level actors as they relate to science education, and 16 of these 23 dealt with actors outside of the United States (NARST, 2008). Peter Fensham's (2008) keynote address at NARST was the fullest statement focused on the conference theme, and he remarked on the lack of a policy research focus within the NARST program. Additionally, he noted that when his fellow Australian, Richard White, was writing his 2001 summary article on science education for the *Handbook of Research on Teaching* he created 45 categories of studies by science education researchers and not a single one dealt directly with policy. Similarly, he noted that Kenneth Tobin and Barry Fraser's edited *International Handbook of Science Education* (1998) had *no* chapters or sections on policy. The same is true for the more recent *Handbook of Research on Science Education* co-edited by Sandra Abell and Norman Lederman (2007).

Our search of the past 5 years of the journal, *Science Education*, one of the leading journals in the field, found not a single article focused on state education agencies and no articles on national education agencies directly as actors in science education policy. It should be pointed out, however, that for a number of years *Science Education* has had a separate section of the journal devoted to policy. There were, of course, articles on PISA and other national and international assessment results, but the role of state agencies as actors in these events was not part of the data collection, analysis, or discussion. Even the Statewide Systemic Initiatives (SSIs), major science education reform efforts carried out at the state level, as well as the Urban Systemic Initiatives focused on urban areas, and Local Systemic Initiatives focused on local schools, funded by the National Science Foundation (NSF) during the 1990s, resulted in little, if any, research on the role of state education agencies in science education reform beyond the federally required evaluations of the programs This is a shortcoming that Jane Kahle (2007), former division director within the Education and Human Resources directorate at NSF thought was due to the fact that only three SSIs were housed at research universities. If more of the SSIs had been housed at research universities, she speculated, research faculty would have been likely to have carried out that basic policy research (p. 928). (See also Kahle & Woodruff, Chapter 3, this volume.)

Due to the paucity of research on state education agencies and K–12 science education policy research to draw on, this chapter can make but a modest contribution to our understanding of that role. We will center much of our narrative on the state of New York for which there is, at least, extensive documentation of programs and practices. New York typifies the roles that states across the nation have played, or could play, in science education policy development. We will consider various actions of the state within a historical context. Policy always is formulated by real actors functioning within the opportunities, constraints, and challenges of a given milieu, and these conditions must be understood to appreciate why policy decisions were made. We also will make reference to activities in other state education agencies as appropriate. It is hoped that this chapter will stimulate science education researchers and others to pursue more grounded research into matters that can inform our understanding of the role of state education agencies in science education policy development.

STATE RESPONSIBILITY FOR EDUCATION

The Constitution of the United States makes no mention of education; therefore, since its inception, education has been under the authority of the respective states and territories. Within the states, most of the day-to-day responsibility for schools is left to local school districts. This "local control" of education originated with the Massachusetts Law of 1789, which granted towns within a state the power to create "districts" that determined who was eligible to attend a particular school and, under later legislation, the authority to arrange for local tax support of the district schools, hire teachers, and choose textbooks. All states in New England quickly adopted this system of organization, and many other states subsequently followed this pattern in the early nineteenth century as well (Spring, 2008). Heavily influenced by the ideals of Jacksonian Democracy, which sought to expand participation of average citizens in government, it was believed that schools could help to democratize and assimilate the immigrants who were arriving from Europe and beyond and prepare them for increased civic participation. For this reason, the early nineteenth century witnessed a large increase in the number of schools and school districts.

In a tradition going back to Horace Mann and the Commonwealth of Massachusetts in 1826, school districts were typically governed by their own school board (or school committee) independent of the rest of municipal government. In the northeast, school districts usually conformed to the boundaries of towns or contiguous towns; southern states tended to have districts organized by counties; and school districts in the southwest and west presented a confusing array of size, shape, and association with municipal governments because the boundaries of the geopolitical units tended to grow by annexation over time with school districts not undergoing the same reconfiguration to match those changes.

The number of school districts in the United States has changed significantly over the years. Because schools were originally established by local communities, it is not surprising that by 1917 there were 195,000 recognized school districts in the nation (Kirst, 2004), a number that decreased steadily during the twentieth century as the school consolidation movement took place. The trend toward fewer school districts can be seen in Table 7.1 (drawn from Snyder 1993, p. 56).

Impetus for consolidation came from a number of factors including the use of gas-powered school buses, which meant that students could attend schools farther away from home, urbanization and its attendant population losses in rural America, an efficiency emphasis by an increasingly professionalized school administration (especially from 1890–1920), and incentives to consolidate in the form of increased state aid for transportation and for the construction of new schools (Horner, 1954; Rury, 2009; Steffes, 2008). More recently, an expanding population, coupled with the growth of charter schools, has resulted in an upward trend in the number of school districts. In most states, charter schools are counted as individual school districts. The number of districts grew from 15,358 in 1990–91

Table 7.1. Number of Public School Districts in the United States, 1937–1991

School Year	*Number of Public School Districts*
1937–38	119,001
1939–40	117,108
1949–50	83,718
1959–60	40,520
1970–71	17,995
1980–81	15,912
1990–91	15,358

to 17,662 by 2004-2005 with a mean district enrollment of 518 pupils located across 98,579 schools (National Center for Education Statistics [NCES], 2008a).

Although, during the early years of the country, many of the day-to-day responsibilities for education were handled at the local level, from the beginning it was the states that had the ultimate authority for educating the youth in a state. The way in which state authority was manifested varied from place to place but, in each case, some form of state-level agency was created to oversee education. Massachusetts holds the distinction of being the first state to appoint a state school board for common schools in 1837 (Dunn, 2006), and New York has the oldest continuous state education entity in the country, the Regents of the University of the State of New York. The Board of Regents was established on May 1, 1784 to oversee the former King's College in New York City (later renamed Columbia College) and to establish other colleges and academies in the state (Horner, 1954). Erasmus Hall Academy in Brooklyn, New York was the first secondary school to be chartered by the Regents in 1786 (Horner, 1954). In 1812, to oversee the common schools, the New York state legislature created an Office of Superintendent of Common Schools outside the control of the Regents (Horner, 1954), which still retained control of the academies and colleges.

The first schools in America were the district elementary schools, where parents paid fees for their children to attend. These were followed by the publicly supported common schools in the early nineteenth century. At the secondary level, both private grammar schools and academies were present in the early years. Many of the academies served a dual role, often by offering both a "Latin" program and an "English" program of study. The Latin program was focused on a classical education typical of private grammar schools and colleges of the seventeenth and eighteenth centuries, and the English program focused on teaching practical skills and knowledge of the contemporary world rather than exclusively on classical studies. The academies were privately funded, nonsectarian, and emphasized science and politics as well as freedom of inquiry. In many respects these private academies (400 were established during the early and mid-nineteenth century) were precursors to the publicly funded high schools that came later in the nineteenth century (Sizer, 1964; Tolley & Beadie, 2001).

The Federal Role in Education

As noted earlier, education in the United States is a function reserved to the states. But, this does not mean that there is no federal participation in education. That participation began in the eighteenth century, and has grown steadily over the years, along with efforts to have the federal government share in the financing of public education. The first federal funding of K–12 education was through the use of revenues from land sales. This practice eventually was formalized in the Northwest Ordinance of 1787, with monies directly turned over to school districts where the sales had taken place. After the Civil War, a federal bureau of education was established within the U.S. Department of the Interior, but the role of the bureau was restricted to collecting educational data and providing advice to schools about organizational and instructional matters. Repeated attempts to have the federal government provide unrestricted aid to public schools were introduced in 36 different bills between 1862 and 1963 to no avail, as opponents constantly pointed out that there was no mention of education in the U.S. Constitution (Kirst, 2004).

Civil rights eventually became the rallying cry under which federal aid to states materialized. From the 1960s onward, the federal role in education increased substantially. Due to this enhanced federal role and the attendant growth of federal legislation, regulation, data collection, and supervision tied to federal dollars, state education agency functions have become far more complex, and the agencies' size has grown considerably larger. State education agencies today conduct an array of functions linked to this increased involvement of the federal government in education. These include administration of federal aid to schools, federally mandated assessment and accountability programs, and oversight of federal programs in such areas as career and technical education, special education, programs for students with limited English proficiency, and bilingual education.

THE NEW YORK STATE STORY

State Education Activity in New York During the Nineteenth Century

The nineteenth century was a period of rapid growth of public education in New York, as it was across the country. Initially, schooling was organized at the district level with each local community having its own school system, usually a single school building. At first, all of these schools were fee-based. The first state monies to support local schools in New York were appropriated in 1795 when the state legislature passed "An act for the encouragement of schools" (New York State Legislature, Chap. 75, Laws of 1795). Under this legislation, local school districts were required to raise an equal amount to match the state contribution (Horner, 1954). This statute was allowed to lapse after only five years. But then, in 1805, the state established a common school fund from proceeds of the sale of 500,000 acres of state land. This fund was enshrined in the state's constitution as a standing source of support for schools. In 1812, the legislature created the position of a Superintendent of Common Schools, who was to have responsibility for managing the common school fund and improving the organization of the schools. The decision to establish a state superintendent of common schools outside the purview of the Board of Regents (previously established in 1784 to oversee the colleges, universities, and academies) was not without controversy. There was considerable consternation on the part of the Regents that supervision for the common schools was not placed under their control given that they had already been overseeing the establishment of academies across the state since 1787 (Horner, 1854). This decision by the New York State legislature to have a split administration of education lasted until the so-called "Unification Act" of 1904 (The Consolidated Laws of the State of New York, 1904, Chapter 40). This 1904 law created the State Education Department, which was headed by a commissioner who also was the chief executive officer of the Board of Regents.

The Preparation, Testing, and Licensing of Teachers

Before 1812, decisions about who was qualified to teach were made at the local district level. In 1812, due to the perceived unevenness of the teaching force across the state, the state began to require local districts to design and administer examinations for teachers, although persons already holding a teaching license prior to the establishment of the examination requirement were exempt. Each district was expected to design its own testing and licensing "system." Then, in 1841, the Deputy Superintendent of Common Schools was granted authority to issue licenses to teachers in schools statewide. Although the tests were still designed and administered locally, the state tried to achieve some degree of uniformity in the procedures and the general scope of the examinations. In 1875, because some jurisdictions were succumbing to local political pressures to allow teachers to teach who had not passed the tests, the state legislature decreed that anyone wishing to hold a state teaching certificate for life must pass the required locally-administered teacher examination. The elements of the examination itself were not standardized statewide until 1888, and the state education agency did not prepare the examination questions until 1904. These state teachers examinations were given until 1926, at which time the process of determining who would be certified was changed so that certification was granted based on the recommendation of an approved program of teacher certification located in a college or university, not by state test (Folts, 1996).

Teacher Preparation

Teachers for the common schools initially were recruited from the academies, which we have already noted were overseen by the Regents of the University of the State of New York. Because the demand for teachers for the common schools became so great in the early nineteenth century, the Regents developed a policy of encouraging the academies to offer courses to prepare teachers. In 1834, the legislature specified that the Regents should take a portion of the Literature Fund that was set aside for the academies and devote it specifically to the training of elementary teachers. Then, in 1844, following in the footsteps of Massachusetts under Horace Mann's influence, the New York state legislature established state normal schools exclusively to prepare teachers for schools, first in Albany and then at 11 additional locations throughout the state over the next 50 years. The mission of these normal schools was later broadened, and they became comprehensive colleges under the State University of New York system (Horner, 1954; Folts, 1996).

In the normal school, science was seen as an integral part of teacher preparation, and the graduates of normal schools were exposed to such subjects as botany, chemistry, physiology, mineralogy, geology, and, less often, physics and astronomy. Both field and laboratory experiences (in the form of demonstrations by the teacher) were emphasized, all under the regulations and inspections of the state. These laboratory and field experiences were guided by the state-issued syllabus for the required teacher certification tests, which were in place by the end of the nineteenth century (Kirst, 2004; Montgomery, 1994; Ogren, 2005).

The Regents Initiate a Student Testing Program

Even though the Regents did not have direct control over the common schools until the Unification Act of 1904, they substantially increased their influence over the common schools in 1865 when they introduced a test that could be used to determine which elementary pupils were qualified to go on to secondary education in the academies which were under the supervision of the Regents. Each student who passed a Regents Examination resulted in that school receiving a monetary sum as a "reward" for the student's successful preparation. Ultimately, the Regents expanded this voluntary system in 1878 to encompass high school examinations. These high school tests greatly influenced admissions decisions to colleges within the state (Kandel, 1936).

The first Regents Examinations for high school students were offered in algebra, Latin, American history, natural philosophy, and natural geography. By 1891, the Regents program in science had grown to include examinations and accompanying syllabi for astronomy, physics, advanced physics, chemistry, advanced chemistry, geology, physical geography, botany, zoology, and physiology and hygiene (University of the State of New York, 1891). In addition to administering the tests, state officials also offered teachers advice on how to prepare students for the tests. The state advised teachers that:

> Instruction should commence in the primary school with simple object-lessons and, always objective, should become more and more systematic as the education of the child progresses. In preparing for these examinations, students thus initiated will avoid the mechanical work of those who begin the study by memorizing definitions and laws, and will form the habit of confirming the knowledge gained from the text-book or teacher by reference, as far as possible, to the objects themselves. (p. 63)

Each school, before it could be incorporated into the Regents' system as a registered entity with eligibility for state funds based on successful passes on the Regents examinations, also had to furnish concrete evidence to inspectors of the possession of scientific apparatus suited to the Regents syllabus along with relevant charts and specimens. Astronomy, for example, required the "actual study of the sky during different parts of the year" with "the aid of suitable instruments" (University of the State of New York, 1891, p. 65). The examinations in physics and chemistry called for:

> discussions, illustrations and proofs of facts and theories. Reason, judgment and common sense will be considered more important than memory ... [along with] a course of experiments which is calculated to develop the explanation of such phenomena as daily come under their observation. Special attention should be given to the construction of simple illustrative apparatus and to the correlation of subjects. (p. 67)

The emphasis on "reason and judgment" reflected one of the many influential recommendations made by the science conference members of the National Education Association's Committee of Ten in 1893 (DeBoer 1991). But, despite these high sounding intents, a report on science teaching by the University of the State of New York in 1900 lamented that, "While the laboratory method is almost universally approved by the science teachers, the text-book method prevails in the schools, to such an extent that laboratory work is incidental, inefficient, and in many cases excluded altogether" (as cited in Smith & Hall, 1902; also cited in DeBoer, 1991, p. 55f.).

A sense of what the Regents examinations entailed can be gained by looking at some of the questions from the Astronomy examination given January 28, 1904. Students were presented with 15 questions of which they had to answer any ten of their choosing. A pass was awarded for grades of 75% and above. Sample questions from the exam included:

1. Explain by aid of a diagram why the sun is a longer time in passing from the vernal to the autumnal equinox than in passing from the autumnal to the vernal equinox.
2. State the form of the earth. Describe a scientific method of determining the precise form of the earth.
3. Mention a proof that the sun (a) rotates on its axis, (b) has a real motion in space.
4. Describe how each of the following may be found by an observation of the sun: (a) the latitude of the place, (b) the longitude of a place. (University of the State of New York, 1904, p. 212)

The Regents examination program began as an effort to judge the suitability of elementary school students for the academies and, later, the suitability of secondary school students for college. Then, in an effort to ensure student success in the academies and colleges, the Regents expanded their influence by advising schools on the kinds of programs that were most suitable for preparing students for the next level of the educational system. As we saw above, sometimes that advice reflected educational ideals that were part of the national conversation, but the *implementation* of those recommended approaches was very much dependent on the effectiveness of the particular administrative structures existing in New York at the time. And, as we also noted, advice about the most effective ways to teach (reasoning about observations made in the laboratory) were often ignored at the local level, suggesting that even in highly centralized states, something more than recommendations from the state are needed if implementation is to be successful.

The Early Twentieth Century

Consolidation of the State's Control Over Education

The New York State Department of Public Instruction, which supervised the common schools, and the Board of Regents, which supervised the academies, colleges, and universities, finally were brought together into a single entity for the administration of education with the passage of the Unification Act of 1904. This act abolished the Department of Public Instruction and the office of State Superintendent of Public Instruction and placed all education functions under the University of the State of New York and its 11 Regents. The Board of Regents also was empowered to provide for a Department of Education. The first Commissioner of Education was elected by the legislature, and all subsequent Commissioners were to be appointed by the Board of Regents and would serve "at their pleasure." Andrew Sloan Draper was elected by the legislature as the first Commissioner of Education, a position he occupied from 1904–1913. He had formerly served as New York State Superintendent of Public Instruction from 1886–1892, and at the time of his election was President of the University of Illinois (Horner, 1954; Folts, 1996). This centralization of authority under the Regents resulted in much greater coordination across the many interactive systems of the state.

State Control Over Curriculum and Instruction

As early as 1914, New York state instituted a series of visitations to school districts across the state on a systematic basis, modeled after the school inspectorate system of Great Britain (Wiley, 1937). In one example, a team of visitors organized by the state visited the Lockport School System in the spring of 1921. Their initial report was submitted in draft form to the superintendent and the school board, and subsequently was revised and published in light of the feedback received. Then, a follow-up visit occurred in 1923 to gauge progress, with a report published the following year after again incorporating feedback from the district. Detailed observations were made, photographs taken, and diagrams created. The section of the final report focused on biology, physiology, health, and nature study noted that two teachers of biology together handled 189 students in seven classes. During the actual visit, the teachers were engaged in reviewing for the upcoming Regents examination in biology. The inspectors noted that:

> Students did not give as accurate and complete definitions as they should have given of important terms commonly used in studying the subject. The notebooks of the students indicated careful laboratory instruction; yet pupils were often not able to relate readily the principles illustrated by the experiments to the explanation of life phenomena they had observed. This was undoubtedly due to the doing of the laboratory exercises in a too technical manner. The teacher expected to remedy these apparent deficiencies in the course of the review. (University of the State of New York, 1924, p. 119)

The Regents, as well as the Commissioner, had the authority to demand remedies deemed suitable for any observed deficiencies. Local superintendents and school boards could be removed by the state under the state's broad executive powers. This early activity by New York is a historical precursor to processes instituted in many states over more recent decades to engage state officials in school site visits encouraging improvements and/or corrective action through mediated processes between the school district and various state-designated parties. It is an excellent example of how oversight and accountability, which might be difficult to implement, can be facilitated through policies that encourage a collaborative spirit between the state and the local schools.

The 30s and 40s

Growth and Success of the Regents Examination Program

As noted earlier, New York was a pioneer in statewide secondary examinations used to award high school diplomas and provide colleges with information about how well students were prepared for higher education. By 1936, a wide array of states administered some form of secondary education examinations including Alabama, Colorado, Georgia, Indiana, Iowa, Kansas, Michigan, Minnesota, Montana, New Hampshire, Ohio, Pennsylvania, South Carolina, Texas, and Wisconsin. Although most of these tests were focused on mathematics, reading, writing, and general ability testing, Iowa, following New York State's lead, offered tests in general science, biology, and physics as well (Kandel, 1936). In most cases, the examinations were given to assess the students' preparation for college admission. Controversy surrounding these early statewide testing programs, especially in regard to the format of the examinations, is discussed in this volume, by Cheek and Quiriconi, Chapter 4.

One complaint made by teachers in New York state over the years has been that the Regents examinations stifled individual teacher creativity and student learning, a refrain now heard frequently as statewide testing has become even more pervasive under the federal No Child Left Behind

Act (e.g., Hurley, 2007). But, not all teachers found the testing requirement to be so onerous. For example, J. Myron Atkin of Stanford University (Atkin & Black, 2003) recalls that, as a New York state high school science teacher in the late 1940s, he readily incorporated topics that were not on the state syllabus but fit with both his own professional knowledge and local community issues. Like many other teachers with whom he was familiar, the end of each course was spent preparing students for the Regents examinations, but never at the expense of professional judgment or locally tailored teaching during the rest of the school year. Sherman Tinkelman, Assistant Commissioner for Examinations and Scholarships at the New York State Education Department, when surveying the history of the Regents Examinations system and its status for a 1965 Educational Testing Service (ETS) Invitational Conference on Testing Problems (Tinkelman, 1966), demonstrated a clear awareness of both the power of the Regents examinations and their inevitable limitations and trade-offs. He observed that:

> A test that does not influence pupils and teachers does not exist. The mere administration of the test will, of course, inevitably have some effect on pupils and teachers.... Now any testing person worth his salt realizes full well the dangers inherent in such a position. Use such power recklessly and the test technician supplants the teacher and the curriculum specialists in determining what will be taught in the classroom. But if one keeps in mind the old adage that testing must follow the curriculum, not determine it, and if one uses this tool judiciously, to achieve curricular objectives that have previously been validated and accepted as State policy, then the examination becomes a powerful instrument for good. (pp. 12–13)

Tinkelman (1966) noted, for example, that when the state began testing the effects of certain curriculum reform practices, schools cooperated in implementing those practices knowing their students would be tested. Although he observed that some teachers taught to the tests, he felt most did not. In addition, the Board of Regents always allowed for local examinations, approved by the State Education Department, to substitute for Regents examinations so that school districts wishing to experiment with new curriculum approaches could do so without the added pressure of having to prepare students for the Regents examinations.

An attitude the State Education Department embraced in regard to its Regents Examinations was cogently expressed by George Martin Wiley, an Associate Commissioner within the Department responsible for the examinations. Writing in 1937, he said:

> A testing program must serve the needs of all pupils. It must be constructive in the service which it renders to the individual pupil. It must help pupils to

> find out where they are, in what direction they are going and what progress they are making. In other words, if examinations and tests are to serve in any large degree as an aid to guidance, they can not be regarded solely as an *ex post facto* device. Furthermore testing materials must bear some relation to the units of instruction, and the interpretation of test results have significance only in relation to the groups under instruction.
>
> There is much traditionalism in much of our school procedure that is overcome with difficulty. Only in an occasional school does one find a system of individual pupil records that will picture a reasonably accurate profile of a high school pupil as he advances through the six years covered by grades 7 to 12. The usual file record shows, for instance, that at the end of the tenth year, Pupil X (and too often he is only *pupil x*) passed English II with a rating of 70, or a letter indicating that bracket. But *why* he passed or *why* he failed, what background of experiences, interests or achievements were his, what progress he had been making, what attitudes he had been developing, what influences were at work outside of the classroom, these and many other factors in his *education* were not a matter of school record or even of information to the teacher. If the point of departure in any social study dealing with a typical community is the community itself, no less significant is it that any program for the growth and development of the adolescent boy or girl must begin with the individual. (p. 28)

These remarks by Tinkelman and Wiley reflect the desire of many educators to use student assessment as a tool for improving student learning. Obvious ways to do that are to make sure the tests match the content of the curriculum and produce data that can diagnose and track student learning.

During the 1930s and 1940s, the policy of testing high school students to determine their suitability for college became entrenched in the thinking of educators throughout the state. Although there was some concern that the tests would drive instruction and limit the creativity of teachers, in the end, most teachers were able to adapt their teaching and scheduling so that they could introduce topics of local interest while at the same time preparing their students for the Regents examinations. It is a remarkable story of policy implementation in what often is a controversial area that has survived with only minor changes for nearly 150 years.

The Mid to Late Twentieth Century

Education policies and practices at the state and local levels shift over time as educational researchers, theorists, and practitioners change their ideas about what is most important in terms of both educational goals to pursue and means with which to achieve them. This shift in thinking about what is the best way to educate also is driven by chief state schools officers who come and go, shifting priorities of the state board of education and governor, pressures and constraints due to shortages of human and fiscal resources, and a host of other factors. This sociopolitical context was even more complicated in the late twentieth century because of the increased role of the federal government in education and the increased size and complexity of state education departments.

As part of this increased complexity of the educational system, the last 50 years of the twentieth century saw proposals for basic skills instruction, criterion-referenced testing, enhanced graduation requirements, site-based school management, open classrooms, charter schools, effective schools, environmentalism, additional consolidation of school districts, school-business partnerships, expansion of student rights, performance-based assessment, global competitiveness, outcomes-based education, revising the scope and sequence of the science curriculum (layered-cake versus every science every year), alphabet soup science curricula (NSF funded programs of the 50s through 70s), bilingual education, multicultural education, improved equity and access for minority group students, the development of a National Board for Professional Teaching Standards (NBPTS), interstate reciprocity agreements for teachers and administrators, preschool through age 20 (P–20) councils, and improved career and technical education. There also was an increased focus on the SAT Reasoning Test or "SAT" (formerly known as the Scholastic Assessment Test and the Scholastic Aptitude Test) and the ACT Assessment or "ACT" (formerly known as the American College test examinations, introduction of the National Assessment of Educational Progress (NAEP), growth of Head Start and community colleges, and the initial publication of *Education Week* (the so-called newspaper of record for K–12 education). The late twentieth-century also saw various reports and legislative acts that commanded the attention of educators including the Coleman report (Coleman, 1966), *A Nation at Risk* (National Commission on Excellence in Education, 1983), *The Goals 2000: Educate America Act* (Public Law 103-227), *Rising Above the Gathering Storm* (Committe on Prospering in the Global Economy of the 21st Century, 2007), *Family Educational Rights and Privacy Act of 1975* (Public Law 108-446), *Akin with Disabilities Act of 1990 as amended in 2008* (Public Law 110-325), *Improving America's Schools Act of 1994* (Public Law 103-382), and the *No Child Left Behind Act* (Public Law 107-110).

How much attention was paid to these various proposals and reports depended on chief state school officers who came and went, shifting priorities of the state board of education and governor, pressures and constraints due to shortages of human and fiscal resources, and various fiscal crises in the state and elsewhere. And, although most of the issues referred to here pertained to education broadly, not specifically to science education, it was in the context of these proposals, reports, and legislative acts that tremendous growth in programs devoted to science education took place during the last half of the twentieth century.

An Era of National Curriculum Reform

The 1950s witnessed the start of the Cold War and with it a massive effort by the U.S. government to improve mathematics and science education, especially in the form of the 1957 National Defense Education Act during the Eisenhower administration (Kirst, 2004; Rudolph, 2002). The recently established National Science Foundation (NSF) poured large sums of federal money into supporting the creation of new mathematics and science curricula and professional development programs for science and mathematics teachers, following the lead of various philanthropic organizations (see Cheek & Quiriconisee, Chapter 4, this volume). New York state was a leader in these reform efforts as well. The Regents syllabus for physics, for example, with its focus on the fundamental concepts of science, which was approvingly described in the *Physics in Your High School* publication of the American Institute of Physics in 1960 (Rudolph, 2002), played a very influential role in the deliberations of the Physical Science Study Committee (PSSC) at the Massachusetts Institute of Technology about how physics should be taught. As one of the few states with high school syllabi in the sciences, the Regents syllabi in chemistry, biology, and earth science served as resources for national reformers during this period of curriculum reform.

Increase in Science Enrollment

Science education had been a priority for many years within the New York State Education Department, long before the national emphasis on science during the 1950s and 60s. The number of students taking Regents courses in physics rose by 73% between 1945 and 1960, with 23% of students taking physics by 1960. The number of students taking the Regents chemistry exams increased nearly 250% over the same period, with a total of 40% of students taking chemistry by 1960. In 1958, to support the increased number of students taking these courses, the state initiated and supported refresher courses for teachers of mathematics and science (University of the State of New York, 1962). Not only did enrollments increase, but there was also evidence that the quality of the programs was high. For example, from the inception of the Westinghouse Science Talent Search in 1942 through 1962, New York state pupils, who represented about 10% of the nation's enrollment, were awarded an average of 31% of the total winner and honorable mention awards.

Differentiated Science Curriculum

By 1960, the state was recommending four general pathways through the science curriculum, although districts had considerable freedom to adopt, adapt, or ignore these suggestions. The "accelerated program" was recommended to start with Regents Earth Science in Grade 9 and then continue through a sequence of biology and/or chemistry and/or physics, and an advanced or college level science course. The "regular academic" track commenced in Grade 10 and followed the sequence of biology and/or earth science and/or chemistry and/or physics. The "general" track recommended one biological science and one physical science course sometime during the Grades 10–12 continuum. Finally, a "vocational" track recommended a general science or related science in Grade 9 followed by "related and applied sciences in vocational and prevocational programs" throughout Grades 10–12 (University of the State of New York, 1960). Providing different programs for students with differing ability and interest in science was consistent with the recommendations of the report by James Conant on high schools in the United States that had been funded by the Carnegie Corporation (see Cheek & Quiriconi, Chapter 4, this volume). The main argument in favor of a differentiated curriculum was that it would ensure that students of superior talent would be available for needed scientific and technical careers even as all students were educated under the same roof in the comprehensive high school (see also Lynch, Chapter 11, this volume, for a further discussion of the equity issues that this raised).

The Impact of Federal Education Legislation on State Programs

The Civil Rights movement of the 1960s, including President Johnson's efforts to remedy educational inequalities as part of his Great Society programs, brought considerable changes to the operations of the New York State Education Department and its counterparts across the nation. The 1965 Elementary and Secondary Education Act (ESEA) was enacted to realize Johnson's ambitions that federal aid to schools should be used to address inequities in the educational system and that disadvantaged students should be the beneficiaries of such aid if we were to create a more just society (Spring, 2008). Initial language for what became ESEA was drafted by a number of senior members of the New York State Education Department, especially its Chief Legal Counsel, the late Al McKinnon (McKinnon, personal communication, April 3, 1991). Rather than providing general unrestricted school aid, ESEA focused on providing funding for certain specific needs categories. The launch of ESEA and its successor federal education acts—Improving America's Schools Act of 1994 and the No Child Left Behind Act of 2001—coupled with a tremendous increase in federal funding for education, required a corresponding increase in staff within state education agencies. In 1995, the General Accounting Office (GAO) found that up to 70% of personnel in many state education agencies were fully supported by federal funds (Kirst, 2004). Many of these personnel were involved in administering the federal programs, with their attendant monitoring and federal reporting requirements. The government also placed an emphasis on science and mathematics and authorized states to use federal dollars to support these functions and to hire the staff needed to do so. Sometimes, this meant federal money was used to create new programs, and in other cases the funds were used to expand programs that New York already had in place.

Largely because of the increased federal funding, the State Education Department was able to create a separate Bureau of Science Education of about five to ten science specialists, headed by a Chief of the Bureau. The specialists all had advanced degrees, and many held doctorates. Each was an expert in some area of science and often had extensive teaching experience and expertise at a particular level in the K–12 system (elementary, middle, and/or high school). Some of the staff were fully or partially supported by federal funds; others were supported by various competitive grants the department obtained from NSF or other federal or state agencies (usually in partnership with corporations and/or universities within the state). Active in professional societies and often writing for various publications, these state education department professionals conducted workshops throughout the state; led and supervised the annual writing of Regents examinations in the sciences; supervised revisions of Regents science syllabi; engaged in site visits to various districts or schools; served on various national and state committees; testified before the state legislature regarding science education issues; interacted with the Commissioner, Deputy and Associate Commissioners, and the Board of Regents as required; responded to the media about matters related to science education; solicited corporate support and involvement for science education activities across the state; wrote and submitted competitive proposals for funding; advised science teacher preparation programs; responded to members of the general public regarding inquiries related to science education; promulgated science safety regulations and ensured that adequate science safety training was provided on a continuing basis across the state; and supervised and supported science competitions for young people and engaged in site visits of those programs.

In many respects, this period was the "golden age" of science education within the State Education Department as there were abundant federal resources (direct aid as well as competitive grant awards) as well as state support that enabled innovations in professional development, curriculum and assessment, and public accountability. These innovations often were accomplished at a very high level of quality that drew admirers and imitators across the nation and outside the United States.

Innovations in Professional Development

Due to the large size of the state, the state education department drew upon a wide variety of resources to provide professional development to administrators and teachers. This included working with union-sponsored teacher centers as well as with colleges and universities. It also included developing a "trainer of trainers" model and designating lead teachers from across the state to conduct workshops on the state's behalf. State education department personnel also worked with the Boards of Cooperative Educational Services (BOCES), which had been created to support local school districts as official arms of the state education department. The BOCES originally were created in 1950–51 to assist rural school districts in obtaining needed services, but they gradually evolved into 37 regional boards that ultimately embraced all but nine of the state's 721 school districts. Since their formation, the New York State Education Department has employed the BOCES extensively to provide professional development, curriculum and assessment assistance, and science safety help for all levels of K–12 education. Many BOCES have highly qualified science supervisors who work with the schools and districts within their region and sometimes with districts outside of their region on broader science education initiatives. Similarly, the Teacher Resource and Computer Training Centers (Teacher Centers) were established by the state legislature due to advocacy by the New York State affiliates of the American Federation of Teachers (AFT) and the National Education Association (NEA) to provide professional development services to teachers. Run by boards comprised of at least 51% teachers, the 44 teacher centers that were initially established have now grown to 126 centers that serve 660 public districts and BOCES. Some of the teacher centers are housed in school districts, some at the BOCES offices, and some are independent of these two entities. Many teacher centers possess considerable expertise related to science education, and both the BOCES and the teacher centers have obtained numerous federal and foundation competitive grants over the years to expand their repertoire of science education services. The structure and role of the BOCES as an intermediate level in the organization of the administration of the state's education system, would make another excellent research study.

The Regents Action Plan and Its Impact on Science Education

A number of major policy initiatives in education were introduced in New York State following the 1983 publication of *A Nation at Risk* (National Commission on Excellence in Education, 1983). That report was part of a 1980s national back-to-basics and public accountability movement. In keeping with that report, the New York State Board of Regents approved in March 1984 a new comprehensive reform agenda called the "Regents Action Plan to Improve Elementary and Secondary Education Results in New York State." The Regents had laid the foundations for this work before *A Nation at Risk* appeared, and many of the recommendations in the Regents Action Plan were mirrored in *A Nation at Risk*. The new regulations began to be phased in during the fall of 1985 and ran for about a decade. Among the many provisions and requirements of the Regents Action Plan were increased requirements for high school graduation, upgrades in course content and objectives, competency and knowledge examinations at certain key points, new personnel standards, revised procedures for school assessment, new procedures for school guidance and discipline, requirements for a school-by-school assessment report for the public, and the creation and administration of program evaluation instruments.

The New York State Education Department, partly in response to the requirements of the Regents Action Plan, then launched a Regents Competency Testing Program. The first administration of the science portion of the program took place in June 1988. The test was a survey of core science knowledge and process skills from the state science syllabus for middle and junior high schools. The test was administered statewide to all ninth grade students. Its aim was to monitor science programs at these grades, not to measure individual student achievement (University of the State of New York, 1987). Districts whose results were viewed as inadequate had to create plans to address problems highlighted by the science test, and the state provided various forms of assistance to help districts and schools realign their programs with the science that was emphasized on the test.

Another program in the science arena that resulted from the Regents Action Plan was the Elementary Grades Program Evaluation Test (ESPET), to be administered in grade four. It included an objective test and a manipulative skills test, which consisted of a series of stations through which students would rotate to determine the extent to which laboratory type skills and scientific processes were being cultivated among the state's elementary school students. Once again, the test was introduced as an indicator of the quality of the science program, not as a measure of individual student performance (University of the State of New York, 1990). Unique for its time, the impact of the program across the state in terms of elementary science teaching was evident to even a casual observer. Science, a frequently ignored or greatly devalued subject within elementary schools suddenly became a top priority for districts as the initial wave of results were quite poor statewide. The success of the program led state education personnel to be invited to train a group of their peers in California for what became a subsequently modified and highly touted California elementary science program evaluation test.

Performance-Based Assessment

As indicated above, the ninth grade and fourth grade program evaluation tests that were developed as part of the Regents Action Plan included a performance component. At the same time, performance assessments to measure individual students also were being introduced in a number of states throughout the country. New York was part of those experiments in the early nineties. Through an NSF-funded project, New York developed and field-tested performance-based assessments for Regents Biology, Regents Earth Science, and Regents Grades four and eight examinations. But after field testing the performance-based assessments, they found the cost too prohibitive to go to scale. The same conclusion was reached in a field trial of over 2,000 fifth and sixth graders by Stecher and Klein (1997), who found hands-on assessment activities were about three times more expensive than open-ended writing assessments if similar reliability was sought. While proponents of these forms of assessment argued that there are many other benefits that accrue from widespread use of performance-based assessments, these types of assessments have been difficult to maintain over the long term because of their increased costs. In addition, some research suggests that their positive impacts on educational outcomes may be exaggerated (Firestone, Mayrowetz, & Fairman, 1998). The performance-based assessment experience is a good example of how a state can test an idea that is being promoted by researchers and policymakers across the country and then decide to abandon the idea for its own good reasons. In this case, the state was not convinced of the merits of performance-based testing and, in addition, they had to face the very practical problem of excessive costs.

Changes in the Regents Examinations and State Syllabi

Just as with professional development, responsibility for the development of new Regents examinations and the syllabi that guided instruction in both the Regents and the local, non-Regents courses, was distributed across a wide range of constituencies throughout the state. Syllabi and examinations were developed by a Commissioner-appointed committee with the involvement of state education department representatives who had expertise in science education and in assessment. The state education department personnel were involved in both the nomination process and in the committee's work. The syllabi and examinations were piloted, revised, field tested, revised again, and then disseminated through a network of teacher trainers so that professional development could be deployed in a coordinated manner along with new program changes. This distributed approach ensured participation by a wide range of constituents and contributed to the successful implementation of science education policy.

The New York State Education Department Today

Today, under the state constitution, the New York State Board of Regents heads both the State Education Department and the University of the State of New York, with an appointed chief executive who simultaneously holds the positions of Commissioner of Education and President of the University of the State of New York. In addition to more than 7,000 public and independent elementary and secondary schools, the Regents system embraces 270 public and independent colleges and universities, 7,000 libraries, 900 museums, 25 public broadcasting facilities, 3,000 historical repositories, 436 proprietary schools, 48 professions encompassing more than 761,000 licensees and 240,000 certified educators, and a wide range of services for children and adults with disabilities. Its implied powers regarding K–12 science education policy within this overarching authority are immense and perhaps unparalleled in the nation.

The distributed BOCES and Teacher Resource and Computer Centers across the state, as described above, provide continuity of support for districts and schools and a means by which even the appointment of superintendents for districts statewide is a shared responsibility. (The BOCES conducts the superintendent searches and vets candidates, and final selection is made by the local district.) Various teacher and supervisor networks organized by the state education department serve as a means to both disseminate new information and approaches and a means for the department to receive continuous feedback from the public.

The state remains a national leader in policies and procedures relating to student learning, standards, and assessments. In fact, the state was already employing many approaches that were center pieces in the No Child Left Behind legislation, and the state already conformed to all the features that the legislation required. For example, even before NCLB, all students had to pass Regents examinations in English, mathematics, science, global history, and U.S. history in order to graduate from high school. Students who fail Regents exams are required to receive remediation. Every 2 years, districts must show evidence that they have such programs in place or risk state censure. Schools throughout the state are held accountable by means of a report to the public on their performance. Schools can be placed under registration review and are subject to state takeover or closure if improvements are not realized. The state also has raised teacher certification standards and has removed the ability of any district to grant a local teaching credential of any form outside of the Regents' authorization. New standards have been promulgated within the past few years strengthening requirements for teacher preparation programs, school leader certification, and professional development (Vergari, 2009). In summary, New York state is clearly an example of how a large governmental entity can successfully assert control over a large and highly diverse system of schooling.

REFLECTIONS ON THE ROLE OF STATES IN SCIENCE EDUCATION POLICY

In this final section of the chapter, we discuss more generally the role of state education departments in science education policy, particularly the ways that state education agencies develop and implement policies to try to improve the teaching of science in their state. In this section we review the functions state education agencies perform, the status of research on the role and functions of states in science education policy, and factors affecting the success of states in improving student learning in science.

The roles of state education departments in science education can be organized into two major interrelated categories. The first category includes the basic regulatory and accountability functions they perform. This encompasses such responsibilities as setting graduation requirements, establishing standards for certification and licensing of school professionals (including alternative routes to certification), coordinating professional development, creating content standards and assessing students with respect to those standards and, in some states, approving textbooks (see also DeLucchi & Malone, Chapter 12, this volume). The second major role is to improve science education so that students are well prepared for science and technology related jobs and well prepared to understand and participate in science-related issues that affect their lives. It should be clear from the previous discussion of New York state that states can be very active when it comes to issues of regulation and accountability. Throughout the nineteenth and twentieth centuries, New York along with many other states, took control of those functions that once were the responsibility of local school districts. Some "local control" states still leave many of these functions to local communities, but even in local control states there is a considerable amount of oversight at the state level. But, what about state involvement in science education reform efforts? How effective have states been in initiating effective policies to improve science education for the students of that state? We have already seen that New York often was at the forefront of national-level reform initiatives. Given that reform ideas must ultimately be operationalized through the states, what does research say about the effectiveness of states in implementing reform?

One study of the involvement of states in education reform was conducted by Margaret Goertz (2001). In her study, she analyzed instruments states employ to influence teaching and learning in schools. She sorted them into three types: (1) performance standards (focused on the individual), (2) program standards (focused on curriculum requirements, staffing, time in school, class size, etc.), and (3) behavior standards (attendance, discipline codes, homework, etc.). Her study of the fifty states showed dramatic increases in the use of all three types of reform initiatives between 1983 and 2001.The impact of these various approaches over this time period is mixed. In one of the most well studied examples, David Cohen and colleagues spent a decade looking at the impact of standards, assessment, and professional development among California teachers of mathematics (the first two categories of Goertz's taxonomy) (Cuban, 2004). They found that even when state and local policies were aligned with each other and sustained over time, only about one in 10 teachers actually changed their practices in a way so that their students' learning improved (Cuban, 2004). One reason for such disappointing results may be that the policies put in place may not have been thought through systematically. That is, when policies are introduced that involve changing one part of the system without thinking through and carefully monitoring how the change intersects with other parts of the system, those policies may not produce the desired outcomes. For example, a study of high school graduates from 1992 regarding their mathematics and science course taking in high school in light of increased requirements found that "increasing the number of credits students have to earn in mathematics and science to graduate from high school by itself may not be sufficient to improve student proficiency in these subjects" (Teitelbaum, 2003, p. 31). Policymakers also need to take into account such considerations as the difficulty level of the new courses, the possibility that teachers may relax requirements in existing courses to accommodate the influx of additional students, student interest in the new courses, and the extent to which the policy is enforced.

To help think about policy development and implementation from a systems perspective, Rogan (2007), in a view from South Africa, borrowed Vygotsky's idea of a zone of proximal development (ZPD) (1978) to create the idea of a zone of feasible innovation (ZFI). He argues that educational reformers should think through more carefully how much change can be realistically expected in schools given the practicalities of human change potential, instructional contexts, and other factors. He argues that, too often, political and social factors (e.g., concerns about economic competitiveness or rankings in international assessments) rather than implementation factors drive policy development, which invariably imposes insurmountable hurdles to implementers. This is an interesting variation of a well-articulated theory in political science and policy studies known as the principal-agent problem. This problem is the inherent disconnect occurring between the chief policy agent (the "principal"), who articulates the policy, and the chief policy implementer (the "agent"), who is tasked with its implementation in the real world that exists outside the realm of policymakers. In addition to usually being from different institutions, "principals" and "agents" often have different degrees of power, operate on different business cycles, and have access to different amounts of information, all of which makes policy implementation difficult. (See also Halverson, Feinstein, & Meshoulam, Chapter 13, this volume.) These differences combine to result in delaying tactics, shifting priorities, information asymmetry, and a host of other problems that delay, impede, undermine, defer, or alter policy decisions and their intended impacts (Braun, 1993; Waterman & Meier, 1998). Echoing the theme of the irrelevance of many policy initiatives to local agents, Larry Cuban of Stanford University, an experienced former superintendent of schools, observed

> Today's education reforms are curious. They seek to slay demons that no longer exist. They apply uniform approaches to dissimilar problems. They take power away from local school boards and educators, the only people who can improve what happens in classrooms, and give it to distant officials, who have little capacity to achieve results. (Cuban, 2004, pp. 104, 112–114; cf. D'Agostino, Welsh, & Corson, 2007)

Although New York is a state that exercises considerable control over curricula, examinations, certification, and a host of other matters, state education department officials in New York have learned that, to be successful, the manner in which these requirements are determined and set must be a highly collaborative process. In New York, this has meant involving hundreds, and sometimes thousands, of players in each major decision taken. The desirability of this collaborative approach also can be seen in changes made by the state of Kentucky as part of the Kentucky Education Reform Act of 1990 (KERA). In addition to equalizing funding across the state and providing substantial support for districts and schools, the act placed more emphasis on collaborative decision making. Prior to the reform act, the state education department tended to be aloof and focused on regulation and accountability but, under the reform act, there is now more collaborative decision making, partnership, and the creation of new levels of trust between districts and the state (Alston, 1996; Keedy & McDonald, 2007; Lusi, 1997). This increased level of shared decision making is one way in which states can improve the likelihood that their reform efforts will be well designed, well received, and result in desired educational outcomes.

The Role of Intermediary Organizations in Assisting States in Policymaking

A discussion of how states influence science education policy would not be complete without mentioning the role that intermediary organizations play in that process. Organizations such as the Education Commission of the States, the National Governors Association, Achieve, the National Association of State Boards of Education, and the Council of Chief State Schools Officers (CCSSO) provide resources to states as they consider which policies will be most likely to improve education in their state. The CCSSO, established in 1928, has been especially active in a sustained manner over decades in both mathematics and science education, much of it under the able leadership of Dr. Rolf Blank. They have produced an enormous amount of helpful literature for state education agencies to use, dealing with science standards and curricula (Blank, Porter, & Smithson, 2001; Blank, 2005; Council of Chief State School Officers [CCSSO], 2007; Webb, 1999, 2002), science assessments and other state science indicators (Blank, Langesen, & Petermann, 2007), science teacher certification and professional development (Blank & de las Alas, 2008; Blank, de las Alas, & Smith, 2008; Interstate New Teachers Assessment and Support Consortium, 2002; Smithson & Blank, 2007), and advice to the federal government regarding how to work with state education agencies in science, technology, engineering, and mathematics (STEM) education reform (e.g., Blank, 2000). Because the presence and influence of science specialists in state education departments varies from state to state and from decade to decade, providing up-to-date summaries of what other states are doing often is invaluable to educators working at the state level. Compilations of state policies, procedures, and future intentions also have served as useful starting points for many educational and policy researchers and think tanks for deeper exploration of issues of common interest. State education agencies have worked under the auspices of these various intermediary organizations to influence federal legislation, regulations, and funding related to science education as the federal government's role in education has increased. With the help of these intermediary organizations, many chief state school officers and other representatives of the states frequently testify before Congressional committees on matters regarding K–12 science education as well as to federal agencies and quasi-federal agencies who support and/or study science education.

A Limited Research Base

As indicated at the beginning of this chapter, the amount of research conducted on the role of state education agencies in science education policy development is meager. The role of state education agencies in state decision making in any area of the curriculum has been the subject of very few comprehensive studies. An analysis of 20 case studies of state decision making in Minnesota between 1971 and 1991 suggests there is no existing overarching analytical framework adequate to explain state education policymaking (Mazzoni, 1993). In addition, none of the cases Mazzoni examined dealt with science education in the state, also suggesting that science education did not rank high in the state's priorities. In another study, a Rasch model was applied to 43 educational policies collected from across 50 states (Lee, 1997). This study concluded that "low-cost, easy-to-implement and conventional policies were more likely to be adopted by the states" (p. 41). This is one reason why increased testing is a policy vehicle of choice for both the federal government and for state governments. Because testing is relatively cheap compared to alternatives such as expanded professional development for teachers, increased pay for teachers, longer school years, longer school days, or substantially smaller class sizes, it tends to be the preferred policy vehicle. Raising standards for beginning teachers and requiring externally administered teacher examinations before issuing teaching licenses also are popular legislative remedies for similar reasons, despite the paucity of evidence that such procedures predict teacher effectiveness (Angrist & Guryan, 2008).

Chatterji (2002) in his study of 63 published reports that purported to be studies of systemic reform found little to commend and much to criticize, and a major review by the National Research Council of research on science teacher preparation found the research so unreliable and spotty that nothing of any substance could be concluded (see also Kahle & Woodruff, Chapter 3, this volume). Even basic questions could not be answered from the research literature (Windschitl, 2005), such as:

What kinds of people make good science teachers?

In what ways is current teacher education effective?

What are the costs and benefits of alternative models of teacher preparation?

Does preservice teacher preparation matter in the life of a professional?

A rare longitudinal study of a sample of 22 secondary science teachers over 17 years in Australia by Richard White and colleagues (e.g., Arzi & White, 2008) is currently investigating how teacher subject matter knowledge changes over time. It argues for career-long support of teachers, even in areas where it is assumed that teachers possess expertise, and it proposes a three-phase model for such support. However, we must note that 22 teachers is a very small sample size, which is consistent with one of Windschitl's criticisms of policy research in science education.

One effort that would improve the quality and potency of policy research in science education would be the construction of better state data systems. A 2008 survey of all 50 states and the District of Columbia in terms of state data systems found 48 states had 5 or more out of 10 essential elements of a "robust state longitudinal data system." But, only 21 states had a teacher identifier to match teachers to students, 17 states collected student-level course completion and transcript information (with another 9 planning to do so), and only 29 states can collect college readiness test scores, with 12 states having no plans to do so (Guidera, Smith, Kowalski, Laird, & Wiggins, 2008).

One problem that could be addressed if better data systems were in place is the problem of attrition among public school mathematics and science teachers. This is a complex issue and a perennial problem in American education going back to the 1950s when it was first raised as a critical issue at the national level. State education agencies have convened numerous panels on this issue, have analyzed their data to determine patterns of attrition, and have implemented policies to address perceived deficiencies. Despite these efforts, nationally, the percent of science teachers who left teaching between 1988–89 and 2004–05 has not changed measurably according to the School and Staffing Survey conducted by the National Center for Education Statistics of the U.S. Department of Education (NCES, 2008b). The story is, of course, different in various locations, where supply and demand problems for science teachers can vary considerably, but nationally, the attrition rate has stayed relatively constant. There also is variation across science disciplines, with physics being a field with perennial shortages. Until we have much better data systems at the state and district level, the reasons for these shortages will remain a puzzle to both researchers and policymakers.

Is More Centralized Authority Over Science Education the Answer?

Despite calls over the years for educational change, and despite substantial funding efforts by various private and public foundations, by NSF, and by numerous other government agencies, the anticipated gains in educational outcomes remain elusive (cf. Clune, 1993; CCSSO, 2006). Frustrated with the pace and effects of educational reform, many experts have advocated a more systemic and centralized approach to education policy development at local, state, and national levels. Such a systemic approach requires centralization of power and authority. We already have noted that highly centralized models already exist in the United States in states like Hawaii and New York, but, curiously, these organizational systems rarely are cited as models in policy debates, and there is a lack of systematic and sustained study on which approach is most effective. As already noted several times before in this chapter, there is almost no study of the role of state education agencies as policy actors except in highly focused areas such as state assessment and teacher certification. We rarely have in-depth evaluations of large-scale change efforts, which are needed to better understand the various problems that arise from the creation of these reform efforts and their implementation, as one study of eleven statewide systemic initiatives (SSIs) noted (Hamilton, McCaffrey, Stecher, Klein, Robyn, & Bugliari, 2003). The SSIs were 10 million dollar investments by NSF intended to serve as levers to substantially transform STEM education in the respective states, but they had, at best, only modest success in achieving their intended outcomes (Kahle, 2007).

Before concluding that more highly centralized power and authority over education is the best route to science education reform, it is instructive to consider the reasons why success is often elusive, regardless of whether the system is highly centralized or decentralized. Research over many years about complex systems applied to K–12 science education suggests many factors that make lasting and positive change difficult in a dynamic and chaotic environment. According to Knapp (1997), these include:

1. *Unbridled enthusiasm of the innovators*. Americans have always been fans of the technological fix. For every problem there is believed to be a technological solution. Policymakers as well as the public at large continue to believe that a quick and effective solution can be found to problems in American science education. This means that new innovations are frequently held up to have more promise than is realistically achievable.
2. *Reification of the current materials, approaches, and systems*. Systems have a strong tendency to be conservative and to resist change by slowly changing the innovation back into that which it was designed to replace.
3. *Skepticism*. Since change is threatening, skepticism naturally arises about the reliability of the evidence favoring the innovation. This situation is not helped by the fact that most "third-party" evaluations are controlled and financed by the developers themselves, so that it is very rare to find a critical evaluation of an intervention. The lack of controlled experiments or even good quasi-experiments,

combined with poor methodological rigor provides further support for a skeptic's position.

4. *Time, fiscal, and emotional commitments.* Rarely is an educational innovation funded sufficiently or for long enough to achieve its effects. Conversely, emotional commitments can inhibit necessary adaptations of the innovations to new contexts for which it is ill suited.
5. *Loosely-coupled system.* America's loosely-coupled system of education does not allow innovations to spread quickly because each jurisdiction must be convinced independently to adopt the new innovation.
6. *Disorganized, fragmented federal R & D system.* The federal government generally favors university-controlled efforts, short time horizons rather than long-term studies, and evaluation of process rather than research on learning. The federal government also provides little or no money to support implementation, usually has no institutional memory with respect to the organization receiving the funds, duplicates development efforts within the same agency and across federal agencies, fails to pay attention to existing systems and existing markets and distribution systems, has little understanding of how K–12 education "actually" works, has a dismissive attitude toward practitioners and administrators, and supports very few of the longitudinal studies needed to measure gains over time in a controlled manner.
7. *Constantly changing assemblage of state, regional, and local initiatives/reforms.* A patchwork quilt of innovations requires ongoing local attention and response, which affects the speed with which reforms and materials are accessed, evaluated, adopted, implemented, and monitored.
8. *Sociopolitical nature of schooling.* Education takes place within a sociopolitical context so that decisions are rarely made solely on the basis of what is good for learners, the availability of research, or sound logic and reasoning.

With such a long list of factors that contribute to making change difficult, it is surprising that any change at all can be made or be effective. But, we know that change is possible, as the New York story indicates many reforms have been successful. What we do not have is a body of systematic research to answer the question of which organizational structures and approaches to policy development and implementation lead to success and which ones do not.

SUMMARY

Our goal in writing this chapter was to help the reader develop an appreciation for the various roles that state education agencies play in the formulation and dissemination of science education policy, drawing principally on the experience of the New York State Education Department. A number of observations emerge from our review that are relevant to the theme of this volume. The first is that during the past two centuries, there has been a progressively greater degree of state and federal control over what happens at the district and local school level. Officially, education has been the responsibility of each individual state even though the states initially granted much of that responsibility to the local districts. But, as the educational system has become larger and more complex over the years, states have chosen to exercise more and more of that authority through centralized state bureaucracies. The increase in federal control is a more recent phenomenon, brought on largely by civil rights legislation in the 1960s and concerns about national security during the Cold War. In the case of science education, the involvement of the federal government grew rapidly following the establishment of the National Science Foundation in 1950. This has created a relationship between the states and the federal government that has greatly affected the way states administer their educational programs.

Related to this federal involvement is the question of where the funds will come from that are needed to make things happen. Clearly, money matters. From the earliest allocations of funding for the common schools to the golden age of the New York State Education Department with large sums provided by the federal government, money has driven innovation. A research question of considerable interest is what is the most appropriate and effective means of funding education. At the time of this writing, the U.S. Secretary of Education is about to begin distributing $4.35 billion to states to improve education. This money is needed because states and local school districts simply do not have the resources to undergo the scope of innovation that the federal government can promote, both through grants from agencies such as the Department of Education and NSF, and through direct programmatic support. Without such funding, states are limited to low-cost reforms.

We also observed that very little research has been conducted, or is currently being conducted, on the role that state education departments play in helping to set a national policy agenda in science education or in how states are organized to carry out various policy mandates. There are many opportunities to conduct policy research of this kind, but little has been done. In the case of New York, for example, the state has carried out a program of statewide accountability testing for nearly 150 years. It is a remarkable story of policy implementation in a controversial area that has survived with only minor changes for a very long time. How has the state been able to carry out such a program for so long in the face of concerns that the program would lead to a narrowing of the curriculum and reduced creativity and innovation in teaching? With such a massive and comprehensive testing program in place for so long, the Regents examination program could provide a wealth of information to policy researchers about the effect of a testing program like this on student learning, teacher morale and attrition, the efficiency of educational systems that utilize such an accountability system, and more. New York also has an intermediate level of organization in the administration of its education system. The Boards of Cooperative Educational Services (BOCES) serve primarily to provide services to smaller, rural school districts, but also function to distribute a wide range of services more efficiently throughout the state. How successful have they and other similar intermediate-level organizational structures that are used by other states been? This, too, would make an excellent subject for study.

Given the increased involvement of the federal government in education, both through increased funding but also through the direction-setting that the federal government engages in, from a research perspective the states could be considered as test beds for policy implementation. In the case of federal legislative mandates, states must operationalize what the federal government creates as policy. The mandates have to be implemented through the organizational and regulatory structures of the states. There is no other way. One promising area of research could be a comparison of how successful the various state systems are at effectively implementing these national policies. Although there are certainly many ways in which states differ in how they conduct their educational affairs, the broad outlines of their responsibilities and policymaking actions regarding science education are reasonably consistent with what has been witnessed in the state of New York. It is hoped that this skeletal outline of the relatively large influence that state education agencies play in the K–12 science education landscape of their own state will stimulate deeper and fuller research about their roles and influence and greater consideration by policymakers of the unique roles that state education agencies play nationally. Despite what anyone outside of the state education agencies thinks about the matter, state education agencies can and will be vital participants in science education policy making and policy implementation as they have from the inception of public schools in the United States.

REFERENCES

Abell, S. K., & Lederman, N. G. (Eds.). (2007). *Handbook of research on science education*. New York, NY: Routledge.

Alston, E. (Ed.). (1996). *Kentucky Education Reform Act*. Frankfort, KY: Legislative Research Commission.

Angrist, J. D., & Guryan, J. (2008). Does teacher testing raise teacher quality? Evidence from state certification requirements. *Economics of Education Review, 27*(5), 483–503.

Atkin, J. M., & Black, P. (2003). *Inside science education reform: A history of curricular and policy change*. New York, NY: Teachers College Press.

Azri, H. J., & White, R. T. (2008). Change in teachers' knowledge of subject matter: A 17-year longitudinal study. *Science Education, 92*(2), 221–251.

Blank, R. K. (2000). *Summary of findings from SSI and recommendations for NSF's role with states: How NSF can encourage state leadership in improvement of science and mathematics education*. Washington, DC: Council of Chief State School Officers.

Blank, R. K. (2005). *Surveys of enacted curriculum: A guide for SEC state and local coordinators*. Washington, DC: Council of Chief State School Officers, CCSSO SEC Collaborative.

Blank, R. K., Porter, A., & Smithson, J. (2001). *New tools for analyzing teaching, curriculum and standards in mathematics & science. Results from Survey of Enacted Curriculum Project. Final report*. Washington, DC: Council of Chief State School Officers.

Blank, R., Langesen, D., & Petermann, A. (2007). *State indicators of science and mathematics education*. Washington, DC: Council of Chief State School Officers.

Blank, R. K., & de las Alas, N. (2008). *Current models for evaluating effectiveness of teacher professional development. Recommendations to state leaders from leading experts. Summary report of a CCSSO conference*. Washington, DC: Council of Chief State School Officers.

Blank, R. K., de las Alas, N., & Smith, C. (2008). *Does teacher professional development have effects on teaching and learning? Analysis of evaluation findings from programs for mathematics and science teachers in 14 states*. Washington, DC: Council of Chief State School Officers.

Braun, D. (1993). Who governs intermediary agencies? Principal-agent relations in research policy making. *Journal of Public Policy, 13*(2), 135–162.

Council of Chief State School Officers. (2006). *Mathematics and science education task force: Policy statement executive summary*. Washington, DC: Author.

Council of Chief State School Officers. (2007). *Surveys of the enacted curriculum interactive CD-ROM*. Washington, DC: Author.

Chatterji, M. (2002). Models and methods for examining standards-based reforms and accountability initiatives: Have the tools of inquiry answered pressing questions on improving schools? *Review of Educational Research, 72*(3), 345–386.

Clune, W. H. (1993). The best path to systemic educational policy: Standard/centralized or differentiated/decentralized? *Educational Evaluation and Policy Analysis, 15*(3), 233–254.

Coleman, J. S. (1966). *Equality of educational opportunity.* Washington, DC: U.S. Department of Health, Education, and Welfare.

Committe on Prospering in the Global Economy in the 21st Century. (2001). *Rising above the gathering storm: Energizing and employing America for a brighter economic future.* Washington, DC: National Academy Press. (Revised and updated 2008)

Cuban, L. (2004). A solution that lost its problem: Centralized policymaking and classroom gains. In N. Epstein (Ed.), *Who's in charge here? The tangled web of school governance and policy* (pp. 104–130). Washington, DC: Brookings Institution Press.

D'Agostino, J. V., Welsh, M. E., & Corson, N. M. (2007). Instructional sensitivity of a state's standards-based assessment. *Educational Assessment, 12*(1), 1–22.

DeBoer, G. E. (1991). *A history of ideas in science education: Implications for practice.* New York, NY: Teachers College Press.

Dunn, R. J. (2006). State departments of education. In F. W. English (Ed.), *Encyclopedia of educational leadership and administration* (Vol. 2, pp. 960–962). Thousand Oaks, CA: SAGE.

Family Educational Rights and Privacy Act of 1975 (Public Law 108-446).

Fensham, P. (2008). The link between policy and practice in science education: The role of research. *Science Education, 93*(6), 1076–1095.

Firestone, W. A., Mayrowetz, D., & Fairman, J. (1998). Performance-based assessment and instructional change: The effects of testing in Maine and Maryland. *Educational Evaluation and Policy Analysis, 20*(2), 95–113.

Folts, J. D. (1996). *History of the university of the state of New York and the state education department, 1784–1996.* Albany: New York State Education Department.

Fusarelli, B. C., & Cooper, B. S. (Eds.). (2009). *The rising state: How state power is transforming our nation's schools.* Albany, NY: State University of New York Press.

The Goals 2000: Educate America Act (Pub. L. 103-227). 103rd U.S. Congress.

Goertz, M. E. (2001). *Assessment and accountability systems: 50 state profiles.* Philadelphia, PA: Consortium for Policy Research in Education, University of Pennsylvania.

Guidera A. R., Smith, N. J., Kowalski, P., Laird, E., & Wiggins, E. (2008). *Data quality campaign: Using data to improve student achievement.* Austin, TX: Data Quality Campaign.

Hamilton, L. S., McCaffrey, D. F., Stecher, B. M., Klein, S. P., Robyn, A., & Bugliari, D. (2003). Studying large-scale reforms of instructional practice: An example from mathematics and science. *Educational Evaluation and Policy Analysis, 25*(1), 1–29.

Horner, H. H. (1954). *Education in New York State, 1784–1954.* Albany, NY: The University of the State of New York, The State Education Department.

Hurley, M. (2007). School reform and high stakes testing: Their impact on students in the United States and England. *International Journal of Learning, 14*(2), 239–246.

Improving America's Schools Act of 1994 (Public Law 103-382)

Indiviuals with Disabilities Act of 1990 as amended in 2008 (Public Law 110-325).

Interstate New Teachers Assessment and Support Consortium (2002). *Model standards in science for beginning teacher licensing and development: A resource for state dialogue*. Washington, DC: Council of Chief State School Officers, Interstate New Teacher Assessment and Support Consortium, Science Standards Drafting Committee.

Kahle, J. B. (2007). Systemic reform: Research, vision, and politics. In S. K. Abell & N. G. Lederman (Eds.), *Handbook of research on science education* (pp. 911–942). New York, NY: Routledge.

Kandel, I. L. (1936). *Examinations and their substitutes in the United States*. New York, NY: The Carnegie Foundation for the Advancement of Teaching.

Keedy, J. L., & McDonald, D. H. (2007). The instructional capacity building role of the state education agency: Lessons learned in Kentucky with implications for No Child Left Behind. *Planning & Changing, 38*(3/4) 131–147.

Kirst, M. W. (2004). Turning points: A history of American school governance. In N. Epstein (Ed.), *Who's in charge here? The tangled web of school governance and policy* (pp. 14–41). Washington, DC: Brookings Institution Press.

Knapp, M. S. (1997). Between systemic reform and the mathematics and science classroom: The dynamics of innovation, implementation, and professional learning. *Review of Educational Research, 67*(2), 227–266.

LaSpina, J. A. (2009). *California in a time of excellence: School reform at the crossroads of the American dream*. Albany, NY: State University of New York Press.

Lee, J. (1997). State activism in education reform: Applying the Rasch model to measure trends and examine policy coherence. *Educational Evaluation and Policy Analysis, 19*(1), 29–43.

Lusi, S. F. (1997). *The role of state departments of education in complex school reform*. New York, NY: Teachers College Press.

Mazzoni, T. L. (1993). The changing politics of state education policy making: A 20-year Minnesota perspective. *Educational Evaluation and Policy Analysis, 15*(4), 357–379.

Montgomery, S. L. (1994). *Minds for the making: The role of science in American education, 1750–1990*. New York, NY: The Guilford Press.

National Association for Research in Science Teaching. (2008). *2008 NARST annual international conference: Impact of science education research on public policy. March 30–April 2, Baltimore. Program and Abstracts.* Reston, VA: National Association for Research in Science Teaching.

National Center for Education Statistics. (2008a, April). *Characteristics of the 100 largest public elementary and secondary school districts in the United States: 2004–05*. Washington, DC: Author.

National Center for Education Statistics. (2008b). *Attrition of public school mathematics and science teachers*. Washington, DC: Author.

National Commission on Excellence in Education. (1983). *A Nation at risk: The imperative for educational reform*. Washington, DC: U.S. Department of Education.

New York State Legislature. (1795). *An act for the encouragement of schools. Chapter 75, Law of 1795*. Albany, NY: Author.

No Child Left Behind Act of 2001 (Public Law 107-110)

Ogren, C. A. (2005). *The American state normal school*. New York, NY: Palgrave Macmillan.

Pankratz, R. S., & Petrosko, J. M., (Eds.). (2000). *All children can learn: Lessons from the Kentucky reform experience*. San Francisco, CA: Jossey-Bass.

Rogan, J. M. (2007). How much curriculum change is appropriate? Defining a zone of feasible innovation. *Science Education, 91*(3), 439–460.

Rudolph, J. L. (2002). *Scientists in the classroom: The Cold War reconstruction of American science education*. New York, NY: Palgrave.

Rury, J. L. (2009). *Education and social change: Contours in the history of American schooling*. New York, NY: Routledge.

Sizer, T. (1964). *Secondary schools at the turn of the century*. New Haven, CT: Yale University Press.

Smith, A., & Hall, E. (1902). *The teaching of chemistry and physics in the secondary school*. New York, NY: Longmans, Green.

Smithson, J., & Blank, R. (2007). *Indicators of quality of teacher professional development and instructional change using data from surveys of enacted curriculum: Findings from NSF MSP-RETA project*. Washington, DC: Council of Chief State School Officers.

Snyder, T. D., Ed. (1993). *120 years of American education: A statistical portrait*. Washington, DC: National Center for Education Statistics, U.S. Department of Education.

Spring, J. (2008). *The American school: From the Puritans to No Child Left Behind* (7th ed.). Boston, MA: McGraw Hill.

Stecher, B. M., & Klein, S. P. (1997). The cost of science performance assessments in large-scale testing programs. *Educational Evaluation and Policy Analysis, 19*(1), 1–14.

Steffes, T. L. (2008). Solving the "rural school problem": New state aid, standards, and supervision of local schools, 1900-1933. *History of Education Quarterly, 48*(2), 181–220.

Teitelbaum, P. (2003). The influence of high school graduation requirement policies in mathematics and science on student course-taking patterns and achievement. *Educational Evaluation and Policy Analysis, 25*(1), 31–57.

Tinkelman, S. N. (1966). *Regents examinations in New York State after 100 years*. Albany, NY: University of the State of New York.

Tobin, K. G., & Fraser, B. J. (Eds.). (1998). *International handbook of science education*. Boston, MA: Kluwer Academic.

Tolley, K., & Beadie, N. (2001). Reappraisals of the Academy Movement. *History of Education Quarterly, 41*(2), 216–224.

University of the State of New York (1891). *Regents Bulletin, No. 5. Academic Syllabus*. Albany, NY: Author.

University of the State of New York. (1904). *High school department. Bulletin 26: Academic examination papers for the academic year 1904*. Albany, NY: Author.

University of the State of New York. (1924). *A report of the survey of the Lockport school system by the State Department of Education*. Albany, NY: Author.

University of the State of New York. (1960). *Administrator's handbook of the secondary school curriculum of New York State*. Albany, NY: Bureau of Secondary Curriculum Development. New York State Education Department.

University of the State of New York. (1962). *Quality in education in New York State–1962*. Albany, NY: Author.

University of the State of New York. (1987). *Information bulletin: Regents competency testing program in science. A guide to the core process skills and content understandings*. Albany, NY: Author.

University of the State of New York. (1990). *New York State program evaluation test in science. Grade 4. Directions for administering and scoring*. Albany, NY: Author.

Vergari, S. (2009). New York. In B. C. Fusarelli & B. S. Cooper (Eds.), *The rising state: How state power is transforming our nation's schools* (pp. 65–87). Albany, NY: State University of New York Press.

Vygotsky, L. S. (1978). *Mind in society: The development of higher psychological processes.* Cambridge, MA: Harvard University Press.

Waterman, R. W., & Meier, K. J. (1998). Principal-agent models: An expansion? *Journal of Public Administration Research and Theory, 8*(2), 173–202.

Webb, N. L. (1999). *Alignment of science and mathematics standards and assessments in four states. Research Monograph No. 18*. Washington, DC: Council of Chief State School Officers & National Institute for Science Education, University of Wisconsin-Madison.

Webb, N. L. (2002). *Alignment study in language arts, mathematics, science, and social studies of state standards and assessments for four states.* Washington, DC: Council of Chief State School Officers.

White, R. (2001). The revolution in research in science teaching. In V. Richardson (Ed.), *Handbook of research on teaching*. Washington, DC: American Educational Research Association.

Wiley, G. M. (1937, April 15). *The changing function of Regents Examinations (*Bulletin No. 1114) Albany, NY: University of the State of New York Press, University of the State of New York.

Windschitl, M. (2005). Guest editorial: The future of science teacher preparation in America: Where is the evidence to inform program design and guide responsible policy decisions? *Science Education, 89*(4), 525–534.

CHAPTER 8

SCIENCE EDUCATION POLICY AND STUDENT ASSESSMENT

Rodger W. Bybee

INTRODUCTION

In recent decades, science educators increasingly have recognized the influence of local, state, and federal policy on science education programs and practices. The United States, for example, has a federal policy that all students be tested to measure their knowledge of science as defined by their state's content standards. States have comprehensive examinations used for accountability and graduation requirements. States also have policies about how many science classes are required for graduation and whether students experience laboratories. What is the origin of policies for science education? How are science education policies developed and implemented?

To further confuse our understanding of policies in science education, some formal policies have unintended consequences for science education, for example, the emphasis in the federal No Child Left Behind Act (2001) on literacy and math has reduced emphasis on science. (See also DeLucchi & Malone, Chapter 12, this volume.) In some cases, there is a

The Role of Public Policy in K–12 Science Education, pp. 211–239

clear and directly negative influence of policies on school science programs, e.g., local boards implementing policies supporting teaching intelligent design and "alternative theories" (DeBoer, Chapter 10, this volume). In other cases, there seems to be a policy, but there is no official authority for the policy, for example, many in the science education community embrace a "policy" for inquiry-based instruction. Finally, there are formal national documents recommending policies. These national documents are voluntary but still directly influence state and local policies, e.g., the *National Science Education Standards* (National Research Council [NRC], 1996) and *Benchmarks for Science Literacy* (American Association for the Advancement of Science [AAAS], 1993).

Although some in the science education community recognize the role of policies for guiding the practice of science education, many do not. Indeed, one often hears comments from science educators indicating very little understanding about the processes by which policies that affect their work are made. Because of the increasing role state and federal agencies are playing in education, the importance of having scientists, science educators, and science teachers who understand the intersection of science education and policy is increasing.

Diverse examples of science education policies introduce the general purpose of this chapter. The chapter sheds light on the complex world of science education policy using assessment as a specific example. Several questions are explored in the chapter.

- Where do science education policies originate?
- How are science education policies formulated?
- Who influences the formulation of science education policies?
- What influences the formulation of science education policy?
- How are science education policies implemented?
- What are the intended and unintended effects of science education policies?

The discussion begins with an assumption that formal policies are formal rules or guidelines established by an authoritative body and that those policies direct the actions of a larger group of people when the policies are implemented. The number of laboratory courses required for graduation serves as an example of a formal policy. This discussion does not address informal "policies" that may be widely accepted preferences or even orthodoxies within the field of science education often not specifically authorized by any official body.

After a discussion that clarifies my perspective on policies in science education, this chapter presents four cases exploring the relationships between policy and assessments. The four cases represent policies at the international, national, and state levels. These cases and the aforementioned questions are the basis for detailed discussion of how the policies actually were formulated.

A PERSPECTIVE ON POLICIES IN SCIENCE EDUCATION

This section presents a framework for describing various reform initiatives in science education. I suggest four basic forms of initiatives for reforming science education, which can be summarized as four Ps: changing emphasis among the *purposes*; stating new *policies*; developing new *programs*; and implementing new *practices*. In general, such changes can be described at local, state, national, and even international levels. The following section introduces these perspectives. The four perspectives are described in greater detail in *Achieving Scientific Literacy: from Purposes to Practices* (Bybee, 1997).

Locating Policies in Science Education

Discussion of policies as one of the steps from purposes to practices is relatively new for the science education community. For decades, science educators have focused on goals or purposes, curriculum programs, and instructional practices but have ignored the question of how policy regarding purposes, programs, and practice is developed and implemented. Typical discussions have been framed by themes of "research to practice" or "theory to practice" as if purposes, programs, and practices exist independently of the influence of policy. Science education requires alignment among goals, plans, programs, and instruction; stated another way; coherence among the 4Ps and perhaps more. Following is a brief introduction to the four perspectives and the particular role that policy plays in educational reform. In the view presented here, policy is one of the four perspectives on reform.

Purposes of Science Education

Purpose refers to the aims and goals of science education. While goals such as scientific knowledge, methods, and applications to personal and social issues have been maintained, the emphasis and priorities of the goals have changed. Very importantly, statements of purpose are universal, applying to the larger science education community (Bybee & DeBoer, 1994). Achieving scientific literacy for all students is an example of a purpose statement. An example of a comprehensive statement of purpose in science education is *Science for All Americans* (AAAS, 1989), which describes what all citizens should know about science to be science literate.

Policies for Science Education

Policies are more specific statements of the purposes for science education. Policy statements translate the more abstract purposes into concrete plans and directions to those responsible for different domains of science education such as teacher education and curriculum and assessment programs. They help clarify and make concrete the various aspects of programs, for example, content for different disciplines, what is appropriate for different ages and stages of students, and programs for school science. This definition places the *Benchmarks for Scientific Literacy* (AAAS, 1993) and *National Science Education Standards* (NRC, 1996) in the policy category, specifically describing science content for school programs. Because these documents provide specific descriptions of content for grade level bands, for example, K–2, 3–4, and so forth, they represent policies that can guide the development of school science programs and science assessments. Other policies, for example, NCLB and standards for teacher education, influence programs for assessments and teacher certification, respectively.

For this discussion, science education policies can be thought of as a plan by some group(s), based on a purpose and designed for a more concrete set of actions. The Program for International Student Assessment (PISA), for example, uses a plan formulated by the Organization for Economic Cooperation and Development (OECD) for science assessment based on a perspective of scientific literacy. The actual policy document is *Assessing Scientific, Reading and Mathematical Literacy: A Framework for PISA 2006* (OECD, 2006).

Programs for Science Education

The translation of policies into actual courses, curriculum materials, software, equipment, and assessments defines the difference between policies and programs. Programs, for example, are unique to grade levels, disciplines, and domains such as teacher education and certification. The actual assessment system, which is based on content standards for the state of Washington—the WASL or Washington Assessment of Student Learning—is an example of a program. The PISA 2006 Science test and questionnaires for students and parents is a second example.

Practices for Science Education

Here, use of the term practices refers to the specific actions of science educators. Practice is the unique and most fundamental application of the more universal and abstract purpose statement. What teachers emphasize when they interact with students, how they help their students achieve the purpose of scientific literacy, serves as an example of practices in science education.

Although one may be tempted to think of the 4Ps as a sequence of stages such that the ideal would be to begin with purposes and develop policies followed by programs and practices, history suggests that, generally, the reform of science education has not represented the ideal. For the most part, the ideal sequence does not happen in reform movements. In the 1960s, science education witnessed reform based on new curriculum programs with subsequent attention to teacher education and other components. In recent history, policies in the form of standards have been a major initiative in reform movements. Using the 4Ps as different perspectives on a reform, one can analyze the changes in science education by considering which P was emphasized at the origin and how well the other Ps were completed as part of the overall structure, function, and complete reform.

Here is the essential issue: science education requires a consistent and coherent system that has a clear and coordinated set of purposes, policies, programs, and practices. If science education had such a system, we would reduce the need for continually addressing the problem of inconsistency of, for example, purposes with programs. Such a view can account for changing purposes without addressing the other components; policies, programs, or practices. One can use this view to analyze reform from pipeline concerns in the 1950s and 1960s to scientific literacy for all in the 1980s and 1990s to science, technology, engineering, and mathematics (STEM) and workforce competencies in the early 2000s.

In the four cases that follow, I discuss assessment as a particular example of formulating policies from purposes that did or will become assessment programs and ultimately affect practice. The cases are based on my own personal experiences with assessment programs at the international, national, and state levels. I begin at the international level with a discussion of PISA 2006 Science.

POLICY FORMATION FOR THE PROGRAM FOR INTERNATIONAL STUDENT ASSESSMENT (PISA): AN INTERNATIONAL CASE

An Overview of PISA

The Program for International Student Assessment (PISA) is an international comparative educational survey carried out for the Organization for Economic Cooperation and Development (OECD). Established in the late 1990s, PISA surveys the mathematics, science, and reading competencies of 15-year-old students in OECD member countries and selected non-member countries. These students are approaching the end of compulsory schooling in most participating countries.

There are a number of purposes for the PISA program. One purpose is to enable each country to assess how well its educational system is providing students with the reading, mathematical, and scientific literacy that they will need in adult life; a second is to monitor trends in performance over time; and a third is to determine the relationship between student performance and various demographic, social, economic, and educational variables in each country (OECD, 2006).

PISA surveys student competencies every 3 years. The first survey took place in 2000 and the second in 2003. For each survey, mathematics, science, or reading is chosen as the major assessment domain and is given greater emphasis. Specifically, a larger proportion of the test is devoted to the subject designated as the major domain for a particular cycle. The remaining two areas are assessed as minor domains. In 2000, the major domain was reading, in 2003 it was mathematics, and in 2006 it was science. Testing is done in four 30-minute blocks, with 5,000 to 10,000 students taking the test in each country. Multiple test forms are used and the forms are linked so that a wider range of ideas can be tested. The results of these surveys have been published in international reports (OECD, 2001, 2003a, 2004a, 2004b, 2007a, 2007b).

Student Tests

The PISA assessments center on the reading, mathematical, and scientific *literacy* of students. This literacy perspective focuses on the extent to which students can use the knowledge and skills they have acquired when confronted with life situations and challenges for which that knowledge may be relevant. In the context of this chapter's focus on science education, PISA seeks to assess the extent to which students can use their scientific knowledge and skills to understand, interpret, and resolve various kinds of scientific situations and challenges. In science, literacy is defined as:

- Scientific knowledge and use of that knowledge to identify questions, acquire new knowledge, explain scientific phenomena and draw evidence-based conclusions about science-related issues
- Understanding of the characteristic features of science as a form of human knowledge and enquiry
- Awareness of how science and technology shape our material, intellectual, and cultural environments
- Willingness to engage in science-related issues and with the ideas of science, as a reflective citizen (OECD, 2006).

This definition of scientific literacy involves knowledge of scientific principles, knowledge of how science is done, and dispositions to both use scientific knowledge and to engage in discussions of science-related issues.

Student and School Questionnaires

In each survey, students also complete a questionnaire designed to gather relevant background data about their personal characteristics, opinions, preferences and aspirations, characteristics of their home and family environments, and selected features of their school environment. In addition, school staff, usually an administrator, completes a short questionnaire about various aspects of organization and educational provision in schools. This background and contextual information is collected to enable a detailed study of factors within and between countries that are associated with varying levels of reading, mathematical, and scientific literacy among the 15-year-old students of each country.

The Structure and Function of the PISA Project

The Structure of PISA

The PISA project is overseen by the General Secretariat within the OECD in Paris. Policies guiding the project are set by the PISA Governing Board, an OECD committee comprising delegates and observers who are largely senior educational administrators from the participating countries.

Each participating country has a national project implementation center. A National Project Manager coordinates activities at the national level. Typically, the National Project Manager works closely with the country's PISA Governing Board Member to establish a national perspective on policy matters, on issues related to the implementation of PISA, and on the analysis and reporting of outcomes that may be of particular relevance to the country. In the United States, PISA is administered by the National Center for Educational Statistics of the U.S. Department of Education.

The Function of PISA

The project is implemented internationally by a contractor appointed by the OECD. An international consortium led by the Melbourne-based Australian Council for Educational Research (ACER) was the sole contractor for each of the first three PISA survey cycles and, along with Cito of the Netherlands, was one of two contractors for PISA 2009. The ACER consortium partners for PISA 2006 were the National Institute for Educational Measurement (CITO) in the Netherlands, Westat and the Educational Testing Service (ETS) in the United States, and the National Institute for Education Policy Research (NIER) in Japan.

Two groups playing important roles in developing the PISA frameworks in science and in establishing policies for project implementation were the Science Forum and the Science Expert Group. The Science Forum was comprised of 30 representatives from OECD countries. The Science Forum provided the opportunity for participating countries to nominate national experts who could directly represent the interests and views of the country in considering certain detailed technical aspects of the science assessment. The Science Forum allowed for a wider base of expert input than was possible through the Science Expert Group which consisted of 12 individuals. The Science Forum considered priorities and issues at the time the science framework was being conceptualized, and it provided important input with respect to the development of survey material related to the assessment of science and the assessment of student attitudes to science. The Science Expert Group, on the other hand, had responsibilities that included: development of the assessment framework (OECD, 2006); review of items for the assessment and questions for the questionnaires; analysis of results from the field test; selection of trend items; and review of final results from the assessment and questionnaires. The Science Expert Group had representatives from Australia, Canada, France, Germany, Italy, Japan, Norway, Poland, Slovak Republic, United Kingdom, and the United States.

Policies for PISA 2006 Science

The Assessment Policy frameworks (OECD, 2000, 2003b, 2006) negotiated and agreed among participating countries form the basis for PISA assessments. The policies directly influence what is to be assessed, they defining the constructs that underpin the assessments in each of the different domains that are tested, and they give direction to the test item developers and thus the final test.

Where Did Policies for PISA Science 2006 Originate?

Policies for the 2006 test originated with the work of earlier Science Expert Groups (SEG) (e.g., OECD, 2000) and were approved by the PISA Governing Board. In late 2002, discussions between committee members and OECD and ACER leaders made it clear that the policy framework for 2006 should acknowledge the original frameworks for science, but the fact that science was to be the major domain in 2006 required an enhanced and more elaborated framework. Specifically, the new framework was to

- clarify and elaborate the concept of scientific literacy,
- include an attitudinal dimension of the assessment,
- identify scientific competencies,

- describe the distribution of items (score points) for scientific competencies and scientific knowledge, and
- recommend reporting scales for PISA Science 2006.

The result of these deliberations was a 25 page science framework document describing what should be tested in science (OECD, 2006).

Who Influenced the Formulation of Policies for PISA Science 2006?

The 30 individuals on the Science Forum representing participating countries certainly influenced what was to be tested, but the greater power resided in the 12-member Science Expert Group. Understandably, the opinions and recommendations of more senior individuals and those with a greater depth and breadth of experience in science education and assessment carried more weight and prevailed in the formulation of the policy framework.

The more experienced individuals also had a vocabulary and the capacity to present arguments that persuaded other members of the Science Expert Group. All suggestions for content to be tested and contexts to be used, however, had to be consistent with the overarching purpose of assessing scientific literacy. The Science Expert Group maintained the position that this was not to be primarily a test of science content as taught in the curricula of the schools in the various countries. Instead, this was to be a test of the uses of knowledge in the context of real-world problems that students might encounter in their adult lives. This sets PISA apart from other international tests that assess students' knowledge.

How Were Policies Formulated for PISA Science 2006?

The formal process consisted of meetings of the Science Expert Group, subsequent drafting of a document, and review by the Science Forum and the PISA Governing Board. The process was iterative and ultimately the proposed framework had to be approved by the PISA Governing Board. The informal process included discussions with OECD and ACER staff, and individual discussions with Science Forum and Science Expert Group members.

What Are the Intended (and Unintended) Effects of Policies for PISA Science 2006?

The science framework for PISA 2006 set parameters for the test and for the questionnaires on student background and attitudes. The policies provided guidance in the development of test items and direction for the assessment psychometrics, for example, reporting scales.

The unintended consequences included, among other things, a misinterpretation by some people that the assessment framework was meant to be a curriculum framework for countries and a misconception of what was meant by scientific literacy. Representatives from several countries saw the framework as a statement, that is, policies, of what should be taught in their school science programs. There was a perception that OECD and the PISA Governing Board were violating the country's sovereignty by suggesting what should be taught. Because most countries respond to the emphasis represented in the assessment framework, if the teaching of science was consistent with the PISA framework, it would not be an altogether negative way to approach science programs and practices. This would be an unintended positive consequence for those interested in scientific literacy. The mission of PISA, however, is not to influence how science is taught; it is instead to assess how well students are achieving the goal of science literacy in the various participating countries. A second misconception centered on the emphasis of the test. PISA explicitly emphasized scientific literacy (see Roberts, 2007). Because PISA 2006 reports scientific competencies (i.e., scientific literacy) and scientific knowledge (i.e., knowledge of science and knowledge about science), some have interpreted PISA Science 2006 primarily a test of scientific knowledge. The issue here rested on the perceived emphasis of the assessment, that is, discipline-based knowledge of science versus the actual emphasis, that is, the application of knowledge of and about science, scientific literacy (Loveless, 2009).

How Were Policies for PISA Science 2006 Implemented?

The policies were implemented as a program that included a student test and questionnaires for students and schools. The policies were implemented through education agencies in each country that were responsible for the sampling of students, the administration of the tests, the collection of the data, scoring of results, and submission of raw data to ACER.

POLICY FORMATION FOR THE NATIONAL ASSESSMENT OF EDUCATIONAL PROGRESS (NAEP): TWO CASES

An Overview of NAEP

The National Assessment of Educational Progress (NAEP) measures student science achievement nationally, state-by-state, and most recently across selected urban school districts. The NAEP is administered by the National Center for Education Statistics of the U.S. Department of Education. The program was authorized by Congress in 1969. Before 1988, results were reported at the national level only. In 1988, Congress passed legislation authorizing the reporting of results at the state level on a trial basis. Beginning with the 1996 assessment, the state-level testing was no longer considered to be a "trial" program. Currently, under federal No Child Left Behind legislation, all states must participate in the mathematics and reading tests, but the science test is voluntary. Students are sampled from participating states and results are reported at the state level, not the individual student or school level. There also is some below-state level reporting on request. The science test is given every 4 years; the mathematics and reading tests are given every 2 years. In 2005, forty-four of 50 states volunteered to have their students take the science test. The NAEP is the only systematically administered national level test that can provide a comparative measure of student knowledge in science across states. (See Cheek & Quiriconi, Chapter 4, this volume, for further discussion of the history of NAEP.)

Periodically, the framework underlying the science assessment (as well as the frameworks for the other subject areas) is revised or updated. Discussion in this chapter is based on two separate and different initiatives. The first is a revision of the science framework for the main NAEP. Development of the *Science Framework for the 2009 NAEP* contains a new set of recommendations for the NAEP Science Assessment to be administered beginning in 2009 and continuing until 2019. The framework provides guidance on the science content to be assessed, the types of assessment questions to be asked, and the administration of the assessment. A second initiative discussed here is the development of a framework for assessing technological literacy which likely will be implemented in 2014.

Any NAEP framework must be guided by NAEP purposes as well as the policies and procedures of the National Assessment Governing Board (NAGB), which oversees NAEP. The NAGB is a bipartisan, independent federal board that sets policy for the National Assessment of Educational Progress (NAEP). For the NAEP Science Assessment and Assessment of Technological Literacy, the main purpose is to establish guidelines regarding what students should know and be able to do in science for the 2009 test in science and for the technological literacy test in 2014. Meeting these purposes requires frameworks built by communities involved in science and technology education, respectively. Those communities must describe a rigorous body of science and technology knowledge and skills that are most important for NAEP to assess.

Developing the Science Framework

In September 2004, NAGB awarded a contract to WestEd and the Council of Chief State School Officers (CCSSO) to develop frameworks and assessment specifications for the 2009 NAEP exam. Working with representatives from the American Association for the Advancement of Science (AAAS) and the National Science Teachers Association (NSTA), the contractors developed a process for identifying what would be tested. A Steering Committee identified the overall emphasis, and a Planning Committee provided the needed expertise to develop the frameworks and specifications. In establishing the relative priority of the different areas of science content, the NAEP Steering Committee recommended the two national documents, *National Science Education Standards* (NRC, 1996) and *Benchmarks for Science Literacy* (AAAS, 1993), as representative of the leading science communities and their expectations for what students should know and be able to do in science. The inclusive nature of both these documents demonstrates the difficulty of identifying a key body of knowledge for students to learn in science and, therefore, what should be assessed. Neither document limits or prioritizes content as is necessary for developing an assessment, posing a considerable challenge to the framework developers. The development of the framework also was informed by research in science and science education, international assessment frameworks, and state standards. In the end, the NAEP 2009 science framework document represents a consensus of senior level science educators about what is most important for students to know about science. The consensus was reached over months of deliberation using the existing *National Science Education Standards* and *Benchmarks for Science Literacy* as a starting point.

Developing the Technological Literacy Framework

In 2014, there also will be a national assessment of technological literacy as part of the NAEP. The process leading up to this assessment began in October 2008 when a contract was awarded to WestEd for development of a framework and test specifications. Similar to the NAEP Science committee structure, a Planning Committee and a Steering Committee will make recommendations about what content to include in the test. The framework includes definitions relevant terms and a recommendation regarding how much of the test should address students' knowledge, skills, and understandings relative to technology literacy. After development of the framework, NAGB will recommend which grade level— 4th, 8th, or 12th—will be assessed in a 2014 probe. The Steering Committee already has recommended Grade 8 based on the rationale that most technology classes (and enrollment) are in middle grades, students have not dropped out, there still is time to address deficiencies, and students will take the test seriously.

The NAEP Technological Literacy Assessment will be the country's first nationwide assessment of student achievement in this area. The work comes at a time when neither nationwide requirements nor common definitions for technological literacy are clear. Few states have adopted separate assessments of technology. All of this happens as business and industry leaders and policymakers voice concern about American students' abilities to compete in a global marketplace.

Several groups are contributing to the assessment framework: the Council of Chief State School Officers, the International Technology Education Association (ITEA), the International Society for Technology in Education, Partnership for 21st Century Skills, and the State Educational Technology Directors Association. Both the Steering Committee and Planning Committee include technology experts, engineers, teachers, scientists, business representatives, state and local policymakers, and employers from across the country. The committees' work involved recommendations to the Board about the content and design of the assessment in the form of the framework and specifications for the 2014 NAEP Technological Literacy Assessment.

The process of developing the framework also includes public hearings and reviews by hundreds of experts in various fields and the general public. The collaboration will reflect the perspectives of many individuals and diverse groups. However, as with the NAEP 2009 Science test, the ultimate approval of, and decisions about, a 2014 assessment of technological literacy resides with NAGB.

Where Did Policies for NAEP 2009 Science and NAEP 2014 Technological Literacy Originate?

The National Assessment Governing Board (NAGB) has determined that the NAEP exams should be stable for at least 10 years. In the case of science, the previous framework (1996–2005) was developed about 15 years ago. A new framework was called for that would acknowledge changes that had taken place since the mid-1990s. These changes included the publication of national standards in science (*Benchmarks for Science Literacy* of AAAS and the *National Science Education Standards* of the NRC), advances in science research, advances in cognitive research, and growth in the prevalence of science assessments nationally and internationally.

As work began on the 2009 science framework, a number of committee members argued that technology should be included in the NAEP 2009 science framework. In the end it was determined that the 2009 science framework should include technology and that a special study of technological literacy be conducted for testing in 2014.

Policies for the NAEP 2014 Technology Literacy Framework were built on initial introductions of technology in national documents such as *Science For All Americans* (Rutherford & Ahlgren, 1989), *Benchmarks for Science*

Literacy (AAAS, 1993), *National Science Education Standards* (NRC, 1996), and *Standards for Technological Literacy* (ITEA, 2000). In addition, The National Academy of Engineering completed two reports that contributed to this initiative. The first report was *Technically Speaking* (Pearson & Young, 2002) which was an exploration of technological literacy addressing the issue, "Why All Americans Need to Know More About Technology." A more direct influence on the national assessment initiative was *Tech Tally: Approaches to Assessing Technological Literacy* (Garmire & Pearson, 2006), the report of a 2-year study that examined the most viable approaches to assessing technological literacy of K–12 students, K–12 teachers, and out-of-school adults. PISA 2006 Science also recognized technological systems as one aspect of knowledge of science. This introduction of technology as a part of an international assessment of science was complemented by the introduction of technology as part of the NAEP 2009 science framework. In other words, there were several relatively independent policy documents that recognized technology as an important component of science and ultimately a domain worthy of an independent national assessment. This had a very significant effect on convincing NAGB of the wisdom of this effort.

Who Influenced the Formation of Policies for NAEP 2009 Science and NAEP 2012 Technological Literacy?

As with many policies, identifying single individuals who influenced these initiatives is difficult. Development of policies for technological literacy serves as an excellent example. The participation of Dr. Sharif Shakrani, now at Michigan State University and formerly Deputy Executive Director of the National Assessment Governing Board (NAGB), directly influenced NAGB's decision to include technology in the 2009 NAEP science framework and develop a probe of technological literacy in 2014. Several NAEP Steering Committee members argued that technology should be included in the 2009 national assessment of science and recommended that NAEP 2009 science include "Using Technological Design" as part of "Scientific Practices" and that NAGB consider a special study of technological literacy.

How Were the Policies for NAEP 2009 Science and NAEP 2014 Technological Literacy Formulated?

A two-tiered committee structure, consisting of a Steering Committee and a Planning Committee, provided the expertise to develop both frameworks specified by NAGB. The two committees were composed of members who were diverse in terms of role, gender, race/ethnicity, region of the country, perspective, and expertise regarding the content of the assessment to be developed.

Comprised of leaders in science and technology, general education, assessment, and various public constituencies, the Steering Committees set the course for the projects. Functioning as oversight bodies, these groups developed a charge that outlined what the Planning Committee should attend to in the development of the frameworks. The steering committees also reviewed and provided feedback on drafts of the respective frameworks and related materials.

The Planning Committees, supported by the project staff, were the development and production group responsible for drafting the frameworks, the specifications, recommendations for background variables, designs for one or more small-scale studies, and preliminary science and technology achievement-level descriptions. These committees were made up of educators at both K–12 and collegiate levels, district and state personnel, scientists, engineers, and assessment experts. The Planning Committee's work was guided by goals and principles identified by the Steering Committees. In addition, the Planning Committees used a number of resources to facilitate their work.

What Are the Intended (and Unintended) Consequences of Policies for NAEP 2009 Science and NAEP 2014 Technological Literacy?

Because NAEP is the only nationally administered standardized test in science, the NAEP 2009 science framework and test will most likely influence science teaching in the United States for the next decade or more. The NAEP frameworks, for example, will influence development of the new common core standards for science education. Currently, NAEP mathematics and reading tests are mandated for all students on a biannual basis. The NAEP science is administered by states every 4 years on a voluntary basis, but as noted earlier, in 2005 all but six states participated. It also is possible that with the reauthorization of federal education legislation, science could be placed on the same footing as mathematics and reading, making NAEP mandatory in science as well. This would mean that NAEP 2009 science would have an even greater influence on what is taught in science in the United States. But perhaps the most important consequence for this assessment will be the recognition of technological literacy as an important outcome of education. There are few domains with such influence on personal and professional life and about which the general public knows so little as evidenced by the general confusion of equating technology with digital devices such as computers. This confusion is even enshrined in legislative policies.

How Will Policies for NAEP 2009 Science and NAEP 2014 Technological Literacy be Implemented?

The 2009 NAEP Science test will be based on the new framework. In 2014 there also will be a National Assessment of Education Progress and subsequent "Report Card" on the nation's technological literacy.

A POLICY FORMATION FOR STATEWIDE END-OF-COURSE ASSESSMENTS FOR SCIENCE: A STATE CASE OF STATE POLICIES

An Overview of the End-of-Course Policy

In 2008, legislators in the state of Washington asked the state school board to consider the role and place of end-of-course assessments (EOCs) in science. The request came in the context of implementing new state standards for science education, a proposed increase in graduation requirements from 2 to 3 years of science (with 2 years of laboratory science), and continuing criticisms of the state's current comprehensive assessment, the Washington Assessment of Student Learning, referred to as the WASL.

In the past 5 years, statewide EOCs have gained the interest of the education community. Although single comprehensive assessments such as the WASL remain the prominent statewide test, the use of EOCs is increasing. In 2008, sixteen states included EOCs in their high school assessment system and another 11 states plan to implement EOCs in the near future. By 2012, twenty-six states will have exit exams and 13 of these states will use EOCs as a graduation requirement. In addition, 12 states use some or all of their EOCs to meet the testing and accountability requirements of No Child Left Behind legislation.

Recognizing the aforementioned emergence of statewide EOC assessments, understanding the development of new science education standards for the state, and in the face of changing graduation requirements, as part of their policy deliberations regarding science assessment, the Washington State Board of Education asked for a review and policy recommendations addressing the implications of using EOCs for science. The following discussion is based on that review. As of January 2009, the State Board of Education had not taken action on EOCs for science.

The review and policy recommendations addressed the question of "How Well Would Comprehensive and EOCs Provide Evidence About Students' Attainment of the Major Purposes of Science Education as outlined in the *K–12 Science Education Standards for the State of Washington*?" Although the review was prepared as background prior to a decision about implementation of EOCs, it was not intended to present formal recommendations. Rather, it provided a deeper understanding and insightful perspectives on issues associated with the possible implementation of EOCs in the state of Washington.

End-of-Course Assessments

For purposes of this discussion, it is assumed that the EOCs being considered by the state of Washington are based on state standards and address course-specific learning outcomes. The assessments generally are administered at the completion of a course (e.g., Physical Science or Integrated Science). The tests generally are prepared by assessment experts in the state agency. Doing so provides uniformity to the test(s) and assures alignment with state standards. Results of the assessments are meant to provide evidence of the degree to which individual students have learned the science content and developed the inquiry abilities identified in the state standards and expected of students who have completed a particular science course. The results may be used to determine a grade for the course, fulfill a course requirement for graduation, and/or receive partial credit toward completion of a required exit exam for graduation. If EOCs are used as exit exams for graduation, satisfactory completion of two or three exams typically are used as the required number. In addition, the EOC results may be used by the school or district to meet NCLB testing and reporting requirements.

EOCs identify what students have learned upon completion of a specific course as opposed to what they learned from multiple courses or sometimes before a course has been completed. In Washington, for example, EOCs might be administered after courses in Grades 9, 10, and 11 (with a 3-year course requirement).

Research Supporting End-of-Course Statewide Assessments

Research conducted by John Bishop and his colleagues suggests a variety of positive benefits of EOCs for attaining higher levels of teaching and learning (Bishop, 1998, 2007; Bishop, Mane, & Bishop, 2001; Bishop, Mane, Bishop, & Moriarity, 2000). Research on EOCs provides preliminary support for several student outcomes including: promotion of learning, increased attention in class, higher levels of engagement in learning, and increased conscientiousness about completing homework. Changes in teachers and teaching included: setting higher standards for students, spending more time teaching cognitively demanding skills, not giving "inflated grades," and improved relationships with students. These results likely are due to the high-stakes value of EOCs and the fact that students must pass all tests as part of the graduation requirement.

The Washington Context and End-of-Course Assessments

In 2009, the State of Washington introduced new standards for K–12 science education. The standards document includes content standards and performance expectations for science content, scientific inquiry, and applications of science to personal and social problems. The standards will serve as the basis for EOCs, a comprehensive assessment, that is, the WASL, or both. Several contextual requirements underlie the policy decision about implementing the EOCs. First, the state assessment should address the full range of science education standards. Second, in the future there will be a 3-year graduation requirement that includes two laboratory courses. Third, graduation will require satisfactory achievement of a high-stakes assessment that may be either some number of EOCs or a comprehensive examination, the WASL. Fourth, school districts in Washington will continue to have local control over many decisions concerning science education programs and practices at the district, school, and classroom levels, including the option to use EOCs. Fifth, the need for a statewide accountability system will continue. Any transition to EOCs must accommodate the need to be accountable at the national, state, district, and school levels.

Policies for End-of-Course Assessment of Science Education

School science programs have several unique features that must be considered when implementing policies for EOCs to assess science education. The most significant feature of school science is that it is organized both by disciplines—Earth science, biology, chemistry, and physics—and as various aggregates of those disciplines in the form of courses such as integrated science, general science, physical science, etc. In addition, in most courses, students are expected to learn facts and principles of science as well as the way in which science is done. The implications of these unique features are discussed in the following sections.

Science Content

Identification of a common core of science concepts that would form the basis of a state exam such as the WASL presents some difficulty due to the continued influence of the separate academic disciplines such as biology, chemistry, physics, geology, and meteorology. It is impossible for a state exam to adequately cover all of the important ideas from each of the separate disciplines. The EOCs could accommodate this feature, but it would require implementing several separate assessments for courses such as Physical Science, Earth Science, Biology, and courses with titles such as

General Science, Coordinated Science, Science I, and Integrated Science. Due largely to economic constraints, a state such as Washington could not prepare EOCs for all science courses offered across the state. Out of economic necessity and the need for consistent alignment with state standards, some authority, most likely the State Board of Education, would have to decide the courses for which EOCs would be developed. In addition, the use of EOCs would require some changes in course offerings at the local level, because the assessments would be aligned with state standards. In some cases, school districts and school science programs would be required to modify courses to accommodate EOCs and standards.

In contrast to a comprehensive assessment such as the WASL, EOCs would provide more feedback on students' depth and breadth of knowledge in specific science content and approach the requirement of assessing all content in the standards. But in some districts and schools, use of EOCs also would have the likely consequence of narrowing the variety of science courses offered. The positive benefit of this would be greater alignment among standards, courses, and assessments.

Scientific Inquiry

The science education community generally agrees on the importance of laboratory experiences as part of school science programs. (See DeBoer, Chapter 10, this volume.) The extension of this perspective suggests the importance of performance-based assessments and introduces the complications and costs of such experiences for comprehensive assessments.

Performance-based assessments center on students' knowledge of scientific inquiry and abilities such as designing investigations, controlling variables, collecting data and using evidence to support a conclusion, recommendation, or decision. The EOCs can incorporate performance-based assessments and present greater opportunities for in-depth and subject specific evaluation of students' knowledge and abilities. These abilities are (or could be) closely related to twenty-first century workforce skills and abilities such as problem solving, critical thinking, complex communication, and systems thinking.

Applications

The Washington science education standards call for the application of science and technology to "real-world" problems. Although comprehensive examinations can include items with contexts such as health, resources, and environments, opportunities in specific disciplines and based on investigations suggest a slight advantage for EOCs in the area of real world applications.

Career Awareness

The development of the important twenty-first century workforce skills such as those listed above can be supported by science instruction and would fit into EOCs. Inquiry and application standards in the new science standards provide many of the necessary outcomes for the twenty-first century workforce skills, but they must be clearly delineated in both the new standards and the assessments.

Obviously, the implications of using EOCs in science would include increased financial costs for new assessments. In addition, EOCs would likely increase costs in some districts and schools due to the need to select new instructional materials and to reform the science program and classroom practices. In short, new policies established by the State Board of Education will create alignment issues with current practice.

Other Considerations for a Policy on End-of-Course Assessments in Science

Implementing EOCs will, of necessity, engage other issues within the educational system. Some of the issues will require additional decisions by policymakers. Following is a brief introduction to several of those policy issues.

Graduation Requirements

Increasing the graduation requirement to three courses (two with laboratory) and maintaining the WASL at 10th grade, both policies proposed by the State of Washington, present a significant alignment issue: most student will not have completed their three science courses by the time they take the WASL. An alternative approach would be to use EOCs for three courses and count satisfactory achievement on those tests as meeting graduation requirements. Another option would be to use EOCs to meet course requirements and administer the WASL at 11th grade instead of 10th. Maintaining the WASL and introducing EOCs would provide the state and local districts with accountability options and maintain a focus on the new standards and purposes of science education.

Development of End-of-Course Assessments

In order to establish and maintain coherence across the state educational systems, the EOCs would be developed and managed by the Office of Superintendent of Public Instruction (OSPI). Such an approach would define the standards and assessments at the state level (with significant input from school personnel and other educators) and leave the local options open relative to instructional materials, teachers' professional development, and course requirements for graduation within the state guidelines.

State Accountability

The legislature has required the State Board of Education to develop a statewide accountability system. Implementing EOCs statewide would align with the legislative mandate and contribute to principles for accountability adopted and under review by the State Board of Education. Results of EOCs would contribute to the multiple indicators of school and district accountability. Further, it would help identify priorities for schools and clarify specific areas for improvement.

Student Accountability

Using EOCs for both recording results on transcripts and as part of the final grade would engage most students and contribute to their motivation to study science. Such an approach would make the EOCs medium stakes. If counted as part of the exit exam, these tests become high stakes. Policymakers need to decide whether or not the EOCs should be used as high stakes exit exams. That decision, in part rests on whether or not the public will have confidence in the results of these tests. Are they valid enough measures of what students know to be used to make decisions about graduation?

Alignment of Standards, Curriculum, Assessments

Implementing EOCs would probably narrow the course options in schools, but clarify the content and performance expectations in each course in which an EOC was available. In 2008 there were approximately 12 courses with titles such as "Science I," "General Science," "Integrated Science," "Freshman Science," and "Essential Science" used by districts in the state of Washington to meet graduation requirements. This wide array of courses could be narrowed by a policy for EOCs if a common EOC were developed for this array of courses.

Making a Policy Decision

The previous discussion represents the various issues that were presented to assist the Washington State Board of Education in comparing EOCs and comprehensive assessment systems' effectiveness in measuring the major goals of science education as outlined in the Washington State Science Standards. The advantages and disadvantages can be summarized as follows:

- As indicated in our comparison of the effectiveness of comprehensive assessments and EOCs in measuring student learning of the core science content, EOCs have an advantage in measuring depth, whereas comprehensive assessments such as the WASL do a more efficient job of measuring the breadth of student knowledge.
- Although EOCs require more time for testing, the tests can be easily administered in the class for which they are designed and within the normal class schedule, thus creating little, if any, disruption to the normal school schedule.
- There are greater costs involved with EOCs, largely as a result of the greater amount of development and testing time. The advantage of the increased testing time is the depth of the measures and their increased validity and reliability, potentially making the EOCs more effective as tools for diagnosis and improvement of instruction, curriculum, and professional support.
- A decision about the new Washington State science education standards, a transition to EOCs, and changes in WASL will affect local decisions about selection of instructional materials, instructional practices, the curriculum, and the courses and exit exams that meet graduation requirements.
- Changes in WASL and/or a transition to EOCs may cause concerns at the local level. Local concerns may be offset by a recognition of the positive effect on learning outcomes for students. While there are positive advantages in achieving higher levels of coherence and alignment among courses and assessments at districts and schools, there may be concerns that arise due to reduced local control.

In conclusion, there are increased costs and risks to implementing EOCs. On balance, it seems there may be increased benefits to students, teachers, schools, districts, and the state of Washington.

Where Did This Policy Originate?

In this case, a request came from the state legislature, and the review was enacted by the State Board of Education (SBE).

Who Influenced the Formulation of the EOC Policy?

The final formulation of the policy resided with the SBE with approval by the state legislature. Details, trade-offs, and benefits of the policy were formulated in the report and subsequent presentation and discussion with the SBE. The policy is under consideration by the SBE.

How Are the Policies for EOC Assessments Formulated?

The policies were formulated by administrative staff for the SBE.

What Are the Intended (and Unintended) Effects of the Policy?

Several of the possible effects of the EOC policy were discussed in the summary. Primary among the positive consequences would be a reduction and narrowing of courses that would count as graduation requirements and accountability for NCLB. The increased coherence within Washington's science education would be a significant benefit.

The most likely negative consequences would center on the increased costs for the assessments and maintaining both the comprehensive assessment and EOCs for a period of transition, if the legislature requires a transition. The political consequences of an apparent reduction of local control also would be significant.

REFLECTIONS ON POLICIES IN SCIENCE EDUCATION

The cases presented above explored various aspects of science education policy with respect to assessment. This section extends that discussion.

The Origin of Science Education Policies

In general, policies for assessment originate with officially sanctioned groups, governing boards, organizations, or agencies. They do not, for example, originate with individual science educators. Although individuals may initiate and influence new policies, it is an official body that signals the process, timeline specifications, and funding for a new policy. The origin of policies does not occur *de novo*. The official body has clear purposes and goals, and the development of policies must be within limits set forth by some variation of a governing board. Examples include the PISA Governing Board (PGB), National Assessment Governing Board (NAGB), state legislatures, and school boards.

One should note that the official governing body also exercises final approval of the policies and assumes responsibility for implementation through financial, legal, and contractual means: for example, NAGB contracted WestEd for the formation of NAEP assessment frameworks. Those frameworks are then used by another contractor, for example, Educational Testing Service (ETS) as the basis for item development. Yet another organization, for example, Westat, may manage administration and scoring of NAEP.

These observations can be generalized to selected policies for science education such as those involving content standards, graduation requirements, textbook adoptions, or teacher certification. Whether at the international, (e.g., TIMSS, PISA), national, (e.g., NCLB, NAEP), or state and local levels, the formal policies that influence science education originate with a formal body. Although policies involving content standards indirectly affect assessments, the origin of national and state standards for science education also rest with a body that had legal responsibility for the recommendations they were making, for example, National Research Council, American Association for the Advancement of Science, state legislatures, and state boards of education. There are, of course, variations and exceptions. Selection of curriculum materials by local committees and policies for teacher education and professional development may not align with the discussions in this chapter.

The Formulation of Science Education Policies

Officially sanctioned groups such as governing boards and agencies typically create committees or commissions of experts to develop polices. In recent times, the expert committees have had diverse representatives and authority to develop policies within the limits set by the governing authority. So, for example, working groups that developed state standards for science education have been represented by scientists, science educators, and science teachers with a balance of disciplines, gender, grade levels, and ethnic groups included. Because the expert committee is set up by the governing board, there is usually a clearly established system of guidelines, checks, and balances. The committee has freedom within limits.

Answering the question "Who influences the formulation of policies?" becomes both a central and complex question to answer. The expert committee working on policies has a chair or a designated leader. Further, there usually is a hierarchy of authority on a committee. All committee members do not have equal influence in all areas involved in the actual formulation of policies. Science teachers, for instance, usually do not have the scientific knowledge as do scientists when defining content for science education standards and assessments. Conversely, scientists do not have the expertise that science teachers have when designating the appropriate ages and grades for content and assessments.

The power to influence the formulation of policies resides in the individual or individuals who understand the broad purposes of the committee; the form and function of policies; and the eventual use of the policies in programs such as curriculum materials, assessment, or teacher education programs. The power and leadership of these individuals is enhanced further by their abilities to persuade and to assume responsibility for the actual statements of policy. These statements, of course, are contingent upon review and revision by the committee and acceptance by the governing body. Because it is at the heart of the matter, I will say it again. There often is a small number of individuals who lead the group because of their vision and persuasive arguments, which are based on their experiences, their knowledge, their ability to listen and communicate, their credibility with all factions of the group, their openness to ideas, their ability to synthesize recommendations, and their willingness to put in the time and effort to actually formulate, present, and defend concrete statement of the policies. Very importantly, these individuals demonstrate these qualities to both the committee members and governing board, as appropriate and required. By the nature of these qualities, these usually are senior individuals in the science education profession.

Next is the question "What influences the formulation of policies?" To repeat what has been mentioned before, one of the most important factors in the formulation of policies is the purpose, the aims and goals of the group originating the policies. Committees and even powerful individuals are not at liberty to go beyond the purposes of the originating group. Policies for NAEP science or technology are for assessments and cannot be designed to reform the school curriculum. In some cases, the formulation of new policies or the revision of extant policies is scheduled. Revisions to NAEP are scheduled every 10 years. Other revisions may include state science education frameworks and criteria for adoption of science textbooks. Some of these revisions are strictly scheduled, while others may vary based on budgets or other factors external to the state or district educational system.

Probably more central to the question about the influences on policies, and most difficult to answer, are internal and external factors. Disappointingly low student achievement on state assessments, particularly low levels of achievement for significant numbers of students or the need for accountability measures, may be internal factors influencing formulation of policies. External factors such as the need for an emphasis on twenty-first century work force skills (which has been identified as a goal by still other policymakers) and other larger, more complex issues such as the economy or national security and the achievement gap may also stimulate changes. After all, education is a social institution and subject to the various changes that affect social institutions.

The Implementation of Science Education Policies

In most cases, the implementation of policies is clear from the beginning. There is a program that makes the abstract policies more concrete for those responsible for their implementation. The PISA and NAEP policy frameworks result in actual assessments. State standards for science education result in assessments, criteria for textbook adoption, and graduation requirements.

What Are the Intended and Unintended Effects of Science Education Policies?

The intended consequences of policies include direction and guidance for various components of the science education system. The formulation of policies has the general goal of improving specific components of science education, in particular school science programs, classroom teaching, and ultimately student learning. As a result of assessments, educators have information, data, and direction for reform and improvement.

The unintended, usually negative, consequences include criticism, controversy, and confusion about the policies. Quite often the negative consequences are based on misunderstandings about policies but sometimes on genuine differences in what people value or think is effective. Some think policies ought to be programs or specific statements of practices; hence, a criticism that the policies are not practical. Others will review the policies and decide they are inadequate and do not cover what they think are important aspects of science education. An emphasis on scientific literacy, such as PISA, spurs controversy because there is not an emphasis on science disciplines such as there is in the TIMSS assessment. Conversely, TIMSS can be criticized for emphasizing discipline-based abstract facts and information at the expense of broader goals having to do with the ability to use knowledge in the consideration of real-world phenomena. There also may be confusion about the policies, precisely because they are policies and not more concrete statements about, for example, test items or teaching strategies. Most frustrating to policymakers are the criticisms based on misunderstanding or on minor parts of the policies that have been extended to the entire document. Stating a criticism about one minor element can be the basis for a recommendation to dismiss the entire set of policies.

CONCLUSION

This chapter explored the development of science education policies for assessments. The exploration was intended to answer questions: "Where do policies originate?" "How are policies formulated?" "Who influences the formulation of policies?" "How are policies implemented?" "What are the intended and unintended effects of science education policies?"

Although the generalization from policies for international, national, and state assessments may be limited, answers to the questions included the following. Policies generally originate with a governing body or responsible agency. Policies are further developed by a working group or committee. Key individuals provide leadership in the development of policies. Their leadership is based on their credibility, knowledge, experience, the ability to persuade, and very importantly, the capacity to do the work. Policies usually are implemented in a predetermined form, for example, an assessment. Although most policies have positive intentions to improve various aspects of science education, there are sometimes unintended negative consequences as well.

Working on various national and state policies reveals how carefully and thoughtfully policymakers go about their work. They invest time and money in deliberating the potential negative consequences and perceived positive benefits of various policy options. They use expert opinion to a large extent, and one of the roles experts play is to filter the research and recommendations of other individuals and groups. One of the roles, then, that researchers can play is to recognize the form, function, and importance of policy research and to present their work in as compelling a form as possible so it is read, understood, and considered by the experts who find themselves on panels giving advice to policymakers or formulating policies.

To conclude, formulating policies is a critical, but often overlooked, component of science education. Greater recognition of the role of policies will contribute to a coherent and integrated science education system that will benefit all concerned, especially the students.

EDITOR'S NOTE

Dr. Bybee played a significant role in each of the cases that he describes in this chapter. For PISA, he served as chair of the Science Expert Group for the 2006 assessment; for NAEP, he served on the NAEP Steering Committee; and for the end-of-course exam study in the State of Washington, he was a member of the committee appointed to develop a report on the policy implications of end-of-course exams.

REFERENCES

American Association for the Advancement of Science. (1989). *Science for all Americans: A Project 2061 report on literacy goals in science, mathematics, and technology.* Washington, DC: Author.

American Association for the Advancement of Science. (1993). *Benchmarks for science literacy.* New York, NY: Oxford University Press.

Bishop, J. (1998). *Do curriculum-based external exit exam systems enhance student achievement?* Philadelphia, PA: Consortium for Policy Research in Education, University of Pennsylvania.

Bishop, J. (2007). *A steeper, better road to graduation*. Hoover Institution, Education Next. Board of Trustees of Leland Stanford Junior University.

Bishop, J., Mane, F., & Bishop, M. (2001). *Is standards-based reform working ... and for whom? Working paper 01–11.* Ithaca, NY: Cornell University Center for Advanced Human Resources Studies.

Bishop, J., Mane, F., Bishop, M., & Moriarity, J. (2000). *The role of end-of-course exams and minimum competency exams in standards-based reforms. Working paper 00-09.* Ithaca, NY: Cornell University Center for Advanced Human Resources Studies.

Bybee, R. (1997). *Achieving scientific literacy: From purposes to practices*. Portsmouth, NH: Heinemann.

Bybee, R., & DeBoer, G. (1994). Research on goals for the science curriculum. In D. Gabel (Ed.), *Handbook of research on science teaching and learning* (pp. 357–387). New York, NY: MacMillan.

Garmire, E., & Pearson, G. (Eds.) (2006). *Tech tally: Approaches to assessing technological literacy.* Washington, DC: National Academies Press.

International Technology Education Association. (2000). *Standards for technological literacy: Content for the study of technology.* Reston, VA: Author.

Loveless, T. (2009). *The 2008 Brown Center report on American education: How well are American students learning?* Washington, DC: Brookings.

National Research Council. (1996). *National science education standards.* Washington, DC: National Academies Press.

Organization for Economic Cooperation and Development. (2000). *Measuring student knowledge and skills. The PISA 2000 assessment of reading, mathematical and scientific literacy.* Paris: Author.

Organization for Economic Cooperation and Development. (2001). *Knowledge and skills for life–First results from PISA 2000.* Paris: Author.

Organization for Economic Cooperation and Development. (2003a). *Literacy skills for the world of tomorrow: Further results from PISA 2000.* Paris: Author.

Organization for Economic Cooperation and Development. (2003b). *The PISA 2003 assessment framework–Mathematics, reading, science and problem solving knowledge and skills.* Paris: Author.

Organization for Economic Cooperation and Development. (2004a). *Learning for tomorrow's world–First results from PISA 2003.* Paris: Author.

Organization for Economic Cooperation and Development. (2004b). *Problem solving for tomorrow's world. First measures of cross-curricular competencies from PISA 2003.* Paris: Author.

Organization for Economic Cooperation and Development. (2006). *Assessing scientific, reading and mathematical literacy. A framework for PISA 2006.* Paris: Author.

Organization for Economic Cooperation and Development. (2007a). *PISA 2006. Science competencies for tomorrow's world, Volume 1: Analysis.* Paris: Author.

Organization for Economic Cooperation and Development. (2007b). *PISA 2006. Science competencies for tomorrow's world, Volume 2: Data*. Paris: Author.

Pearson, G., & Young, T. (Eds.). (2002). *Technically speaking: Why all Americans need to know more about technology*. Washington, DC: National Academies Press.

Roberts, D. (2007). Scientific literacy/Science literacy. In S. Abell & M. Lederman (Eds.), *Handbook of research on science education* (pp. 729–780). Hillsdale, NJ: Lawrence Erlbaum.

Rutherford, F. J., & Ahlgren, A. (1991). *Science for all Americans*. New York, NY: Oxford University Press.

CHAPTER 9

HOW CAN SCIENCE EDUCATORS INFLUENCE LEGISLATION AT THE STATE AND FEDERAL LEVELS?

The Case of the National Science Teachers Association

Jodi Peterson

INTRODUCTION

In this chapter I describe the role that the National Science Teachers Association (NSTA) plays in attempting to influence policy decisions in science education, primarily at the national level but also at the state and local levels as well. The chapter also provides practical advice to individuals on how to lobby members of congress or their state legislators and how to advocate for policy positions in science education.

The Role of Public Policy in K–12 Science Education, pp. 241–271

LOBBYING VERSUS ADVOCACY

In any discussion of policy and policymaking, it is helpful to first understand what is considered "lobbying" and what is considered "advocacy." Although the terms are often used synonymously, the difference is important to nonprofit organizations such as NSTA because lobbying is strictly controlled by the Internal Revenue Service (IRS). At the simplest level, lobbying is an attempt to influence specific legislation, and advocacy more generally involves promoting a cause but not influencing specific legislation. Advocacy activities may include an individual educating a legislator on an issue, or an organization educating members on an issue, as long as the organization members are not encouraged to ask the legislator to vote in a particular way on a particular piece of legislation. Any contact with a nonlegislative body, such as the mayor or governor of a state, is considered to be advocacy and not lobbying.

With respect to influencing legislation, the IRS distinguishes between direct lobbying and grassroots lobbying. For lobbying to be considered "direct," the communication between the person doing the lobbying and the legislator or staff member again must refer to a specific piece of legislation, and it must reflect the person's particular view on that legislation. A conversation with a member of Congress about a bill is not lobbying if you do not state a position on that bill, and it is not lobbying if you state a policy position without referring to a specific bill. Grassroots lobbying includes the same two criteria as for direct lobbying and, in addition, involves direct communication by an organization asking the readers of that communication to take action by contacting their member of Congress and expressing a specific position. The amount of grassroots and direct lobbying a nonprofit organization can engage in is controlled by limits placed by the IRS on the amount of money it can spend on lobbying activities. The amount an organization can spend on lobbying is typically defined in terms of a percent of the organization's total budget, with volunteer (nonpaid) lobbying activities not included in the total. Nonprofit organizations that engage in lobbying must follow certain reporting rules, which can be found on the website of the Center for Lobbying in the Public Interest (http://www.clpi.org/). Unlike lobbying, there are no limitations placed on the nonlobbying advocacy activities that a nonprofit organization can engage in. Nonprofit organizations can spend as much of their budget as they wish on advocacy activities.

Lobbying often is perceived negatively by the public, but lobbying is an essential part of a democratic society. It is a way for people and organizations to have their voices heard and for them to influence policy development. A particularly helpful definition of lobbying comes from political and economics columnist Robert Samuelson (2008) who writes that,

> Lobbyists sharpen debate by providing an outlet for more constituencies and giving government more information. Lobbyists primarily woo lawmakers with facts. If lawmakers see merit in a position and there is a public outcry in its favor, that's the way they tend to vote.... Lobbying is an expression of democracy (para. 6).

Clearly, lobbying and advocacy are distinct activities that (should) aim for the same result, that is, to influence policy. NSTA has been lobbying and advocating for science education as an official activity of the organization since establishing its Office of Legislative and Public Affairs in 1995. The single goal of NSTA's lobbying and advocacy activities is to ensure that federal lawmakers have clear, concise knowledge and information that will help them to make policy decisions in the best interest of science education.

ALL POLITICS IS LOCAL: THE IMPORTANCE OF MEMBER PARTICIPATION

Many science and mathematics education groups, including NSTA, maintain an active lobbying presence on Capitol Hill. Here, legislative liaisons from organizations such as NSTA and their staff members work to advance specific legislation affecting science, technology, engineering, and mathematics (STEM) education. Some of those lobbying activities will be described later in this chapter. Yet, the real strength of an effective legislative affairs program is an organization's ability to mobilize members and help them advocate, or lobby, for a specific policy by bringing their experiences, messages, and stories to their lawmakers.

Politics is all about preserving what you have that you don't want to lose and acquiring what you don't have but would like to gain. With this in mind, who can carry the message to their elected representatives better than practitioners, educators, researchers, students, and other, in the STEM education community? As Tip O'Neill (2009), former Speaker of the House of Representatives said, "All politics is local," which was meant as a rallying cry to involve constituents in advocacy activities. Constituents who are knowledgeable about an issue, passionate about a cause, and know when and how to interact with their members of Congress can be very effective. Any real change to STEM education policy comes from individuals within the STEM education community and the influence they can bring to the policy decisions that are made.

The larger the number of genuine messages of concern that come from constituents, the more likely members of Congress and their staffs are to listen to them. The operating assumption of many congressional staff is that the more time and effort constituents take to communicate, the more passionately they care about the issue. According to the Congressional Management Foundation (2005): "Thoughtful, personalized constituent messages generally have more influence than a large number of identical form messages" (p. 4).

Involving Members of NSTA in the Policy Process

From its beginning, the NSTA Legislative Affairs office sought to involve NSTA members in interacting with their representatives in Congress. NSTA staff from the NSTA traveled to scores of area conferences and signed up teachers for the NSTA Legislative Network, a group of members who wanted to receive legislative information on a regular basis. The network soon grew to over 100,000 individuals who received legislative updates and alerts, as needed, on key science education issues.

A few years ago, the Legislative Network was incorporated into NSTA's e-newsletter, NSTA Express, which now goes to almost 350,000 individuals each week. From 1995 to 2009, more than 300 legislative updates have been sent to subscribers, often asking them to contact members of Congress on issues, including federal funding for science and mathematics education programs at the U.S. Department of Education and the National Science Foundation (NSF), reauthorization of the No Child Left Behind Act of 2001, and passage of the America COMPETES Act of 2007.

During development of the Legislative Affairs initiative, the NSTA Board of Directors and NSTA staff created a policy agenda for the association. Together, they identified key issues and learned more about the existing legislation on Capitol Hill that addressed these issues. In the early years, the key issue for NSTA was funding for professional development for science teachers. From this, NSTA developed a series of messages using available research, and it developed white papers and talking points using these messages. Later, when addressing a specific issue or piece of legislation, these messages were included in the NSTA communiqués sent to key members of Congress with NSTA's concerns and thoughts on federal policy. This process has continued over the years, with the NSTA Board of Directors identifying the policy agenda and providing direction on the key issues to address.

Early on, the association also developed a legislative handbook (NSTA, n.d.) and made it available online to both NSTA members and to state science teacher organizations. The legislative handbook contains information on the related committees in both the House and Senate that have oversight of science education policies, tips for sending effective messages to elected officials, and tips on how to advocate with legislators or other elected officials.

Since the NSTA Legislative Affairs initiative began, the association has sent dozens of letters to members of Congress on science and STEM education-related issues. Over the years, NSTA also has been asked to provide testimony on a number of key issues, including increased funding for the Mathematics and Science Partnerships at the U.S. Department of Education, and for higher funding for STEM education programs at the science-related agencies National Science Foundation (NSF), National Aeronautics and Space Administration (NASA), and National Oceanic and Atmospheric Administration (NOAA) under the jurisdiction of the Commerce, Justice, and Science Appropriations Subcommittee. Testimony also has been provided by NSTA on nanotechnology education, science labs, and the America COMPETES Act of 2007.

CREDIBILITY AND RELIABILITY ARE KEY TO HAVING ONE'S VOICE HEARD

The ability to get current and useful information to policymakers and their staffs is key to an effective government affairs initiative. Members of Congress need accurate and trustworthy information. As a result of its congressional testimony, letters to Congress, member advocacy, and personal interactions, over the years NSTA has established a reliable and valued presence on Capitol Hill that allows many congressional offices to turn to the association as a trusted source of information on science education.

Much of the information NSTA provides to congressional offices is based on research reports from the National Research Council, studies sponsored by the National Science Foundation, evaluation studies undertaken by the Department of Education, and research done by members of the National Association for Research in Science Teaching (NARST) and the Association for Science Teacher Education (ASTE) members. The information is summarized and formatted in a concise and user-friendly way, so that it specifically addresses the policy issue under discussion and is easy to use and understand.

Another effective outreach initiative with Congress that demonstrates NSTA's commitment to high-quality science education is the NSTA Center for Science Education (CSE) Champion Awards. These public service awards were created in 2008 to honor outstanding individuals for their efforts on behalf of science education. The awards acknowledge the contributions of members of Congress to science education, and they help contribute to a positive relationship between members of Congress and the members of NSTA. In 2008, NSTA presented CSE Champion Awards to former Senator John Glenn for his lifelong dedication and commitment to science education. In September 2000, the 25-member National Commission on Mathematics and Science Teaching for the 21st Century, led by Senator Glenn, issued *Before It's Too Late*, the Commission's report to the nation. Members of the influential "Glenn Commission" included key leaders from business, industry, education, and the federal government. *Before It's Too Late* contained straightforward and powerful recommendations to improve mathematics and science teaching. The CSE Champion Award also was presented to Representative Vernon Ehlers (R-MI) and Representative Rush Holt (D-NJ), two "Congressional physicists" who have worked to promote science education in the U.S. House of Representatives. Both Representatives Ehlers and Holt have been tireless advocates for quality science education and have spearheaded long-term efforts to increase funding for science and mathematics education programs at the Department of Education and the National Science Foundation. They have also worked to ensure that high-quality programs for science education are included in authorizing legislation such as the America COMPETES Act of 2007 and the No Child Left Behind Act of 2001.

JOINING TOGETHER FOR A COMMON CAUSE

The STEM Education Caucus

Another way to have one's voice heard on policy issues is to work with congressional caucuses to provide members of Congress with up-to-date information regarding policy issues of concern. There are dozens of congressional caucuses made up of groups of members of Congress who share a particular interest and who wish to influence their colleagues with regard to the issues involved in that policy area. The National Science Teachers' Association is an active member of the steering committee for the STEM Education Caucus, formed in 2004 by Representative Vernon Ehlers from Michigan and then Representative Mark Udall from Colorado. Representative Udall was later elected to the Senate in 2008. The STEM Education Caucus, which is comprised solely of members of Congress, seeks to strengthen STEM education at all levels (K–12, higher education, and workforce) by providing a forum for Congress and the science, education, and business communities to discuss challenges, problems, and solutions related to STEM education. Although the caucus itself is made up of members of Congress, the steering committee of the STEM Education Caucus is made up of representatives from business and industry, nonprofit organizations, professional scientific societies, and education. The STEM Education Caucus serves as a core group for vetting new legislative proposals and circulating letters on budgets, existing legislation, and proposed agency programs that concern STEM education. The work of NSTA and the other steering committee members with the STEM Education Caucus falls under the category of advocacy, not lobbying, because it is a forum for sharing and discussing relevant information rather than for influencing a particular piece of legislation.

The Caucus educates members of Congress and Hill staff by providing information on STEM education issues directly and by sponsoring briefings on STEM concerns. These briefings often frame important STEM education issues. Since 2005, the STEM Education Caucus has collaborated with a number of outside organizations to host congressional briefings on a wide range of STEM related topics. For a list of recent activities, see Table 9.1.

Table 9.1. Forums and Events in Collaboration With the STEM Education Caucus for 2007–2009

2009
• March 17, 2009: New Member of Congress Orientation on STEM Education Issues • March 25, 2009: Innovative STEM Teacher Preparation Programs • April 29, 2009: Science Professionals: Master's Education for a Competitive World A Policy Briefing • May 20, 2009: Bringing Innovative Computing Curriculum Across the Digital Divide • June 4, 2009: Learning Science in Informal Environments (report from the National Research Council) • September 16, 2009: The Role of STEM Education in the Growing Green Collar Economy • October 16, 2009: Engineering in K–12 Education: Understanding the Status & Improving the Prospects
2008
• January 29, 2008: The Importance of STEM Literacy for All Employees • February 25, 2008: STEM in the FY09 Budget Event (Congressional Research staff-only briefing) • April 2, 2008: NanoDays reception and poster session • April 17, 2008: STEM Public and Private Programs that Work • May 13, 2008: Math Education Standards and Curricula Coherence • May 21, 2008: Louis Stokes Institute launch reception • July 16, 2008: STEM Education, Girls, and the Challenges that Follow: From the Classroom to STEM Careers (in conjunction with Congressional Caucus for Women's Issues) • September 24, 2008: STEM at the Elementary level–Engineering is Elementary

Table 9.1 continues on next page.

Table 9.1. Continued

2007
• February 27, 2007: STEM Education in the FY08 budget (staff event)
• March 28, 2007: JASON Foundation–Getting Kids Excited about STEM through Informal Education
• April 26, 2007: National Science Board and National Governors Association–STEM Education and Innovation priorities and reports
• April 24, 2007: The Important Role of School Counselors in Fueling the STEM Pipeline
• May 17, 2007: Effective Interventions in K–12 STEM Education
• May 22, 2007: Innovation Partnerships: The Role of Foundations
• June 5, 2007: Undergraduate Research: What it Means for U.S. Competitiveness
• July 18, 2007: Society of Women's Engineers/Women's Caucus Pathways of Women in STEM education and careers
• July 20, 2007: SAE Foundation Engineering Design Competition Field Trip
• July 24, 2007: AAAS Science Research, Ed. and Leadership reception/poster session
• July 27, 2007: Elementary Science Education
• September 21, 2007: Putting the STEM in No Child Left Behind
• September 24, 2007: Innovative Research in STEM Education Underway at Leading Universities
• November 19, 2007: (NAE) Engineering Education Research and National Competitiveness

Congressional Caucus members also serve as experts on STEM education issues and initiatives and work to educate their congressional colleagues on these important issues. Caucus members often speak on STEM education issues to groups outside of Congress as well, including chambers of commerce and business groups, school boards and PTAs, and professional scientific organizations. In addition to serving on the steering committee for the STEM Education Caucus, NSTA maintains the Caucus website at www.stemedcaucus.org. Steering committee members work to facilitate communication and collaboration between and among members of Congress and the scientific, education, and business communities concerned with STEM education.

The STEM Education Coalition

A general principle when trying to influence policy is that the larger the number of individuals and organizations that feel strongly about an issue, the more attention policymakers will pay to their views.

Policymakers need the support of their constituents to gain reelection, so it is important for them to know what it is that the majority of their constituents care about. This means that strength in numbers can be a key component of effective advocacy. In 2004, NSTA and the American Chemical Society (ACS), also a leading voice in advocating for effective STEM education policy, created the Science, Technology, Engineering, and Mathematics (STEM) Education Coalition.

The STEM Education Coalition works to support STEM programs for teachers and students at the U.S. Department of Education, the National Science Foundation, and other agencies that offer STEM-related programs. The coalition is composed of 1,200 individuals and diverse groups representing all sectors of the technological workforce from educators to scientists, engineers, and technicians. The participating organizations of the STEM Education Coalition are dedicated to ensuring high-quality STEM education at all levels. The Coalition works aggressively to raise awareness in Congress, the Administration, and other organizations about the critical role that STEM education plays in enabling the U.S. to remain the economic and technological leader in the global marketplace of the twenty-first century. Specifically, the Coalition works with Congress and the Administration to meet the following coalition objectives:

1. Strengthen effective STEM education programs at all levels—K–12, undergraduate, graduate, continuing education, vocational, informal—at the National Science Foundation, the U.S. Department of Education, and other federal agencies with STEM related programs.
2. Encourage national elected officials and key opinion leaders to recognize and bring attention to the critical role that STEM education plays in U.S. competitiveness and our future economic prosperity.
3. Support new and innovative initiatives that will help improve the content knowledge skills and professional development of the K–12 STEM teacher workforce and informal educators and improve the resources available in STEM classrooms and other learning environments.
4. Support new and innovative initiatives to recruit and retain highly-skilled STEM teachers.
5. Support new and innovative initiatives to encourage more of our best and brightest students, especially those from underrepresented or disadvantaged groups, to study in STEM fields.
6. Support increased federal investment in educational research to determine effective STEM teaching and learning methods.

7. Encourage better coordination of efforts among federal agencies that provide STEM education programs.
8. Support new and innovative initiatives that encourage partnerships between state and local educators, colleges, universities, museums, science centers and the business, science, and technology communities that will improve STEM education. (STEM Education Coalition, n.d.)

Although the STEM Education Coalition advocates for strengthening STEM-related programs for educators and students and increased federal investments in STEM education, it also supports robust federal investments in basic scientific research to inspire current and future generations of young people to pursue careers in STEM fields. Since 2005, hundreds of organizations have signed the 45 STEM Education Coalition letters delivered to key Congressional leaders on a host of issues, most notably increased funding and support for science and mathematics education programs at the U.S. Department of Education and the National Science Foundation. The Coalition also has commended congressional leaders for passing, and the president for signing, the American Recovery and Reinvestment Act of 2009 and has put forth policy recommendations for the reauthorization of the Elementary and Secondary Education Act (No Child Left Behind), including support for adding science to the Adequate Yearly Progress indicators. In the past year, the STEM Education Coalition provided policy leaders with key recommendations for the Higher Education Act, and specific recommendations were submitted to the President's Council of Advisors on Science and Technology (PCAST) and the National Science Foundation Education and Human Resources Directorate. A sample letter from the Coalition to members of PCAST appears in Appendix A. The Coalition, which is co-chaired by NSTA and ACS, meets monthly in Washington, DC. It also maintains a website at www.stemedcoalition.org, which provides information on the Coalition's positions and activities, STEM-related legislation, and the available research on science and mathematics education. Membership in the Coalition is free, and participation by members is completely voluntary. Many members simply elect to receive the as-needed e-mails sent by the NSTA and ACS co-chairs. A legislative task force of volunteers work together to formulate the Coalition's positions on STEM-related issues and policy, develop white papers and correspondence to Congress, and, in some cases, directly advocate on behalf of science and mathematics education. Appendix B contains a list of policy positions the Coalition took in 2007 and outcomes of the various policy initiatives.

INFLUENCING THE POLICY AGENDA

As stated earlier, any real change to STEM education policy will come from the many individuals within the STEM community and the influence they together can bring to the policy decisions that are made. Also, as noted earlier, there are limits placed on the lobbying activities of nonprofit organizations, but there are generally no restrictions on the lobbying activities of individuals (except that many federal employees and contractors cannot advocate during business hours if their time/salary is paid for with federal dollars). The limitations described earlier apply to non-profit organizations such as NSTA, not to individuals. An individual citizen may contact a member of Congress and ask for support for a particular bill that is pending in Congress. Individuals and groups can work to educate federal (and state) policymakers on the role of science and science education in public policy, and they can advance specific recommendations on issues affecting science and science education. But, before attempting to influence a member of Congress on a particular policy issue, it is important to be well prepared. Here are some ideas to keep in mind:

1. Take some time to learn about the existing policy, legislation, and funding resources that are important to science and science education. Find out more about the political process and what you can do. (Many groups, such as the STEM Education Coalition, can assist with this.)
2. Identify the key issues that you want to address, and develop a succinct message that expresses what you want to convey (Remember to "keep it short and simple." The point is to not provide too much information. Members of Congress have little time to delve into complex analyses or to review pages of data.).
3. Know who your legislators are and how they stand on the key issues. Find out who, besides you, in your organization is interested and committed to these issues, and then work together.
4. Remember to provide real life success stories in your contacts with policymakers. Seek out opportunities to interact with legislators, both in Washington and in the home state/district. Many organizations sponsor "fly-in lobby days," "meet and greets," and legislative briefings that provide their members the chance to interact with legislators.
5. Use some of the tips for a successful contact (e-mail, phone, letter, visit) listed in Figure 9.2 and always follow up a visit or an encounter

with a thank-you and more information. Become comfortable writing or talking to legislators.

Advocating for an issue is essential to legislative success. By cultivating relationships with your elected officials, you are establishing yourself as an authority on a specific issue. These individuals want the best information to do their jobs, and in many cases they will turn to reliable experts who have demonstrated that they have the knowledge needed to assist in the development of policy.

Table 9.2. Guidelines for Communicating With Members of Congress

Include Your Name and Address at the Top of Message.
You should send an e-mail or message only to your elected official. If you are not a constituent, then do not send an e-mail. The first thing your representative will do is to determine if you live in his or her district. Representatives and staff do not have any obligation and little time to read messages from people who are not constituents, so it is vital that you make it clear that you live in the district.
Humanize Your Message.
This is one of the most important things you can do to ensure your e-mail makes an impact. Many people are uncomfortable sharing their feelings or talking about their own experiences, or believe that such information is inappropriate to the legislative process. Yet, it is this information that separates one's message from the standardized, bulk messages drafted by interest groups. These messages are more likely to be read than simply tallied.
Be Brief.
Members of Congress and their staff are extremely busy. Respect their time and try to tell them only what they need to know. Two or three paragraphs should be sufficient. Do not feel that you have to make every single argument that relates to the issue, only the strongest points you can make.
Be Clear About Your Position.
Your request should be stated as a concrete, actionable item, e.g., "I would like you to support H.R. 100."
Make Your Message Timely.
Send your message when the legislation is being considered. Your message is worthless if it arrives after a critical vote.
Don't "Flame."
You are allowed to disagree with your member of Congress, but you will not be effective if you abuse or threaten them. Abusive letters seem more desperate than intimidating to the recipient, and they are seldom taken seriously.
Avoid Attachments.
Congressional offices rarely print or read attachments to e-mail. Offer to provide supporting documents on request, but avoid sending attached files.

Don't Become "Spam."

Do not send Congress a message every single day about every issue you read about or develop an opinion on. An office that receives numerous messages from a single person quickly loses sight of the urgency or expertise that the constituent can bring to a specific issue.

Establish Your Credibility.

Explain if you are an expert in some area. Also, do not shy away from saying that you are either a personal supporter or a party supporter (but never imply that because you voted for somebody or contributed money to their campaign that they owe you a vote).

Don't Lie.

Political professionals are adept at spotting a tall tale. Any story that sounds too perfect or any statistic that is not substantiated will not bolster your position.

Don't cc Everybody.

Resist the urge to send a copy of your message to every member of Congress. You will persuade no one and annoy everybody. A legislative office wants to know that you have appealed to them for specific action, not just sent them a copy of a memo distributed to all.

Crafting an Effective Message to Members of Congress

Most organizations with advocacy initiatives encourage their members to send messages to members of Congress via e-mail. A survey by the Congressional Management Foundation (CMF) (2005) found that the Internet has made it easier for citizens to become involved in public policy, has increased public understanding of what goes on in Washington, and has made members more responsive to their constituents. E-mail is now the preferred means of communication with constituents for members of Congress and their staff.

All correspondence to Congress should be short, targeted and informative. NSTA encourages teachers to follow the guidelines listed in Figure 9.2 when writing to lawmakers. This information is also available on the NSTA website at www.nsta.org. Timing is everything though. Before you send your message, it is critical to understand the process of how a bill becomes law and when and where constituents can have the most impact. It's important to know the status of a bill because the request to a member of Congress will often depend on where the bill is in the process. The best way to stay abreast of the status of a bill is to use the Library of Congress THOMAS website at http://thomas.loc.gov/ , which provides up-to-date information on where a bill stands in the legislative process and the bill's history. Several publications and associations (including NSTA) publish information on how a bill becomes law. Familiarizing yourself with the basics of this process is important, see Appendix C: How a Bill Becomes Law.

Registering your opinion early on, when the bill is at the subcommittee level or earlier, is more effective than, say, before the bill goes to the House floor for a vote. Knowing how the process works, and aligning yourself with such as NSTA or the STEM Education Coalition, is crucial.

Maintaining a Relationship With Your Elected Representative

Sending an e-mail begins a relationship with your elected official, but, then what? The NSTA encourages members to become advocates for science and mathematics by maintaining a relationship with their elected officials, at both the federal and state levels. Staying in touch can be done with little time and effort, and is often as easy as sending an e-mail to a staffer on an issue of concern or to address specific legislation.

When specific legislation is not the focus of this contact, maintaining direct contact with the elected representative and his/her staff can include e-mails that provide local statistics, anecdotes, or information on STEM education in that lawmaker's district. Another suggestion is to send information on national research or new test data relative to U.S. STEM education, or data that shows the link between STEM education and economic growth, and linking this research or data to your community. Providing data and information on key issues that will help that staffer or member of Congress do his/her job will be appreciated and will develop and maintain key relationships. In addition, NSTA also encourages members to develop and maintain direct face-to-face contact with the elected representative and his/her staff, either when the elected official is home in the district or in Washington, DC. (See Appendix D for important ideas to consider when making an appointment with an elected official and during the visit.)

CONCLUSION

Advocacy and the more specific type of advocacy known as lobbying are important activities in a democratic society. By communicating with legislators, whether members of a city council, state legislature, or Congress, policymakers come to know the importance of various policy positions. Legislators are responsible for making policy decisions in a wide range of areas affecting life in our society. The Committee on House Administration, for example, lists dozens of separate caucuses in the House of Representatives that cover areas from STEM education to Alzheimer's disease to public broadcasting. To communicate the importance of STEM education takes a significant effort on the part of individuals, organizations, and alliances of those organizations. The NSTA is committed to improving that communication through its various legislative activities. For more information on those activities and resources to help advocate for high-quality science education, please go to the NSTA Web site at www.nsta.org.

APPENDIX A

LETTER FROM THE STEM EDUCATION COALITION TO MEMBERS OF THE PRESIDENT'S COUNCIL OF ADVISORS ON SCIENCE AND TECHNOLOGY

March 3, 2010

The Honorable John Holdren
Co-Chair, President's Council of Advisors on Science and Technology
Director, Office of Science and Technology Policy
White House
1600 Pennsylvania Avenue, NW
Washington, DC 20500

The Honorable Eric Lander
Co-Chair, President's Council of Advisors on Science and Technology
President and Director,
Broad Institute
7 Cambridge Center
Cambridge, MA 0214

The Honorable Harold Varmus
Co-Chair, President's Council of Advisors on Science and Technology
President, Memorial Sloan-Kettering Cancer Center
1275 York Avenue New York, NY 10065

Dear Drs. Holdren, Lander, and Varmus:

The Science, Technology, Engineering, and Mathematics (STEM) Education Coalition is pleased to provide comment to the President's Council of Advisors on Science and Technology (PCAST) as you move forward with a report on a wide range of policies related to STEM education.

As our country deals with the current economic downturn and prepares for a robust recovery, it is absolutely essential that we pay close attention to the role STEM education plays in ensuring the competitiveness of our workforce. To bolster our nation's STEM education system, we must employ a robust range of policies, solutions, and partnerships.

The STEM Education Coalition has actively promoted positive STEM education reform before Congress and the Executive Branch, and we have been engaged in many of the major legislative debates of the last several years, including the *Higher Education Opportunity Act*, the *America COMPETES Act*, efforts to reauthorize the *Elementary and Secondary Education Act*, and the annual appropriations process.

As you proceed with your work, we respectfully request the following key principles be given strong consideration in your report to President Obama.

1. **The Federal Government Must Provide Strong and Sustained Support for Key STEM Education Priorities**

We strongly urge that PCAST recommends increased funding for NSF's EHR Directorate and the Math and Science Partnership Program at the U.S.Department of Education.

APPENDIX B
2007 STEM EDUCATION COALITION POSITIONS AND OUTCOMES
Updated 8/3/2007

<table>
<tr><th>Letter</th><th>Coalition Position</th><th>Congressional Outcome</th></tr>
<tr><td colspan="3">Appropriations</td></tr>
<tr><td>Coalition on House Labor, Health, and Human Services (LHHS) Appropriation</td><td rowspan="2">"provide $450 million in funding for the Math and Science Partnership (MSP) program at the United States. Department of Education"</td><td>House bill funded Department of Education MSP program at $182 M, floor amendment increased funding by $16 M to $198 M (8.6%), bill passed House</td></tr>
<tr><td>Coalition on Senate LHHS Appropriation</td><td>Senate bill funded MSP program at $184 M, floor action pending</td></tr>
</table>

<table>
<tr><td>Coalition on NSF to House Commerce, Justice, and Science (CJS) Appropriation</td><td rowspan="2">"support of the Administration's request of $6.43 billion in fiscal year 2008 for the National Science Foundation (NSF)"

"...if additional resources are available, they be devoted to bolstering the programs of NSF's EHR Directorate without diminishing essential support for the Foundation's research directorates."</td><td>House bill funded NSF Education and Human Resources Directorate at $822.6 M, a $72 M increase over Administration request, bill passed House</td></tr>
<tr><td>Coalition on NSF to Senate CJS Appropriation</td><td>Senate bill funded EHR Directorate at $850 M, a $100 M increase of Administration request, floor action pending</td></tr>
<tr><td>Coalition to Leadership on CR Appropriation</td><td>"make investments in STEM Education programs at the National Science Foundation (NSF) and the Department of Education (DoEd) a priority as Congress completes action on the FY 2007 appropriation bills."</td><td>Although NSF's research programs received a modest boost in FY07, NSF's Education programs received the same funding level as FY06.</td></tr>
<tr><td colspan="3">Competitiveness</td></tr>
<tr><td>Coalition to Gordon-Hall on Partnerships for Access to Laboratory Science (PALS)</td><td>"support of HR 524, the Partnerships for Access to Laboratory Science (PALS) Act."</td><td>Bill incorporated into House and Senate Competitiveness bills and final Competitiveness bill passed by Congress</td></tr>
<tr><td>Coalition to Miller-McKeon on Competitiveness Priorities</td><td>"support for several key STEM education policy priorities related to a competitiveness bill from the Education and Labor Committee:
• Establish P-16 STEM Councils
• Dedicate Funding for Elementary and Middle School Mathematics
• Strengthen Emphasis on STEM fields in After-School and Summer Programs
• Recruit Highly Qualified STEM Teachers
• Expand Professional Science Master's (PSM) Degree Programs in all the STEM Fields</td><td>All items incorporated into final Competitiveness bill passed by Congress</td></tr>
</table>

Appendix B continues on next page.

APPENDIX B Continued

Letter	*Coalition Position*	*Congressional Outcome*
Competitiveness		
Coalition to Miller-McKeon on Competitiveness Priorities	"support for several key STEM education policy priorities related to a competitiveness bill from the Education and Labor Committee: • Establish P-16 STEM Councils • Dedicate Funding for Elementary and Middle School Mathematics • Strengthen Emphasis on STEM fields in After-School and Summer Programs • Recruit Highly Qualified STEM Teachers • Expand Professional Science Master's (PSM) Degree Programs in all the STEM Fields	All items incorporated into final Competitiveness bill passed by Congress
Coalition in Support of HR 362	"we commend you for your leadership in establishing a new Diversity and Innovation Caucus in the House of Representatives that will focus much-needed attention on the issue of increasing the participation of underrepresented groups in the STEM fields."	Caucus successfully launched in June
No Child Left Behind		
Coalition NCLB Recommendations–House Coalition NCLB Recommendations–Senate	"support for several key STEM education policy priorities related to the reauthorization of the No Child Left Behind (NCLB) Act: • Strengthen Math and Science Partnerships Establish P-16 STEM Councils • Establish K-8 Master Teacher Programs • Dedicate Funding for Teacher Professional Development under Title II A • Strengthen Emphasis on STEM fields in After-School Programs • Promote STEM Specialty High Schools • Dedicate Funding for Elementary and Middle School Mathematics	Draft bill pending, some portions of these recommendations incorporated into Competitiveness bill Draft bill pending, some portions of these recommendations incorporated into Competitiveness bill

Coalition to Representatives Miller and McKeon on Science in AYP	"support for amending the No Child Left Behind Act to require that the results from state science assessments be included in Adequate Yearly Progress."	Draft bill pending

We urge the PCAST to carefully review STEM-focused education initiatives authorized in the Higher Education Act and America Competes Act as a part of your study of potential federal STEM education initiatives.

The PCAST report should include strong language that clarifies the roles and responsibilities of federal R&D mission agencies in STEM Education and calls for the coordination of STEM education programs across the federal agencies.

The PCAST report should also address the portion of the federal STEM portfolio dedicated to K–12 programs

2. **Science, Technology, Engineering, and Mathematics Education Must Be Clearly Defined**

PCAST should include a clear definition of STEM education and define what STEM education means in the context of preparing the next generation to be career or college ready. Federal STEM education initiatives must include technology and engineering educators and programs. Computer science education should also be a major component within the STEM conversation.

3. **Stakeholders Must Work Toward the Alignment of STEM Education**

To ensure that all students have an opportunity to learn 21st-century skills, we encourage PCAST to support the development and implementation of policies that will encourage a vertical alignment of P–20 STEM education that includes these stakeholders: Higher Education/Undergraduate, Community Colleges/CTE, After-School Programs, and Informal Education.

4. **STEM Teaching and Learning Must Be Improved**

A systemic approach to improving teaching and learning in the STEM fields must focus on Standards, Assessments, and Accountability; Teacher Preparation and Professional Development; Increasing Diversity in the STEM Pipeline; Linking Research to Classroom Practice; Increasing Classroom Resources; and Recognizing the Importance of Informal Learning.

We appreciate the strong commitment of the Administration to addressing the challenges facing STEM education and our nation's competitiveness in the global economy and hope that the recommendations offered here will help inform your deliberations on this vitally important subject.

For any additional information on STEM education please do not hesitate to contact Coalition Co Chairs, James Brown (American Chemical Society) at 202-872-6229 or Jodi Peterson (National Science Teachers Association) at 703-312-9214.

Sincerely,

Action Works
Aerospace Industries Association
Alabama Mathematics, Science, and Technology Education Coalition (AMSTEC)
Altshuller Institute for TRIZ Studies
American Association of Colleges for Teacher Education
American Association of Physicists in Medicine
American Association of Physics Teachers
American Association of University Women (AAUW)
American Astronomical Society
American Chemical Society
American Helicopter Museum & Education Center
American Institute of Aeronautics and Astronautics
American Institute of Biological Sciences
American Museum of Natural History
American Society for Engineering Education
American Society for Microbiology
American Society of Agronomy
American Society of Civil Engineers
American Society of Heating, Refrigerating and Air-Conditioning Engineers, Inc.
American Statistical Association
ASME Center for Public Awareness
Association for Computing Machinery
Association of Public and Land-grant Universities–APLU
Association of Science-Technology Centers
ASTRA
Baltimore Washington Corridor Chamber
Battelle
Biophysical Society
BSCS (Biological Sciences Curriculum Study)
Carnegie Corporation of New York
Center for Excellence in Education (CEE)

Center for Minority Achievement in Science and Technology (CMAST)
Computer Science Teachers Association
Computing Research Association
Council on Undergraduate Research
Crop Science Society of America
DEPCO, LLC
Destination ImagiNation, Inc.
EAST Initiative
Education Development Center, Inc.
Engineers Without Borders-USA
Entertainment Industries Council, Inc.
Exploratorium
Falcon School District 49 pre-K-12 STEM Educational Initiative, Colorado Springs, CO
Funutation Tekademy LLC
Hands On Science Partnership
Illinois Mathematics and Science Academy
Institute for Advanced Study
International Technology and Engineering Education Association (ITEEA)
Knowles Science Teaching Foundation
LearnOnLine, Inc
Museum of Science and Industry, Chicago
Museum of Science, Boston
NASA STEM School Administrators Association
National Alliance for Partnerships in Equity
National Alliance for Partnerships in Equity Education Foundation
National Center for Science Education
National Center for Technological Literacy
National Council for Advanced Manufacturing
National Council of Teachers of Mathematics
National Girls Collaborative Project
National Science Teachers Association
National Society of Professional Engineers
National Youth Science Foundation
NDIA
Ohio Mathematics and Science Coalition
Pathways into Science
PBS
Project Exploration
Project Lead The Way
PTC
PTC-MIT Consortium
Real World Design Challenge

REVOLUTIONARY DESIGNS
SAE International
Science Teachers Association of New York State
Society for the Advancement of Chicanos/Hispanics and Native Americans in Science (SACNAS)
Society of Women Engineers (SWE)
Soil Science Society of America
South Carolina's Coalition for Mathematics and Science
SPIE, the International Society for Optics and Photonics
STEMES
Technology Student Association
The Society of Naval Architects and Marine Engineers
Triangle Coalition
Vernier Software & Technology
Water Environment Federation

APPENDIX C

HOW A BILL BECOMES LAW

GOVERNMENT 101: How a Bill Becomes Law

House—Legislation is handed to the clerk of the House or placed in the hopper.

Senate—Members must gain recognition of the presiding officer to announce the introduction of a bill during the morning hour. If any senator objects, the introduction of the bill is postponed until the next day.

- The bill is assigned a number (e.g., HR 1 or S 1)
- The bill is labeled with the sponsor's name.
- The bill is sent to the Government Printing Office (GPO) and copies are made.
- Senate bills can be jointly sponsored.
- Members can cosponsor the piece of Legislation.

B. Committee Action—The bill is referred to the appropriate committee by the Speaker of the House or the presiding officer in the Senate. Most often, the actual referral decision is made by the House or Senate parliamentarian. Bills may be referred to more than one committee and it may be split so that parts are sent to different committees. The Speaker of the House may set time limits on committees. Bills are placed on the calendar of the committee to which they have been assigned. Failure to act on a bill is equivalent to killing it. Bills in the House can only be released from committee without a proper committee vote by a discharge petition signed by a majority of the House membership (218 members).

Committee Steps:

1. Comments about the bill's merit are requested by government agencies.
2. Bill can be assigned to subcommittee by Chairman.
3. Hearings may be held.
4. Subcommittees report their findings to the full committee.
5. Finally there is a vote by the full committee—the bill is "ordered to be reported."
6. A committee will hold a "mark-up" session during which it will make revisions and additions. If substantial amendments are made, the committee can order the introduction of a "clean bill" which will include the proposed amendments. This new bill will have a new number and will be sent to the floor while the old bill is discarded. The chamber must approve, change or reject all committee amendments before conducting a final passage vote.
7. After the bill is reported, the committee staff prepares a written report explaining why they favor the bill and why they wish to see their amendments, if any, adopted. Committee members who oppose a bill sometimes write a dissenting opinion in the report. The report is sent back to the whole chamber and is placed on the calendar.
8. In the House, most bills go to the Rules committee before reaching the floor. The committee adopts rules that will govern the procedures under which the bill will be considered by the House. A "closed rule" sets strict time limits on debate and forbids the introduction of amendments. These rules can have a major impact on whether the bill passes. The rules committee can be bypassed in three ways: (1) members can move rules to be suspended (requires 2/3 vote) (2) a discharge petition can be filed (3) the House can use a Calendar Wednesday procedure.

C. Floor Action

1. Legislation is placed on the Calendar

House: Bills are placed on one of four House Calendars. They are usually placed on the calendars in the order of which they are reported yet they don't usually come to floor in this order - some bills never reach the floor at all. The Speaker of the House and the Majority Leader decide what will reach the floor and when. (Legislation can also be brought to the floor by a discharge petition.)

Senate: Legislation is placed on the Legislative Calendar. There is also an Executive calendar to deal with treaties and nominations. Scheduling of legislation is the job of the Majority Leader. Bills can be brought to the floor whenever a majority of the Senate chooses.

2. Debate

House: Debate is limited by the rules formulated in the Rules Committee. The Committee of the Whole debates and amends the bill but cannot technically pass it. Debate is guided by the Sponsoring Committee and time is divided equally between proponents and opponents. The Committee decides how much time to allot to each person. Amendments must be germane to the subject of a bill—no riders are allowed. The bill is reported back to the House (to itself) and is voted on. A quorum call is a vote to make sure that there are enough members present (218) to have a final vote. If there is not a quorum, the House will adjourn or will send the Sergeant at Arms out to round up missing members.

Senate: debate is unlimited unless cloture is invoked. Members can speak as long as they want and amendments need not be germane—riders are often offered. Entire bills can therefore be offered as amendments to other bills. Unless cloture is invoked, Senators can use a filibuster to defeat a measure by "talking it to death."

3. Vote—the bill is voted on. If passed, it is then sent to the other chamber unless that chamber already has a similar measure under consideration. If either chamber does not pass the bill then it dies. If the House and Senate pass the same bill then it is sent to the President. If the House and Senate pass different bills they are sent to Conference Committee. Most major legislation goes to a Conference Committee.

D. Conference Committee

1. Members from each house form a conference committee and meet to work out the differences. The committee is usually made up of

senior members who are appointed by the presiding officers of the committee that originally dealt with the bill. The representatives from each house work to maintain their version of the bill.

2. If the Conference Committee reaches a compromise, it prepares a written conference report, which is submitted to each chamber.
3. The conference report must be approved by both the House and the Senate.

E. The President—the bill is sent to the President for review.

1. A bill becomes law if signed by the President or if not signed within 10 days and Congress is in session.
2. If Congress adjourns before the 10 days and the President has not signed the bill then it does not become law ("Pocket Veto.")
3. If the President vetoes the bill it is sent back to Congress with a note listing his/her reasons. The chamber that originated the legislation can attempt to override the veto by a vote of two-thirds of those present. If the veto of the bill is overridden in both chambers then it becomes law.

F. The President—the bill is sent to the President for review.

GLOSSARY OF TERMS

House Legislative Calendars

The Union Calendar—A list of all bills that address money and may be considered by the House of Representatives. Generally, bills contained in the Union Calendar can be categorized as appropriations bills or bills raising revenue.

The House Calendar—A list of all the public bills that do not address money and maybe considered by the House of Representatives.

The Corrections Calendar—A list of bills selected by the Speaker of the House in consultation with the Minority leader that will be considered in the House and debated for one hour. Generally, bills are selected because they focus on changing laws, rules and regulations that are judged to be outdated or unnecessary. A 3/5 majority of those present and voting is required to pass bills on the Corrections Calendar.

The Private Calendar—A list of all the private bills that are to be considered by the House. It is called on the first and third Tuesday of every month.

Types of Legislation

Bills—A legislative proposal that if passed by both the House and the Senate and approved by the President becomes law. Each bill is assigned a bill number. HR denotes bills that originate in the House and S denotes bills that originate in the Senate.

Private Bill—A bill that is introduced on behalf of a specific individual that if it is enacted into law only affects the specific person or organization the bill concerns. Often, private bills address immigration or naturalization issues.

Public Bill—A bill that affects the general public if enacted into law.

Simple Resolution—A type of legislation designated by H Res or S Res that is used primarily to express the sense of the chamber where it is introduced or passed. It only has the force of the chamber passing the resolution. A simple resolution is not signed by the President and cannot become Public Law.

Concurrent Resolutions—A type of legislation designated by H Con Res or S Con Res that is often used to express the sense of both chambers, to set annual budget or to fix adjournment dates. Concurrent resolutions are not signed by the President and therefore do not hold the weight of law.

Joint Resolutions—A type of legislation designated by H J Res or S J Res that is treated the same as a bill unless it proposes an amendment to the Constitution. In this case, 2/3 majority of those present and voting in both the House and the Senate and ratification of the states are required for the Constitutional amendment to be adopted.

Other Terms

Calendar Wednesday—A procedure in the House of Representatives during which each standing committees may bring up for consideration any bill that has been reported on the floor on or before the previous day. The procedure also limits debate for each subject matter to 2 hours.

Cloture—A motion generally used in the Senate to end a filibuster. Invoking cloture requires a vote by 3/5 of the full Senate. If cloture is invoked further debate is limited to 30 hours, it is not a vote on the passage of the piece of legislation.

Committee of The Whole—A committee including all members of the House. It allows bills and resolutions to be considered without adhering to all the formal rules of a House session, such as needing a quorum of 218. All measures on the Union Calendar must be considered first by the Committee of the Whole.

Co-Sponsor—A member or members that add his or her name formally in support of another members bill. In the House a member can become a co-sponsor of a bill at any point up to the time the last authorized committee considers it. In the Senate a member can become a co-sponsor of a bill anytime before the vote takes place on the bill. However, a co-sponsor is not required and therefore, not every bill has a co-sponsor or co-sponsors.

Discharge Petition—A petition that if signed by a majority of the House, 218 members, requires a bill to come out of a committee and be moved to the floor of the House.

Filibuster—An informal term for extended debate or other procedures used to prevent a vote on a bill in the Senate.

Germane—Relevant to the bill or business either chamber is addressing. The House requires an amendment to meet a standard of relevance, being germane, unless a special rule has been passed.

Hopper—Box on House Clerk's desk where members deposit bills and resolutions to introduce them.

Morning Hour—A 90-minute period on Mondays and Tuesdays in the House of Representatives set aside for 5 minute speeches by members who have reserved a spot in advance on any topic.

Motion to Recommit—A motion that requests a bill be sent back to committee for further consideration. Normally, the motion is accompanied by instructions concerning what the committee should change in the legislation or general instructions such as that the committee should hold further hearings.

Motion to Table—A motion that is not debatable and that can be made by any Senator or Representative on any pending question. Agreement to the motion is equivalent to defeating the question tabled.

Quorum—The number of Representatives or Senators that must be present before business can begin. In the House 218 members must be present for a quorum. In the Senate 51 members must be present however, Senate can conduct daily business without a quorum unless it is challenged by a point of order.

Rider—An informal term for an amendment or provision that is not relevant to the legislation where it is attached.

Sponsor—The original member who introduces a bill.

Substitute Amendment—An amendment that would replace existing language of a bill or another amendment with its own.

Suspension of the Rules—A procedure in the House that limits debate on a bill to 40 minutes, bars amendments to the legislation and requires a 2/3 majority of those present and voting for the measure to be passed.

Veto—A power that allows the President, a Governor or a Mayor to refuse approval of a piece of legislation. Federally, a President returns a vetoed bill to the Congress, generally with a message. Congress can accept the veto or attempt to override the veto by a 2/3 majority of those present and voting in both the House and the Senate. (Project Vote Smart, n.d.)

APPENDIX D

DOs AND DON'Ts WHEN MAKING PERSONAL VISITS TO MEMBERS OF CONGRESS AND THEIR STAFF

DO these things:

Make an appointment in advance.
Time is always at a premium in legislative offices. Contact the legislator's scheduler in advance to arrange a meeting. It is best to make your meeting request in writing and follow up with a phone call. Be clear about who will be attending the meeting and the specific reason for the meeting. Legislative schedules are unpredictable so don't be put off if your meeting is rescheduled or if you have to meet with staff in lieu of the elected official.

Your homework.
Prepare carefully and thoroughly for your meeting. Take the time to "know" your legislator by reviewing past votes or statements on the issue, his/her party's position, and committee assignments. Develop an agenda that all your participants clearly understand. Know your talking points in advance and be prepared to make your case. Research the opposition's arguments against your position and, if possible, acknowledge and rebut those arguments in your presentation.

Stay "on message."
Effective legislative meetings should be narrow in scope. Stick to a single issue, state only a few key points in support of your position and make a definite request for action. Many meetings are ineffective because a participant brings up other issues or strays from the key arguments supporting your position. Have a message and stick to it.

Go local.
Your effectiveness is based on geography. Legislators want to hear your thoughts and opinions because you are a constituent. One of your most useful strategies is to relate the issue and your position to your community. Legislators have many other avenues to get national

or state analysis, reports, and statistics. Local statistics and stories are important and you may be the only source for such rich information. Don't be afraid to humanize the issue by relating it to your local community or personal experience.

Make a clear, actionable request.
Many people are afraid that it's impolite to make a direct request. But, don't forget that the purpose of your meeting is to secure support for your issue. It is appropriate and expected that you will make a request at your meeting. The key is to make sure that your request is clearly articulated and actionable by the legislator. Keep in mind that your request should be timely and consistent with the legislative process. It is usually not enough to ask for generic support for an issue or cause, rather make a direct and specific request that is tied to pending legislative activity (if possible). For example, ask that a legislator co-sponsor a bill. You should make reference to bill numbers and be knowledgeable about the status of the bill. Making a specific request gives you the opportunity to evaluate the legislator's response.

Cultivate a relationship with staff.
Many grassroots advocates underestimate the important role of legislative staff. A supportive staff person can often make the difference between success and failure. Staff play an invaluable role in shaping a legislator's agenda and position on issues. It is important that you make every effort to cultivate a positive working relationship with staff. Over time, staff may even come to regard you as a helpful resource for information on your issue.

Follow-up.
What happens after a meeting is almost as important as the meeting itself. Send a 'thank you' letter after the meeting that not only expresses appreciation but reinforces your message and any verbal commitment of support made by the legislator or staff. If you promise during the meeting to get back in touch with additional information, be sure that you do so. Failure to follow up on your promise will call your credibility into question.

DON'T do these things:

Go "off-message" or discuss unrelated issues.
You must deliver a unified message during your meeting. Sending different messages or discussing unrelated subjects will only undermine your ability to secure support. Limit your advocacy to a single issue. Legislators meet with many groups and constituents so it is important that your message and request be clear and uniform.

Engage in partisan critiques.
It is best to keep the discussion based on the merits of the policy or issue. Avoid characterizing your position in strictly partisan terms. Worse, do not make snide or disparaging partisan comments. You are working on behalf of an issue, not a party. So, you want legislators of both parties to support your position. Be careful not to alienate legislators or staff based on partisanship.

Use threats.
While it may be tempting to tell a legislator who has rebuffed your request that "you'll never vote for him/her again" or that "you pay his/her salary," such discourtesy only ensures that your arguments will be discounted—now and in the future.

Be late.
Time is a valuable and scare commodity for legislators. Punctuality conveys professionalism and demonstrates your commitment to your issue, which is after all the reason for the meeting. Arrive early and if you are meeting as a group allow time to calm nerves and make a final review of the talking points and message.

Get too comfortable.
Advocates are sometimes surprised by the courteous reception they receive, even from lawmakers who disagree with their position. As a constituent you will be accorded respect by the legislator and staff. Don't mistake this respect for agreement. Don't let the comfortable nature of the exchange deter you from making your request. And, don't mistake "concern" for your issue with support for your position.

Forget to follow-up.
Immediately send a thank you letter. Stay informed on your issue and track how your legislator responds. Did the legislator follow through on his/her promise? If not, request an explanation. If so, express your appreciation.

REFERENCES

America COMPETES Act of 2007, 20 U.S.C. § 9801 *et seq.* (2007).

American Recovery and Reinvestment Act of 2009, 26 U.S.C. § 1 *et seq.* (2009).

Congressional Management Foundation. (2005). *Communicating with Congress: How Capitol Hill is coping with the surge in citizen advocacy.* Retrieved from http://www.cmfweb.org/storage/cmfweb/documents/CMF_Pubs/cwc_capitolhillcoping.pdf

National Commission on Mathematics and Science Teaching for the 21st Century. (2000). *Before it's too late*. Washington, DC: U.S. Department of Education. Retrieved from http://www2.ed.gov/inits/Math/glenn/report.pdf

National Science Teachers Association. (n.d.). *Legislative handbook*. Arlington, VA: Author.

No Child Left Behind Act of 2001, 20 U.S.C. § 6301 *et seq.* (2002).

O'Neill, T. P. (2009). *Thomas P. O'Neill, Jr. Congressional papers 1936-1994*. Boston: John J. Burns Library, Boston College. Retrieved from http://www.bc.edu/bc_org/avp/ulib/oneill_findingaid2.html

Project Vote Smart. (n.d.). *Government 101: How a bill becomes law*. Retrieved from http://www.votesmart.org/resource_govt101_02.php

Samuelson, R. J. (2008, December). Lobbying is democracy in action, *Newsweek*. Retrieved from http://www.newsweek.com/id/174283

STEM Education Coalition. (n.d.). *STEM Education Coalition objectives*. Retrieved from http://www.stemedcoalition.org/content/objectives/Default.aspx

PART II

IMPACT OF POLICY ON CURRICULUM, INSTRUCTION, AND THE EQUITABLE TREATMENT OF ALL STUDENTS

Part II of this volume is concerned with the influence of policy on practice. The three chapters focus on how various policies have influenced curriculum, classroom instruction, and the equitable treatment of all students. George DeBoer begins the section with a discussion of how policies have influenced what is taught and how it is taught in science classrooms, focusing in particular on the role of content standards and the use of assessment for accountability in shaping the content of the curriculum and on the role of the courts in determining that school districts cannot ask teachers to teach theories of evolution that are outside the accepted scientific canon. Sharon Lynch reviews the history of equity legislation in the United States and points out how different interpretations of what equity means have influenced the kind of science instruction various diverse groups of students receive in science classrooms. Linda De Lucchi and Larry Malone use the Lawrence Hall of Science experience to show how over time a curriculum development effort is subject to changing policies both at the state and federal levels.

CHAPTER 10

HOW STATE AND FEDERAL POLICY AFFECTS WHAT IS TAUGHT IN SCIENCE CLASSES

George E. DeBoer

INTRODUCTION

Who determines what will be taught in science classes in American schools? We know that teachers have choice in what they teach, but how far does that freedom extend? What policies, if any, whether formal or informal, constrain and limit the choices that teachers can make? In this chapter, I describe the influence of state and federal legislative bodies, the judiciary, professional societies devoted to science and science education, as well as individual scientists and science educators on the content of the science curriculum.

EVOLUTION OF THE SCIENCE CURRICULUM IN U.S. SCHOOLS

At the broadest level, the content of the science curriculum in U.S. schools is determined by state policies regarding the number and type of

The Role of Public Policy in K–12 Science Education, pp. 275–303

courses that are needed for graduation. Currently, the average number of years of science required for a regular high school diploma in the United States is just under 3 years, although states have been continually increasing that requirement, especially during the past 20 years (Education Commission of the States, 2007; see also Lynch, Chapter 11, this volume). Typically, a state will require all students to take at least one course in life science and one course in physical science to graduate. Actual enrollments, of course, are greater than minimum requirements because many students take more courses than are required. (For a discussion of trends in high school science enrollment during the first half of the twentieth century, see DeBoer, 1991.) At a more specific level, the content of the school science curriculum is described in state syllabi, curriculum frameworks, and standards documents written and approved by each state education department and typically approved by the state legislature. Under current federal law, the teaching of science content is enforced by means of a comprehensive state test or a series of end-of-course exams written at the state level (see Bybee, Chapter 8, this volume). In some cases, passing these tests is required for graduation, and in other cases, the test results are used to identify schools that need improvement, but they are not used as a requirement for student graduation.

Over the years, the content of the science curriculum also has been determined to a large extent by the expectations of the next higher level in the educational system. The colleges and universities have a considerable influence on the curriculum because high school teachers, administrators, and the parents of college-bound students want the students to be well-prepared for college (see also Halverson, Feinstein, & Meshoulam, Chapter 13, this volume). This attitude trickles down to each lower level in the system. But, most schools also offer elective courses where there is more flexibility in what is studied. Within each course that is required, there also is considerable flexibility in what teachers teach as long as the students do well on state tests and parents, students, and school administrators believe that the students are well-prepared for college and the world of work.

The teaching of science also has been influenced by recommendations of national education policy groups. For example, in part because of recommendations concerning the school curriculum such as those by the National Education Association's (NEA) Committee of Ten (NEA, 1894), by about 1920 most schools had settled on the sequence of biology, chemistry, and physics in high school, and general science in junior high school. The biology course was a transitional course between general science and chemistry and physics, which were taken primarily by college-bound students. Biology as a single course emerged between about 1910 and 1920 as a synthesis of physiology (primarily health and hygiene), zoology, and botany. In 1911, neither biology nor general science were listed by the U.S. Bureau of Education as courses offered by schools, but by 1922 the Bureau's report showed that enrollment in biology and general science greatly surpassed enrollment in the separate courses of physiology, zoology, and botany (DeBoer, 1991). It took a little longer for schools to settle the question of whether it was better to study physics or chemistry first, but by 1947, the ratio of seniors to juniors taking physics was about two to one. Because only a very small percentage of students were preparing to go to college during the first half of the twentieth century, this sequence of courses meant that biology often was the last course most students took in science.

Many other policy recommendations were made by the Committee of Ten (NEA, 1894) concerning the amount of time that students should engage in the study of science and the way science should be taught, that is, by direct investigation of the world rather than as a study of the world solely through books (p. 50). Another example of a policy recommendation by a national organization was the statement by the Thirty-First Yearbook Committee of the National Society for the Study of Education (NSSE) in 1932 that the best way to make the subject matter of chemistry meaningful to students was to focus on a limited number of unifying themes of the discipline and by relating application to theory. They proposed two such organizing themes in chemistry, The first was, chemical substances change during chemical reactions to form substances that behave differently than they did before the reaction occurred. The second was, there is a quantitative relation between the amounts of substances that react and the amounts of substances that are produced in a chemical reaction (NSSE, 1932). But, in 1932, these were simply recommendations. Most states did not have examinations that could be used to enforce such recommendations. Therefore, in most states, teachers decided for themselves if those were the most important ideas to teach.

During the early twentieth century, states tended to define the content that should be taught in the biology, chemistry, and physics courses by means of a state syllabus and, in the case of New York, they enforced the teaching of that syllabus through state examinations used to judge the suitability of students for college (see Cheek & Quiriconi, Chapter 7, this volume). The first attempt at the national level to organize the science curriculum came in the 1950s when Congress began to fund, through the National Science Foundation (NSF), the development of curriculum projects focused on the content structure of the disciplines and the way science was conducted. The NSF-funded courses upgraded the content of science, presented narratives that were more coherent and logically compelling, and explicitly linked the ideas of science to the methods that were used in the pursuit of those ideas. Although the NSF funding for those projects was relatively short lived, the decision by the federal government to support those curriculum initiatives had an impact on the nature of the science curriculum for years to come and significantly affected what teachers were expected to teach.

Additional efforts to define the science curriculum at the national level were evident during the 1970s and early 1980s. The National Science Teachers Association (NSTA), for example, identified scientific literacy as the most important goal of science education in a 1971 statement. The Association followed this identification of scientific literacy with a statement in 1982 that again argued for teaching science in the context of its societal applications. "The goal of science education during the 1980s is to develop scientifically literate individuals who understand how science, technology, and society influence one another and are able to use this knowledge in their everyday decision-making" (NSTA, 1983, p. XX).

Then came the release of the report of the National Commission on Excellence in Education (NCEE), *A Nation at Risk*, in 1983. The Commission called for higher standards in each of the major content areas of the curriculum, including science, mathematics, social studies, English, and computer science. The policy strategy the Commission proposed was for states and local schools to establish high standards in these content areas followed by national testing to ensure accountability. Higher standards were being recommended so that the country could regain its international economic competitiveness. The call for higher standards led to an effort by the National Science Board, the advisory board to the National Science Foundation, to identify the topics all students should study in the various science disciplines (National Science Board, 1983). Then, in 1989, the American Association for the Advancement of Science (AAAS) published *Science for All Americans*, which described knowledge, skills, and habits of mind that constitute adult literacy in science, broadly defined to include science, mathematics, and engineering, and technology. *Science for All Americans* was subsequently followed by the publication of *Benchmarks for Science Literacy* in 1993 (AAAS, 1993) and the *National Science Education Standards* in 1996 (National Research Council [NRC], 1996), both of which laid out the ideas that students should know by the end of various grade bands in order to achieve the goal of science literacy by the time they graduated from high school.

Although states were not compelled by federal law to teach these particular science ideas or to test student knowledge of them, many states used those documents to write their own state standards and to create tests to measure student understanding of those science ideas. In addition, the U.S. Department of Education and the National Science Foundation "urged states and local districts to incorporate or demonstrate consistency with Project 2061's vision of science literacy in their proposals for important federal initiatives" (SRI International, 1996, pp. 10–11). These funding agencies had leverage over the states in this regard by using this as a consideration for funding (see Earle, Chapter 5, this volume).

Perhaps one of the most significant implications of the efforts by AAAS and the NRC to define the science that all students should know is that they described not only the science content to be learned but also the systematic ordering of that content by grade band. For the first time, the scientists and science educators who negotiated the writing of these documents had to make decisions about which ideas were appropriate for the elementary grades, the middle grades, and the high school grades so that the ideas would build progressively toward the goal of science literacy. Here we see an example of how the scholarly writings of educational theorists can intersect with efforts of professional organizations like AAAS and the priorities of funding agencies in the policy domain. The idea that there are logical progressions of understanding that could be exploited during instruction was not a new concept. Robert Gagne had written about hierarchical learning in his *Conditions of Learning* in 1977. Gagne said:

> Many subjects taught in schools have an organization that can readily be expressed as a learning hierarchy. The rule, or set of rules, that is the learning objective may be shown to be composed of *prerequisite* rules and concepts. The learning of the intellectual skills which are the "target" of instruction is a matter of combining these prerequisite skills, which have been previously learned. (p. 143)

As an example, Gagne used the idea of calculating the horizontal and vertical components of forces using vector diagrams. He identified three prerequisite ideas and skills, each of which itself could be broken down into another set of prerequisite ideas and skills. In his example, he said that students should already know how to "(1) use rules to verify the conditions for equilibrium, (2) represent the magnitude and directions of forces as parts of triangles, and (3) employ trigonometric rules to represent the relationships in a right triangle (sine, cosine, tangent, etc.") (p. 143). With that knowledge in hand, students would then be ready to grasp the idea of using vector diagrams to calculate the components of forces. Gagne's ideas of a hierarchical ordering of ideas and skills were applied in a number of science curriculum projects, including *Science: A Process Approach (SAPA)* sponsored by the AAAS Commission on Science Education (Livermore, 1966).

In the case of AAAS Project 2061, the process of identifying progressions of understanding began when they attempted to "back map" the ideas in *Science for All Americans* across four grade bands, K–2, 3–5, 6–8, and 9–12 (AAAS, 1993, pp. 305–307). The team members used published research on student learning and their own professional judgment to make decisions about the progress students could make from grade band to grand band toward achieving each of the learning goals identified in *Science for All Americans*. In the case of the structure-of-matter topic, for example, they created four story lines to organize the progression of specific ideas through the grade bands. The four strand labels were *properties, common ingredients, invisibly small pieces, and conservation of matter.* When *Benchmarks* finally was ready for publication, however, a decision was made not to organize the content vertically by fine-grained ideas within these story lines, but rather to organize the content as clusters of ideas under broader chapter headings such as Structure of Matter, Energy Transformations, and so forth. So, instead of being able to trace a progression of understanding about the conservation of matter within and through the grade bands, *Benchmarks* was organized using a set of ideas listed for the topic of Structure of Matter at each of four grade bands, but without any ordering within that topic. This model also was used by the NRC in the *National Science Education Standards*.

But, then in 2001, with funding from the National Science Foundation, AAAS Project 2061 published the *Atlas of Science Literacy*, in which the progress-of-understanding maps that had been started during the writing of *Benchmarks* were completed. Finally, in 2007, AAAS published a second volume of maps, which covered the remaining benchmark statements. The progressions of understanding are mapped in vertical strands, with the vertical strands linked to each other horizontally within topics. There also are "off-map" connections that show how ideas are conceptually related to more distant topics within the curriculum. The *Atlas* provides curriculum developers and school district personnel a model for creating coherent and logically consistent story lines throughout the K–12 curriculum.

As part of this movement toward a more coherent sequencing of science content, in 2005 the Division of Elementary, Secondary, and Informal Education at the National Science Foundation made a decision to make research on such learning progressions one of their top priorities (see Earle, Chapter 5, this volume). In their 2005 Program Solicitation (NSF 05-612), they said:

> Science and technology education in K–12 schools tends to lack coherence and widely accepted course sequences that allow educators and instructional designers to build on prior knowledge.... To build the foundation for more coherent curricula in science and technology, a pilot initiative will support the development of learning progressions (Catley, Lehrer, & Reiser, 2005) of STEM content and processes that build student understanding over time. These would build models of instruction and demonstrate how instructional materials and professional development would look over multiple grade bands (K–5, 6–8, 9–12) when intended to support students building deep

> understanding of a key concept over time. The *Atlas of Science Literacy* (AAAS, 2001) provides examples of strands (learning progressions) for K–12 students to achieve understanding of important science content concepts described in the *Benchmarks for Science Literacy* (AAAS, 1993), *National Science Education Standards* (NRC, 1996) and some of the *Standards for Technological Literacy* (International Technology Education Association, 2000). Less is known about how effective instruction can accelerate the pace at which students are capable of moving along the developmental trajectory of learning processes such as argumentation, design, assessing evidence, experimentation, inquiry, interpretation of data and modeling. (NSF, 2005)

This discussion of learning progressions shows the complexity of policy formation in the United States. The work of AAAS to carefully define science literacy in *Science for All Americans*, the sequencing of science ideas through the grade bands by AAAS in *Benchmark for Science Literacy* and by the NRC in the *National Science Education Standards*, and the theoretical work of Robert Gagne and others, came together to influence the National Science Foundation to create a program to encourage additional research in this area. These efforts also have influenced policymakers at the state and local levels who must create coherent sequences of instruction for students.

The work done by AAAS Project 2061 in organizing various core science topics throughout the grade bands in *Benchmarks* and the *Atlas of Science Literacy*, as well as the NSF-funded research on learning progressions, give curriculum developers and school administrators a sound basis for creating sequences of courses and sequences of ideas within courses that will provide students with a coherent introduction to science. The effort to provide students with a coherent and carefully articulated presentation of content is becoming policy in American schools as states are moving toward the development of grade level (as opposed to grand band) sequences of course content throughout the K–12 grades (see DeLucchi & Malone, Chapter 12, this volume). In addition, textbooks are being evaluated on how well they provide coherent story lines (Kesidou & Roseman, 2002; Roseman, Stern, & Koppal, 2009; Stern & Roseman, 2004), and a number of science education research centers have focused on coherence as a major theme of their work (see Kali, Linn, & Roseman, 2008; Roseman, Linn, & Koppal, 2008).

This trend toward greater specification in the science curriculum, both in terms of what is taught and the way it is sequenced may not seem like a dramatic change to those of us who are in the middle of this transition, but it is important to remember that the move away from a more open and diffuse approach to curriculum organization has occurred in just the past 30 years or so. When compared to the curriculum traditions in other countries, this change is certainly significant. For example, the European tradition of *Didaktiks* does not value linearity in the same way that American curriculum developers do. In fact, the more humanistic liberal arts tradition in Europe, which also tends to dominate much of higher education in the United States, places more faith in the teacher to interpret the culture of the society and to create learning experiences to help students come to understand that culture (Westbury, 1995). Along the same lines, it is interesting to note the recent change in the national curriculum of England toward more flexibility in course selection after a period of greater specification. Jonathan Osborne (Chapter 2, this volume), describing the change, says:

> In short, the view that school science should address not only what we know but also how science works was an argument that had gained significant ground with those responsible for drafting the National Curriculum.
>
> As a consequence, the requirement to teach about "how science works" was increased and the detailed specification of the content was reduced. In addition, the new National Curriculum framework (Qualifications and Curriculum Authority, 2005) was designed to be much more skeletal so that many different curricular offerings could be offered rather than one standard course for all.... Having one standard curriculum for the complete diversity of students at age 14–16 was increasingly seen as no longer justifiable. In addition, the development of new courses such as *Twenty First Century Science* (Millar, 2006) was possible only if the national specification for the curriculum moved more toward defining a required set of outcomes and specifying in much less detail the route by which these were to be attained. Hence this policy change must be seen as a response to demands for a more flexible curriculum framework. (p. 31)

The point of this discussion about the specification of content is that the courses taught, the content of many of those courses, and the sequence in which they are taught in the United States is now determined by state-level policymakers. And, in the future, given current trends, it is possible that district or state-level polices will control not only the content of the courses but also the sequence of content *within* specific courses. As noted throughout this section, for the most part, these policies have come into being at the state level with input from the federal government, the work of national scientific societies such as AAAS and the NRC, and the research findings and theoretical writings of science education scholars.

ENFORCING THE STANDARDS: THE RISE OF PUBLIC ACCOUNTABILITY

It is one thing to describe what all students should know in science through various policy documents, but it is quite another to enforce those policies and ensure that all students are taught these ideas. One way that policymakers have attempted to enforce the teaching of the science specified in the standards documents is through public accountability. Although we think of this as a relatively new phenomenon because of the current federal education legislation requiring public reporting by schools of student performance in mathematics and reading, public accountability at the state-level has been in place in some locations for a long time.

In New York State, for example, the law (Chapter 655 of the Laws of 1987, which amended Section 215-1 of State Education Law) requires the New York State Board of Regents and the State Education Department to submit an annual report card to the governor and the state legislature on the performance of schools. The report card is to reflect "enrollment trends; indicators of student achievement in reading, writing, mathematics, science, and vocational courses; graduation, college attendance and employment rates; ... [and] information concerning teacher and administrator preparation, turnover, in-service education and performance" (New York State Department of Education, 2009). Although that heightened level of public accountability was new in New York in 1987, the Board of Regents of the State of New York had been using statewide standardized tests to assess student understanding in science since the nineteenth century. First available in 1865, by 1891 the state offered Regents exams in the science areas of astronomy, physics, advanced physics, chemistry, advanced chemistry, geology, physical geography, botany, zoology, and physiology and hygiene. The exams were used to encourage high performance on the part of schools and students and to determine the suitability of students for admission to colleges in the state (see Cheek & Quiriconi, Chapter 7, this volume). The main implication of state-level accountability through testing is that it provided a way for the state to maintain a high degree of control over the local schools. Schools were not free to decide themselves what was best for the students in that local community but had to adhere to a curriculum established and monitored by the state. In fact, Spring (2001) argues that the recent accountability movement is in part a reaction to the ideal of "democratic localism" (p. 423) and the community school movement that re-emerged during the 1960s. In other words, the accountability movement was an effort by professional educators to maintain control of the schools in the face of challenges by local lay leaders.

The public accountability movement received the support and encouragement of the federal government during the Reagan administration with the publication of *A Nation at Risk* (NCEE, 1983). That report called for higher standards in education, particularly in science and mathematics; public accountability through standardized assessment; and local responsibility for accomplishing the educational goals. The federal government continued to move toward a policy of standards and accountability when President George H. W. Bush met with state governors in September 1989 in Charlottesville, Virginia, to discuss a national agenda for education. At the summit, the president and the governors agreed to establish clear national performance goals and strategies to ensure U.S. international competitiveness. As part of their policy agenda, they also agreed that there should be annual reporting on progress toward meeting those goals. Then, on April 18, 1991, the president released *America 2000: An Education Strategy* (U.S. Department of Education, 1991), which described a plan for moving the nation toward national goals.

The policies the president wanted to pursue included the development of an accountability package that would encourage schools and communities to measure and compare results and insist on improvement when the results weren't acceptable. The policy package included national standards, national tests, reporting mechanisms, and various incentives. The tests would be national but voluntary and tied to those national standards. The President's policy proposals called for Congress to authorize the National Assessment of Educational Progress—which had been established by Congress in 1969 to provide national level data on educational outcomes—"regularly to collect state-level data in grades four, eight and twelve in all five core subjects, beginning in 1994. Congress will also be asked to permit the use of National Assessment tests at district and school levels by states that wish to do so" (U.S. Department of Education, 1991, p. 22). This move toward state-level, and sometimes district-level, reporting represented a significant increase in public accountability and a significant increase in the role of the federal government in determining education policy.

Many of the policy proposals in President Bush's *America 2000* report became law when President Clinton signed the *Goals 2000: Educate America Act* on March 31, 1994. The act focused on educating workers for productive employment, with special reference to competition in international trade. In addition to stating national goals, the *Goals 2000* legislation also created the National Education Standards Council, which had the authority to approve or reject the states' content standards. This body was subsequently dissolved following the 1994 midterm elections when the Republicans took control of Congress and voiced objections to the increasing intrusion of the federal government in education (National Conference of State Legislatures Report, 2004). Also in 1994, President Clinton signed the *Improving America's Schools Act* (IASA), which reauthorized the *Elementary and Secondary Education Act* of 1965 (ESEA). Under ESEA, states had to: (1) develop challenging content standards describing what students should know in mathematics and language arts; (2) develop performance standards representing three levels of proficiency for each of those content standards—partially proficient, proficient, and advanced; (3) develop and implement assessments aligned with the content and performance standards in at least mathematics and language arts at three grade spans: 3–5, 6–9, and 10–12; (4) use the same standards and assessment system to measure Title I students as the state uses to measure the performance of all other students; and (5) use performance standards to establish a benchmark for improvement referred to as "adequate yearly progress" (AYP). All schools were to show continuous progress or face possible consequences, such as having to offer supplemental services and school choice options to students or replacing the existing staff (National Conference of State Legislatures Report, 2004).

The modern movement toward holding schools accountable for their students' performance through standards setting and assessment, began in the early 1980s and was continued and strengthened with that 1994 legislation. The legislation also provided the basis for the No Child Left Behind Act of 2001 (NCLB) as it moved the focus away from national standards and voluntary national testing to a state-by-state system of standards setting and mandated accountability. In addition, as would be true under NCLB, the Improving America's Schools Act required states to test students in math and language arts but not in science. The pullback from national-level standards and toward state-level accountability was due to a continuing concern among many national policymakers about the nationalization of education, a concern that had been present from the earliest days of the country.

Today, using state content standards to describe what students should know, and measuring students with respect to those standards, has become official policy, the accepted law of the land. A report of the Committee on State Standards in Education by the National Research Council says: "[State education leaders] generally take standards-based reform and accountability for granted, viewing its approach as a 'central framework guiding state education policy and practice' " (NRC, 2008, p. 7). A consensus has developed over the years that public accountability is the best way to achieve an efficient public educational system and that the best way to monitor performance is through standardized testing. Federal policy (currently articulated in NCLB) now requires that all states measure students' performance in mathematics and reading each year from grades three to eight and that each school makes progress toward meeting the goal of universal proficiency in those subjects by 2014. State policy determines what those content standards will be, what the tests will look like, and what the proficiency levels will be. Currently, federal policy does not directly affect the teaching of science because it is not used as part of annual yearly progress (AYP) determination, although many believe that policy has negatively affected the teaching of elementary school science by placing so much attention on reading and math (see DeLucchi & Malone, Chapter 12, this volume).

Even without being part of federal AYP, the combination of specifying what all students should know and then holding schools accountable through various state-mandated policies has given states considerable control over what is taught in science and the mechanisms of enforcement. Because science is not included in AYP determination—even though since 2007–2008 science has to be tested once each in elementary, middle, and high school—a number of national organizations, foundations, legislators, and prominent individuals have argued that science should be treated the same way as math and reading. This is clearly an opportunity for individuals and groups of individuals who are interested in science education to make their arguments heard regarding the teaching and testing of science (see Peterson, Chapter 9, this volume, for advice on how to advocate for a particular point of view or to lobby for a particular bill). For the past several years we have been in the middle of a national policy debate on the issue of whether the same degree of public accountability in science as in mathematics and reading is a good thing or whether this kind of accountability would have negative effects on the teaching of science (see also Lynch, Chapter 11, this volume). That provides each of us an opportunity to take part in the debate and to attempt to influence the outcome of that debate.

TWO SPECIFIC ISSUES REGARDING SCIENCE EDUCATION POLICY AND WHAT IS TAUGHT IN SCHOOLS

Up to this point, we have seen that there has been progressively greater control over the content of the curriculum through policies that specify what should be taught and through the establishment of public accountability through mandated testing. There are many specific parts of the curriculum that standards-based accountability affects, and in the remainder of this chapter, I discuss two examples of how the science curriculum is shaped by standards-setting and enforcement policies.

The first example involves the teaching of evolution. Most state standards documents include some of the basic ideas of the evolution of living organisms by natural selection. But this has been a contentious issue in a number of states and local communities to the point that the judiciary has often had to intervene in determining how this topic is to be addressed in schools. The second example involves debates over the teaching of the nature of science along with the content of science. Today, all states include statements about the importance of teaching about how science is done, but there has been considerable discussion about what is meant by the nature of science and what students should learn about it.

Teaching Content That is Outside the Scientific Canon

One of the most contentious issues facing policymakers regarding the content of the curriculum in the United States has been the teaching of evolution. This issue has engaged a variety of policy actors including local school board members, school district administrators, state education department officials, state legislatures, and the judicial system. Although the judiciary does not make policy, it plays a critical role in determining the legality or constitutionality of policies.

The most recent legal challenge to the teaching of evolution came in a 2005 decision involving the Dover, Pennsylvania school district. In 2004, the school board in Dover created a policy on how they wanted teachers in the school district to approach the teaching of evolution. They voted to require the teachers to read the following statement to students in the ninth grade biology class.

> The Pennsylvania Academic Standards require students to learn about Darwin's Theory of Evolution and eventually to take a standardized test of which evolution is a part.
>
> Because Darwin's Theory is a theory, it continues to be tested as new evidence is discovered. The Theory is not a fact. Gaps in the Theory exist for which there is no evidence. A theory is defined as a well-tested explanation that unifies a broad range of observations.
>
> Intelligent Design is an explanation of the origin of life that differs from Darwin's view. The reference book, *Of Pandas and People*, is available for students who might be interested in gaining an understanding of what Intelligent Design actually involves.
>
> With respect to any theory, students are encouraged to keep an open mind.
>
> The school leaves the discussion of the Origins of Life to individual students and their families. As a Standards-driven district, class instruction focuses upon preparing students to achieve proficiency on Standards-based assessments. (*Kitzmiller v. Dover*, 2005)

The U.S. District Court for the Middle District of Pennsylvania ruled that the Dover school district's Intelligent Design (ID) policy violated the Establishment Clause of the First Amendment of the U.S. constitution, and the court ordered that the school board could not maintain its ID policy in any school within the Dover Area School District. The judge added that the order prohibited the district from "requiring teachers to denigrate or disparage the scientific theory of evolution and from requiring teachers to refer to a religious, alternative theory known as ID" (Kitzmiller v. Dover, 2005, p. 138).

Previously, state legislatures had adopted laws prohibiting teachers from teaching evolution, resulting eventually in the Scopes "monkey trial" of 1925 in Tennessee (*Scopes v. State of Tennessee,* 1927). The trial tested Tennessee's Butler Act (1925), which said that it was "unlawful for any teacher in the Universities, Normals, and all other public schools of the state which are supported in whole or in part by the public school funds of the State, to teach any theory that denies the story of the Divine Creation of man as taught in the Bible, and to teach instead that man has descended from a lower order of animals." The law was immediately challenged by the American Civil Liberties Union, and John Scopes, a high school coach who sometimes taught as a substitute teacher, agreed to teach evolution and be arrested. After his conviction, the decision was appealed to the Tennessee State Supreme Court, which upheld its constitutionality in 1927 (*Scopes v. State of Tennessee,* 1927). At that time, the state Supreme Court also reversed Scopes' original conviction on a technicality. But, the Butler Act remained on the books until 1967 when it was repealed by the Tennessee state legislature following a freedom of speech complaint by a dismissed teacher. Then, in 1968, the U.S. Supreme Court struck down the prohibition against teaching evolution in a case involving the state of Arkansas (*Epperson v. Arkansas,* 1968). Following these decisions, the opponents of teaching evolution began to argue for "balanced treatment" so that equal time would be devoted to teaching the biblical view of creation along with evolution. For example, Chapter 377 of the 1973 Public Acts of Tennessee said:

> Any biology textbook used for teaching in the public schools, which expresses an opinion of or relates a theory about origins or creation of man and his world shall be prohibited from being used as a textbook unless it specifically states that it is a theory as to the origin and creation of man and his world and is not represented to be scientific fact. Any textbooks so used...shall give in the same textbook and under the same subject commensurate attention to, and an equal amount of emphasis on, origins and creation of man and his world as the same is recorded in other theories, including, but not limited to, the Genesis account in the Bible. (Public Acts of Tennessee, 1973)

Plaintiffs in the subsequent case of *Daniel v. Waters* (1975) were Tennessee biology teachers and the National Association of Biology Teachers. The defendants were the members of the Tennessee state board responsible for selecting public school textbooks. The case was heard by the Sixth Circuit Court of Appeals, which held that the statute requiring equal time for Biblical creationism was unconstitutional under the Establishment Clause of the First Amendment. Then in 1987, in *Edwards v. Aguillard,* the U.S. Supreme Court ruled that requiring public schools to teach "creation science" along with evolution violated the Establishment Clause of the First Amendment.

Over the years a number of state legislatures have yielded to pressure from anti-evolution groups to establish anti-evolution policies in the form of state legislation, and school boards have established anti-evolution policies at the local level, but the courts have consistently ruled against these policies. Despite the rulings, teachers themselves have not always been consistent in their teaching of evolution. In a survey conducted by Berkman, Pacheco, and Plutzer (2008), the authors found that 25% of 939 teachers in a nationally representative sample spent at least one or two classroom hours devoted to creationism or intelligent design, and half of them said that they taught creationism as a "valid scientific alternative to Darwinian explanations for the origin of species" (p. 0922). Other teachers who taught creationism or intelligent design said they did so to point out the weaknesses of those ideas and to make it clear that these ideas were not accepted by most scientists. The researchers also found that 17% of teachers said they did not teach human evolution at all in their classrooms and that 60% of the teachers said they spent only one to five hours on it. (It should be noted that, from the Butler Act in 1925 through today, it is the teaching of *human* evolution that is at the center of the controversy.)

In addition, even though anti-evolution groups have consistently lost in the courts, they continue to press to limit the teaching of evolution and to introduce nonscientific content into the science curriculum. From January to March of 2009, for example, the Texas State Board of Education debated the suitability of wording in the state's high school guidelines that said that students should "analyze the strengths and weakness" of different scientific theories. This language was considered by many as an invitation to challenge the validity of the theory of evolution and to introduce creationism or intelligent design as alternatives. At one point during the debate, an amendment was proposed that would have opened the door for even greater opportunities to teach alternatives to the theory of evolution by requiring students to study the "sufficiency and insufficiency" of theories of common ancestry and natural selection in examining the fossil record and cell structure (Stutz, 2009). In the end, the State Board of Education voted 13–2 to add language into the state science standards that students are expected to examine "all sides of arguments" in the process of examining scientific theories and hypotheses (Gulick,

2009). Proposed language that students would "analyze and evaluate scientific explanations using empirical evidence, logical reasoning, and experimental and observational testing was rejected" in favor of that compromise language (Stutz, 2009). Some will see the vague language that is now part of the Texas standards as another opportunity for teachers who wish to do so to engender doubt in their students regarding the validity of the theory of evolution, not on the basis of "empirical evidence, logical reasoning, and experimental and observational testing," but by using nonscientific criteria instead.

So, in addition to continuing to advocate for the teaching of evolution as part of the science curriculum, constant vigilance is being practiced throughout the scientific community to ensure that school districts and individual teachers treat evolution as the important unifying theory in biology that it is. Because states are mandated to have content standards and to test their students with respect to those standards, the exact language in those standards and the questions students are asked on the assessments provide a public record of the attention that evolution is receiving at the state level and should be receiving in each classroom. This makes it somewhat easier for a state's scientific societies or national organizations such as the National Center for Science Education, the National Academy of Sciences, the National Science Teachers Association, or the American Association for the Advancement of Science to monitor what the states include in their state curriculum. As we saw in the Texas case, however, standards documents also provide an opportunity to introduce language that can be interpreted broadly.

Scientific societies have, in fact, been watchful in this area over the years and continue to pay attention to what is happening across the country. For example, the Board of Directors of the American Association for the Advancement of Science (AAAS), in response to reports that a number of states were considering removing evolution from the state standards, passed a resolution on February 16, 2006, in support of the teaching of evolution. In their resolution they said:

> Evolution is one of the most robust and widely accepted principles of modern science. It is the foundation for research in a wide array of scientific fields and, accordingly, a core element in science education. The AAAS Board of Directors is deeply concerned, therefore, about legislation and policies recently introduced in a number of states and localities that would undermine the teaching of evolution and deprive students of the education they need to be informed and productive citizens in an increasingly technological, global community. Although their language and strategy differ, all of these proposals, if passed, would weaken science education. (AAAS, 2006)

The evolution story illustrates that policy decisions often involve disputes over highly contested areas of the curriculum. It also illustrates that there can be many interested parties to the debates, and it shows where ultimate authority for policy formation and policy implementation resides. School boards and state legislatures can establish policies, but the courts determine their legality, and individual teachers carry them out. The evolution story shows that teachers are not free to teach non-scientific alternatives to the scientific canon and states are not free to tell them to do so. It also shows that policy is an ever-shifting terrain and that what is established today can very well be revisited tomorrow.

Teaching Both Science Content and the Processes of Science

Beginning in the nineteenth century and to the present, science teaching has included both the facts and principles of science as well as the practices and processes that scientists use in their work. For the most part, the science laboratory has been designated as the place where students learn how science is practiced. Virtually all states currently include the nature of scientific reasoning in their state standards, and state policy often requires students to take laboratory courses for graduation from high school (Education Commission of the States, 2008). Although the issue is relatively noncontentious compared to some other educational issues, there have been challenges regarding the efficacy of teaching students how science is done, especially when that includes having students actually doing science investigations.

Why is the laboratory used in science instruction? Is it an effective instructional approach? Can teachers choose whether they will use laboratory instruction or not? In the middle of the nineteenth century, scientists began to advocate for science in the school curriculum. Perhaps surprising to us today, they argued the merits of science largely on its value as a way of thinking and only secondarily on its value as a body of useful knowledge. The study of the classics and mathematics were widely believed to impart disciplined thinking through the precise deductive logic they required. Mathematical and linguistic studies were based on rules that, when followed, imparted a mental discipline to the student that could not be obtained any other way. Scientists argued that the study of science offered another kind of intellectual training through observation of the world and inductive logic. They claimed this was a more appropriate kind of mental discipline for those living in a democratic society because it left the individual free to investigate the world as an independent observer and to draw conclusions about the world based on those observations.

Observation, experimentation, and reasoning were the key elements of the nineteenth century scientists' idea of what a quality science education should entail. In addition to introducing the logic of science, the laboratory would also provide students with "a clear and definite conception" of natural phenomena and allow the student to develop "a definite image in his mind" of the names encountered in books (Huxley, 1899, p. 285).

In 1894, the Conference on Geography of the National Education Association's (NEA) Committee of Ten said that students should be led to "an understanding of the origin, the development, and the future history of geographic features" and that the "the evidence leading to the conclusions, and not just the conclusions, should receive careful consideration" (pp. 233–234). The Natural History Conference said that courses in botany and zoology should be laboratory based, with 3 days per week spent in direct observation of the objects of study and 2 days spent in discussions of those observations, lectures related to them, and quizzing (NEA, 1894). The Committee of Ten strongly supported the science conference reports on laboratory instruction and advocated that schools provide double laboratory periods, Saturday morning laboratory exercises, and one afternoon per week to be set aside for "out-of-door instruction in geography, botany, zoology, and geology" (NEA, 1894, p. 50).

At the end of the nineteenth century and well into the twentieth century, the NEA was one of the most prestigious bodies for making policy recommendations regarding education. But, the NEA did not have the authority to establish official education policy. Policy formation was left to individual school districts, colleges and universities, and to the states (see Cheek, Holland, & Quiriconi, Chapter 7, this volume, for a discussion of the consolidation of authority over education by states during the nineteenth and into the twentieth centuries). One official policy that significantly impacted the trend toward laboratory instruction was the decision by Harvard College in 1886 to allow a course of experiments in physics to be used for admission to the college. As Hall and Bergen (1892), professors at Harvard, said:

> The success of the [laboratory-based physics] course has been very gratifying. It is now followed by hundreds of pupils in the schools of New England, and is established in many other places throughout the country. At the examinations for admission to Harvard College, held in 1891, more than half of all the candidates in physics offered this course in place of the alternative text-book course. (p. vi)

By 1920, the focus in science education, both for classroom and laboratory instruction, had shifted. By then, practical studies, student interest, and a form of laboratory work in which students would solve problems that had personal and social relevance had taken hold. Again, the NEA used its considerable influence to bring together educational experts to build consensus around a particular approach to science education, this time with a focus on science as an applied and practical study. The NEA's Commission on the Reorganization of Science included a Committee on Science that was made up of four subcommittees. The physics subcommittee spoke about the use of "problem solving" as an instructional strategy in which the laboratory would be used as a central part of that approach.

> The unit of instruction, instead of consisting of certain sections or pages from the textbook, or of a formal laboratory exercise, should consist of a definite question, proposition, problem, or project, set up by the class or by the teacher. Such a problem demands for its solution recalling facts already known, acquiring new information, formulating and testing hypotheses, and reasoning, both inductive and deductive, in order to arrive at correct generalizations and conclusions. This method calls for an organization in which information, experimental work, and methods of attack, all are organized with reference to their bearings on the solution of the problem.
>
> With a project or a problem as the unit of instruction and its solution as the motive for work, the pupil should go to the laboratory to find out by experiment some facts that are essential to the solution of this problem, and that can not be obtained at first hand by other means. With such a motive he is more nearly in the situation of the real scientist who is working on a problem of original investigation. He is getting real practice in the use of the scientific method. (NEA, 1920, pp. 52, 53)

There were also challenges to the value of the laboratory as an instructional approach. This led to a number of experiments intended to answer the question of whether the added cost of the laboratory was justified. One line of research had to do with the comparative effectiveness of teacher demonstrations and individual laboratory work. In reviewing the various research studies and arguments on the issue, the Thirty-First Yearbook Committee of the National Society for the Study of Education (NSSE) said "in the interests of economy both of time and of money, it seems desirable to perform more laboratory exercises by the demonstration than by the individual method" (NSSE, 1932, p. 106). The Committee also recognized that research had not demonstrated conclusively the relative value of the two approaches:

> It is very probable that experimenters have not yet been able to measure the more valuable outcomes of laboratory instruction, as a number of critics have pointed out. If there are valuable outcomes, the added expense needed to secure them may be justified, but just how valuable the laboratory experience as a whole may be, we do not yet know. (p. 270)

The ambiguity of these results and a lack of consensus among practitioners on the relative value of demonstrations versus individual laboratory experiences provided district-level policymakers a way to justify their decisions to reduce the number of laboratory experiences students received. School administrators around the country, especially in urban schools, quickly replaced the individual laboratory approach with the demonstration approach. Speaking of this policy shift, the Yearbook Committee said that the policy was carried out "with enthusiasm and in some cases with such complete thoroughness that certain large city high schools were constructed with no provisions whatever for individual experimentation by pupils" (NSSE, 1932, p. 98). Other cost-cutting measures included switching from double-period to single-period labs (Krenerick, 1935).

In 1940, Ford said of laboratory work:

> Laboratories have been a constant source of irritation to some administrators because they are expensive and because the double laboratory periods upset an otherwise smooth running curriculum. Further, the charge is made that the laboratory serves to prepare for further college work in this field and few students go on to college. The techniques learned in laboratory manipulation serve no useful purpose. (p. 556)

Also in 1940, Brown noted that many of the experiments used in physics and chemistry were very much the same as those Harvard had listed in 1887 as appropriate for college entrance and should be replaced by problems of genuine interest. But the College Entrance Examination Board and other standardizing bodies continued to use such lists to make judgments about the suitability of students for college admission and the suitability of programs to prepare them for college, so they continued to be used by schools (Brown, 1940).

Teaching the methods of science by means of laboratory instruction again became the focus of attention during the curriculum reform movement of the 1950s and 60s because one of the goals of reform was to provide a more accurate view of the nature of science and scientific activity. The Biological Sciences Curriculum Study (BSCS) group produced a series of "Invitations to Enquiry" that provided teachers with activities to help develop students' abilities to carry out scientific investigations. The BSCS also created a Laboratory Block Program, which devoted 6 weeks of a course to in-depth student investigations of a specific biological topic. The laboratory blocks were intended to have the students make discoveries for themselves (Lee, 1961). The Chemical Education Material Study (CHEM Study) group created a course that integrated laboratory and classroom work in which all laboratory activities would fit into a 45–50 minute classroom period. The course was intended to give students a better idea of the nature of scientific investigation and the way that scientific knowledge was generated. George Pimentel, editor-in-chief of the project, said:

> The laboratory was designed (1) to help students gain a better idea of the nature of scientific investigation by emphasizing the "discovery approach," and (2) to give students an opportunity to observe chemical systems and to gather data useful for the development of principles subsequently discussed in the text and classwork. (as cited in Merrill & Ridgway, 1969, pp. 33–34)

It should be noted that the use of the "discovery approach" in laboratory instruction had been controversial since the nineteenth century, and numerous debates laid out its merits and limitations, although it was generally discredited as ineffective in its most extreme form. Typically, the arguments led to recommendations for some form of modified discovery approach as the ideal. For example, Edwin Hall said that "learning by experience is a plodding method, and the student who aspires to any great height or breadth of intellectual reach must not confine himself to it" (Smith & Hall, 1902, p. 305). Alexander Smith said that "heuristic" work (in which students are placed in the role of discoverers) took too much time and did not furnish the knowledge of chemistry needed at the secondary level. He recommended a method that carefully guided the student through the discovery process for most of the laboratory activities and only occasionally left the students free to devise their own approach to attach the problems (Smith & Hall, 1902). Later, Jerome Bruner provided tentative support for discovery and inductive approaches to learning but said that more research was needed to know if it was an effective way to teach the fundamental ideas of a field and what the proper balance was between having students discover principles for themselves and having them learn from direct statements from the teacher (Bruner, 1960).

Because of the decision by the National Science Foundation (NSF) to fund the development of curriculum projects during the 1950s and 60s and to use scientists in their development, the new curriculum materials—which incorporated student investigations and an inductive approach to laboratory instruction—received considerable support from the scientific community. And, throughout the educational community, the impact of the new materials, particularly on the number of teachers who began to teach the processes of science along with the content of science, was impressive. A study funded by NSF in 1977 showed that nearly 50% of school districts were using at least one of the BSCS materials, 20% were using one of the new chemistry materials, and 23% were using one of the new physics materials. New elementary school science materials were being used in 32% of the surveyed districts, and about 50% of districts were using the new junior high school materials. Often, district-level decisions were made for all schools in the district to use the NSF-funded materials and for teachers to teach using the inductive approaches that the materials encouraged. But, the *Modern Biology*, *Modern Chemistry*, and *Modern Physics* textbooks published by Holt, which focused largely on science content, were still used by nearly half of the school districts throughout that period (Helgeson, Blosser, & Howe, 1978; Weiss, 1978).

During the era of curriculum reform of the 1950s and 60s, "inquiry teaching" came to be widely considered to be an ideal worth striving for. Rutherford (1964), for example, said:

> When it comes to the teaching of science it is perfectly clear where we, as science teachers, science educators, or scientists stand: we are unalterably opposed to the rote memorization of the mere facts and minutiae of science. By contrast, we stand foursquare for the teaching of the scientific method, critical thinking, the scientific attitude, the problem-solving approach, the discovery method, and, of special interest here, the inquiry method. (p. 80)

In 1981, Welch, Klopfer, Aikenhead, and Robinson summarized several surveys sponsored by NSF during the 1970s, and concluded that there was considerable discrepancy between belief about the value of an inquiry approach among teachers and actual practice. The authors said that teachers were often reluctant to use the approach because of the difficulty of implementation, which included obtaining equipment and supplies for the laboratory and the confusion that was often created among all but the most capable students. In summary, they said:

> The greatest set of barriers to the teacher support of inquiry seems to be its perceived difficulty. There is legitimate confusion over the meaning of inquiry in the classroom. There is concern over discipline. There is a worry about adequately preparing children for the next level of education. There are problems associated with a teacher's allegiance to teaching facts and to following the role models of the college professors. (p. 40)

Research on the effectiveness of the laboratory, whether it is used to teach students the process of scientific inquiry, to have them make discoveries themselves, or simply to demonstrate scientific principles and give students first-hand experiences with the natural world, has tended to be inconclusive, in part because of the difficulty of defining exactly what those approaches entail. The NRC's 2006 report on laboratory instruction focused its examination on instruction that provided: "opportunities for students to interact directly with the material world (or with data drawn from the material world) using the tools, data collection techniques, models, and theories of science" (NRC, 2006, p. 3). They concluded their report by saying:

> The committee does not recommend any specific policies or programs to enhance the effectiveness of laboratory experiences, because we do not consider the research evidence sufficient to support detailed policy prescriptions. A serious research agenda is required to build knowledge of how various types of laboratory experiences (within the context of science education) may contribute to specific science learning outcomes. (pp. 9–10)

Along these lines, a recent synthesis study by Minner, Levy, and Century (2009) on the effect of various aspects of inquiry teaching on science learning revealed inconclusive but generally positive effects. The researchers identified a number of components of inquiry teaching and then examined their individual and combined effects on student learning. The researchers concluded:

> The evidence of effects of inquiry-based instruction from this synthesis is not overwhelmingly positive, but there is a clear and consistent trend indicating that instruction within the investigation cycle (i.e., generating questions, designing experiments, collecting data, drawing conclusions, and communicating findings), which has some emphasis on student active thinking or responsibility for learning, has been associated with improved student content learning, especially learning scientific concepts.
>
> We did not find, however, that overall high levels of inquiry saturation in instruction were associated with more positive learning outcomes for students. The only learning associations we found with the amount of inquiry saturation were modest. However, future research may be able to further explore these associations both in terms of conceptual learning as well as other kinds of student outcomes that were not addressed in this synthesis. (Minner, Levy, & Century, 2009)

As Minner, Levy, and Century (2009) realized, the problem with research in this area has been that different researchers define the approaches differently and compare different outcome measures. In addition, when curriculum materials that take one of these approaches is compared to more traditional materials, the research is complicated by the question of whether or not the materials were, in fact, implemented as intended. Anderson (2002) pointed out that some of the more productive research being conducted in this area has to do with analyzing the barriers and dilemmas teachers face when attempting to implement such approaches. Barriers include a lack of resources and support, and dilemmas include conflicting attitudes about these approaches to teaching when such a high value is now being placed on the results of testing and preparation for the next level of schooling. Because of this pressure to produce results in the form of student outcomes, teachers are tending to focus more on what works in the classroom rather than on more theoretical and propositional knowledge about teaching (Blumenfeld, Krajcik, Marx, & Soloway, 1994; Duschl & Gitomer, 1997). This practical bent of teachers is becoming more and more apparent as schools are now faced with having to meet the requirements of a federal education law that says all states must test their students with respect to a set of content standards that defines the knowledge and skills that students must have.

As noted throughout this volume (see, e.g., Osborne, Chapter 2, this volume), values are a major determinant of policy development and implementation. Empirical evidence for the success of a particular approach plays a role in policy making, but the personal values of the policymakers and the policy implementers are often more important. Whether supported by research, educational theory, or personal values, throughout the history of American education, science educators have advocated particular ways of teaching about the nature of science, and often those ways include having students engage in the "doing of science." These recommendations have found their way into virtually all state content standards, and many states require students to take laboratory courses to graduate from high school. For the most part, exactly how that is done is still left up to individual teachers, given that state and local policy for science teaching tends to leave decisions about pedagogy to teachers. It is in the area of pedagogy that most teachers still retain control over what they do in the classroom. This often means there is no requirement that teachers use inductive or discovery approaches in their teaching, only that they provide students with opportunities to develop knowledge and skill regarding both the content and practices of science.

Despite differences of opinion regarding the way the nature of science should be taught and the historical controversies over the efficacy of having students engage in their own scientific investigations, there is no question that a broad consensus has been building in the United States over the years that students should know something about the nature of science and should engage with at least some of the practices of science themselves. The nature of science was prominently featured in *Science for All Americans*, *Benchmarks for Science Literacy*, and the *National Science Education Standards* (NSES). NSES begins each grade band with a section on Science as Inquiry in which both abilities necessary to do scientific inquiry and understandings about scientific inquiry are listed. A recent publication by the National Research Council dealing with the teaching of elementary and middle school science, *Taking Science to School* (NRC, 2007), reaffirms the commitment to teach both the content of science and the practices of science.

In each case, the argument is that knowledge of science and the way science is done are important if citizens are to function effectively in a world in which science plays such a large role. This is a consensus that has resulted from the arguments of scientists, professional educators, educational researchers, and policy-influencing documents such as *Science for All Americans* (AAAS, 1990), *Benchmarks for Science Literacy* (AAAS, 1993), and the *National Science Education Standards* (NRC, 1996). The effect of this consensus, achieved over a century of debate, is that students in the U.S.A. are expected to be introduced to both the facts of science and the way science is done, and teachers are expected to teach each of these in accordance with their state standards and using pedagogical strategies that they believe to be most effective. In principle, they are not free to do otherwise.

SUMMARY

In this chapter, we have seen that over the past century there has been a progressive specification of what should be taught in science classes in the United States. This means that individual teachers have had progressively less freedom to select which content to teach. This specification has been accomplished through federal legislation that mandates states to have clear statements of what students should know and to assess them on these those ideas. The public reporting of assessment results produces a form of accountability that makes it difficult for schools to wander too far from what is specified in the state standards. This means that some ideas get taught and others do not, and that processes and practices of science get taught as well as science concepts. In addition, it has been clearly determined that it is not acceptable to teach outside the established scientific canon, at least not when matters of religious faith are used to make judgments about the world instead of empirical observations. All of this has been established through policy debates that have evolved over many decades. Ultimately, policy decisions fall within the purview of state and federal legislators, but those policymakers are influenced by the opinions of individual scientists and educational researchers, and by the positions taken by professional scientific societies as well as school professionals and the general public. And, as we have seen in this historical overview of key issues regarding the content of the science curriculum, those opinions evolve over long periods of time until a consensus of opinion forms and policies can be effectively implemented.

It is also important to note that it is possible for two major policy directions to be fundamentally at odds with each other for a long period of time. As Minner, Levy, and Century (2009) concluded in their study:

> The implications of this study are somewhat at odds with current educational policy, which encourages coverage of a large number of scientific concepts to be tested at various stages in a child's educational experience. Since the assessments used by states largely test knowledge or recall of discrete science facts, concepts, and theories, teachers are constrained by the need for wide topical coverage within an already crowded daily schedule.

This brings us to a final point. As we have seen throughout this volume, policy statements do not always guarantee policy implementation. The teachers responsible for implementing policy must believe in the value of a policy in order for its potential to be realized. If two broad policy directions are at odds with each other, the teachers will make decisions about the relative emphasis to place on each one of them. And in that way teachers continue to influence what is taught.

REFERENCES

American Association for the Advancement of Science. (1990). *Science for all Americans*. New York, NY: Oxford University Press.

American Association for the Advancement of Science. (1993). *Benchmarks for science literacy.* New York, NY: Oxford University Press.

American Association for the Advancement of Science. (2006, February). *Statement on the teaching of evolution*. Retrieved on September 18, 2009, from http://www.aaas.org/news/releases/2006/pdf/0219boardstatement.pdf

American Association for the Advancement of Science. (2001). *Atlas of science literacy* (Vol. 1). Washington, DC: Author.

American Association for the Advancement of Science. (2007). *Atlas of science literacy* (Vol. 2). Washington, DC: Author.

Anderson, R. D. (2002). Reforming science teaching: What research says about inquiry. *Journal of Science Teacher Education*, *13*(1), 1–12.

Berkman, M. B., Pacheco, J. S., & Plutzer, E (2008). Evolution and creationism in America's classrooms: A national portrait. *PLoS Biology*, *6*(5), e124. doi:10.1371/journal.pbio.0060124. Retrieved on August 21, 2009, from http://www.plosbiology.org/article/info:doi/10.1371/journal.pbio.0060124

Blumenfeld, P. C., Krajcik, J. S., Marx, R. W., & Soloway, E. (1994). Lessons learned: How collaboration helped middle grade science teachers learn project-based instruction. *The Elementary School Journal*, *94*(5), 539–551.

Brown, H.E. (1940). The plight of high school physics, IV. The languishing laboratory. *School Science and Mathematics*, *40*, 457–462.

Bruner, J. S. (1960). *The process of education*. Cambridge, MA: Harvard University Press.

Butler Act, 17 Tenn. Code Ann. (1925).

Catley, K., Lehrer, R., & Reiser, B. (2005). *Tracing a prospective learning progression for developing understanding of evolution.* Paper Commissioned by the National Academies Committee on Test Design for K–12 Science Achievement.

Retrieved on August 31, 2009 from http://www7.nationalacademies.org/bota/Evolution.pdf

Daniel v. Waters, 515 F.2d 485 (6th Cir. 1975).

DeBoer, G. (1991). *A history of ideas in science education: Implications for practice*. New York, NY: Columbia University Teachers College Press.

Duschl, R. A., & Gitomer, D. H. (1997). Strategies and challenges to changing the focus of assessment and instruction in science classrooms. *Educational Assessment, 4*(1), 37–73.

Education Commission of the States. (2007). *Standard high school graduation requirements* (50-state). Retrieved on August 21, 2009, from http://mb2.ecs.org/reports/Report.aspx?id=735

Edwards v. Aguillard, 482 U.S. 578 (1987).

Epperson v. Arkansas, 393 U.S. 97 (1968).

Ford, L. A. (1940). Laboratory science. *School Science and Mathematics, 40*, 556–557.

Gagne, R. (1977). *The conditions of learning* (3rd ed.). New York, NY: Holt, Rinehart & Winston.

Goals 2000: Educate America Act of 1994, 20 U. S. C. § 5801

Gulick, J. (2009, April 11). Compromise changes Texas teaching of evolution [Electronic version]. *Lubbock Avalanche-Journal*. Retrieved from http://lubbockonline.com/stories/041109/loc_428039905.shtml

Hall, E. H., & Bergen, J. Y. (1892). *A text-book of physics, largely experimental on the basis of the Harvard Descriptive List of Elementary Physical Experiments*. New York, NY: Henry Holt and Company.

Helgeson, S., Blosser, P., & Howe, R. (1978). *The status of pre-college science, mathematics, and social science education: 1955–1975: Vol. I. Science Education*. Washington, DC: U.S. Government Printing Office.

Huxley, T. (1899). *Science and education*. New York, NY: Appleton.

International Technology Education Association. (2000). *Standards for technological literacy: Content for the study of technology*. Reston, VA: Author.

Kali, Y., Linn, M., & Roseman, J.E. (Eds.). (2008). *Designing coherent science education: Implications for curriculum, instruction, and policy*. New York, NY: Columbia University Teachers College Press.

Kesidou, S., & Roseman, J. E. (2002). How well do middle school science programs measure up? Findings from Project 2061's curriculum review. *Journal of Research in Science Teaching, 39*, 522–549.

Kitzmiller et al. v Dover Area School District, et al., 400F Suppl 2d 707 (M. D. Pa. 2005).

Krenerick, H. C. (1935). A single laboratory period, a demonstrated success. *School Science and Mathematics, 35*, 468–476.

Lee, A. (1961). An introduction to the BSCS laboratory block program. *The American Biology Teacher, 23*(7), 409–411.

Livermore, A. H. (1966). AAAS Commission on Science Education: Elementary science program. *Journal of Chemical Education, 43*(5), 270.

Merrill, R., & Ridgeway, D. (1969). *The CHEM Study story*. San Francisco, CA: Freeman.

Millar, R. (2006). Twenty First Century Science: Insights from the Design and Implementation of a Scientific Literacy Approach in School Science. *International Journal of Science Education, 28*(13), 1499–1521.

Minner, D., Levy, A., & Century, J. (2009) Inquiry-based science instruction - what is it and does it matter? Results from a research synthesis years 1984 to 2002. *Journal of Research in Science Teaching*. Retrieved December 4, 2009, from www3.interscience.wiley.com/journal/123205106/abstract

National Science Board Commission on Precollege Education in Mathematics, Science and Technology. (1983). *Educating Americans for the 21st century: a report to the American people and the national science board.* Washington, DC: National Science Foundation.

National Commission on Excellence in Education. (1983). *A nation at risk: The imperative for educational reform*. Washington, DC: U.S. Department of Education.

National Conference of State Legislatures. (2004). *No Child Left Behind: History.* Retrieved on March 29, 2004, from http://www.ncsl.org/programs/educ/NCLBHistory.htm

National Education Association. (1894). *Report of the committee on secondary school studies*. Washington, DC: U.S. Government Printing Office.

National Education Association. (1920). *Reorganization of science in secondary schools: A report of the commission on the reorganization of secondary education* (U.S. Bureau of Education, Bulletin No. 26). Washington, DC: U.S. Government Printing Office.

National Research Council. (1996). *National science education standards*. Washington, DC: National Academy Press.

National Research Council. (2006). *America's lab report: Investigations in high school science*. Washington, DC: National Academy Press.

National Research Council. (2007). Taking *science to school.* Washington, DC: National Academies Press.

National Research Council. (2008). Common *standards for K-12 education?* Washington, DC: National Academies Press.

National Science Teachers Association (1971). NSTA position statement on school science for the 70s. *The Science Teacher, 38*, 46-51.

National Science Teachers Association (1983). *Science-technology-society: Science education for the 1980s.* Washington, DC: Author.

National Science Foundation (2005). *Solicitation for Instructional Materials Development* (IMD). Retrieved on August 31, 2009 from http://www.nsf.gov/pubs/2005/nsf05612/nsf05612.txt

National Society for the Study of Education. (1932). *A program for teaching science: Thirty-First Yearbook of the NSSE*. Chicago: University of Chicago Press.

New York State Education of Department (2009). *The Chapter 655 Report.* Retrieved on June 28, 2009, from http://www.emsc.nysed.gov/irts/chapter655/

No Child Left Behind Act of 2001, 20 U.S.C. § 6301 *et seq.* (2002).

Public Acts of Tennessee, 377 (1973).

Qualifications and Curriculum Authority. (2005). *Programme of Study for KS4 from 2006*. London: Author.

Roseman, J. E., Stern, L., & Koppal, M. (2010). A method for analyzing the coherence of high school biology textbooks. *Journal of Research in Science Teaching, 47*(1), 47–70.

Roseman, J. E., Linn, M. C., & Koppal, M. (2008). Characterizing curriculum coherence. In Y. Kali, M. C. Linn, & J. E Roseman, (Eds.), *Designing coherent science education*. New York: Teachers College Press.

Rutherford, F. J. (1964). The role of inquiry in science teaching. *Journal of Research in Science Teaching, 2*, 80–84.

Scopes v. State of Tennessee, 154 Tenn. 105, 289 S. W. 363, 367 (1927).

Smith, A. & Hall, E. (1902). *The teaching of chemistry and physics in the secondary school*. New York, NY: Longmans, Green.

Spring, J. (2001). *The American School* (5th ed.). New York, NY: McGraw-Hill.

SRI International. (1996). *Evaluation of the American Association for the Advancement of Science's Project 2061, Volume I: Technical report*. Menlo Park, CA: Author.

Stern, L., & Roseman, J. E. (2004). Can middle-school science textbooks help students learn important ideas? Findings from Project 2061's curriculum evaluation study: Life science. *Journal of Research in Science Teaching, 41*(6), 538-568.

Stutz, T. (2009, March 27). Texas education board cuts provisions question evolution from science curriculum [Electronic version]. *Dallas Morning News*. Retrieved from http://www.dallasnews.com/sharedcontent/dws/dn/education/stories/032809dntexevolution.78a4720b.html

U.S. Department of Education. (1991). *America 2000: An education strategy sourcebook*. Washington, DC: Author.

Weiss, I. (1978). 1977 national survey of science, mathematics and social studies education highlights report. In *The status of pre-college science, mathematics, and social studies educational practices in U.S. schools: An overview and summaries of three studies* (pp. 1–25). Washington, DC: U.S. Government Printing Office.

Welch, W., Klopfer, L., Aikenhead, G., & Robinson, J. (1981). The role of inquiry in science education: Analysis and recommendations. *Science Education, 65*, 33–50.

Westbury, I. (1995). Didaktik and curriculum theory: Are they the two sides of the same coin? In S. Hopmann & K. Riquarts (Eds.), *Didaktik and/or curriculum*. Kiel, Germany: Institute for Science Education.

CHAPTER 11

EQUITY AND U.S. SCIENCE EDUCATION POLICY FROM THE GI BILL TO NCLB

From Opportunity Denied to Mandated Outcomes

Sharon J. Lynch

INTRODUCTION

The purpose of this chapter is to examine the intersection of education policies and equity issues in science education in U.S. Kindergarten through Grade 12 (K–12) schools. Some policies are broad-based and affect many aspects of life in our society (e.g., The Civil Rights Act of 1964); others are directed specifically toward science teaching (e.g., policies regarding the qualifications of science teachers); and still others are policies in science education that have an indirect effect on the equitable treatment of students in science classrooms (e.g., the curriculum reforms of the 1960s). Policies are defined in this chapter as the collection of

The Role of Public Policy in K–12 Science Education, pp. 305–354

established laws and rules that govern the operation of social systems, but may also include recommendations of influential individuals or organizations that affect practice, although they may not have a governmental mandate. This chapter concentrates on the laws and rules that have affected science education, as well as documents and events that have influenced its direction, especially as they relate to equity issues. The chapter takes an historical approach, focusing on policies and events taking place in American society from post-World War II to present. Key policy landmarks are listed in the Appendix at the end of this chapter.

When viewing policy initiatives historically, it is important to examine them in the context of time, with awareness of the values of those who held power. Public policy often is seen as created in the best interests of the society, but looking back from the current vantage point, it is possible to see that many policies have had negative consequences. For example, *Plessy v. Ferguson* (1896), allowed "separate but equal" schools, when the Supreme Court, by a vote of 7–1 (with one abstention), ruled that state laws requiring separation of the races were within the bounds of the Constitution as long as equal accommodations were made for African Americans. This decision, which justified segregationist policies, was later overturned in the 1954 Supreme Court *Brown v. Board of Education* decision. A second example is from the late 1800s to the mid-1900s, when some policy makers believed it was in the best interest of the country for Native American children to be removed from their families and forced to attend boarding schools so they could be assimilated into white society (Expanding the Circle Resources, 2010). Both of these policies reflected the mores of dominant social groups at the time, but seem unconscionable today. They serve as cautionary examples as we consider changes to science education policies. The chapter consists of four major sections. It begins with an historical overview of science education and equity policy, interweaving a discussion of major civil rights landmarks with changes in U.S. education policy and practice over time. The next three sections are based on three different views of equity: equality of inputs, equality of outcomes, and equity as mandated in the accountability movement. Each of these sections provides examples of current education practices and policies designed to remedy problems. But, as will be seen, well-intentioned education solutions often are met with varied success and may give rise to new concerns and problems. Finally, the chapter concludes with a discussion of the implications for the equitable treatment of students in science classes given the new vectors of change currently piercing the landscape of K–12 public education. But, first, we discuss a variety of ways in which the concept of equity is used.

Definitions of Equity

The term, equity, has many meanings. Apple (1995) has described it as a sliding signifier, and Secada (1994) as a moving target. In this chapter, we use the term to mean justice, fairness, and impartiality, both in a legal sense and in the broader sense of fairness that exists in the unwritten rules of social arrangements. Rawls (1971) pointed out that justice is the first virtue of social institutions, as truth is to systems of thought. To Rawls, a just institution is one that distributes social goods—rights, liberties, access to power and opportunity—equally among its participants. In education policy, equity can take a legal turn or the more subtle, sometimes elusive, notion of "fairness." Because notions of fairness can differ, it is useful to return to Rawls' powerful idea of fairness that asks one to imagine what would be fair if one were unsure of one's position in society, with rights and opportunities distributed completely by chance rather than by social or economic station. Under such circumstances, inequalities in education might not be tolerated. How would one countenance the decrepit science facilities often found in urban schools, the sparse resources available to some rural students, or the substandard instruction sometimes provided to students in low track science classrooms? (See Lynch, 2000 for an extended scenario derived from Rawls' construction of equity.)

Even when justice and fairness are seen to be at the heart of equity, there is still considerable room to interpret what equity actually means in the context of educational opportunity and outcomes. Below are a number of ways in which equity has been interpreted. They have been drawn from previous work by Secada (1994) and Kahle (1996), and rearranged by Lynch (2000):

1. Equity involves maximum return on the minimum investment: resources go to those most likely to succeed (post-Sputnik, c. 1957).
2. Equity is the same treatment for everyone, equal access or equality of inputs (civil rights era, c. 1960s).
3. All students have an equal opportunity to meet and master standards, resulting in an equality of outcomes (or similar distributions of outcomes no matter the groupings of learners based on social class, ethnicity, etc.), rather than large achievement gaps (civil rights era, c. 1960s).
4. Equity is concern for the whole child as an individual with unique educational, socioemotional and physical needs (women's movement, c. 1970s).

5. Equity is seen as triage, or investing in students whose success or failure depends on their school experience (women's movement, c. 1970s).
6. Equity compensates for social injustice for specific groups who have not received fair treatment (affirmative action, c. 1980s).
7. Equity is a safety net for individual differences, including alternative programs or resources; if one program is ineffective for an individual student, other options are available (persons with disabilities movement, c. 1980s).
8. Equity is accountability; education standards are used to equalize opportunities for all students (further identified by demographic subgroup in the No Child Left Behind Act of 2001 (2002) abbreviated in print as NCLB), regardless of where they live or their socioeconomic status (SES), by spelling out and making public what all students should learn and providing additional funding to schools that need more resources (The Education Trust, 2009) (accountability movement c. 1990s to present).

Especially significant is the change in the perception of equity from the perspective of an identifiable *group's access* to opportunity, to an emphasis on an *individual's right* to equal opportunity. Moreover, in these eight equity perspectives, equity is not always synonymous with "equality," equal access, or equal outcomes.

AN HISTORICAL VIEW OF EQUITY POLICY LANDMARKS

Equity issues in education are embedded in a more encompassing, deep, and hard-fought struggle for civil rights in this country. It is difficult to understand equity policy development in education without also paying attention to the struggle for civil rights that has taken place in the larger society. Tracing equity issues in U.S. science education policy is often inseparable from other aspects of schools and schooling that affect students' opportunities to learn. As Darling-Hammond (1998) points out:

> the U.S. educational system is one of the most unequal in the industrialized world, and students routinely receive dramatically different learning opportunities based on their social status. In contrast to European and Asian nations that fund schools centrally and equally, the wealthiest 10 percent of U.S. school districts spend nearly 10 times more than the poorest 10 percent, and spending ratios of 3 to 1 are common within states. Despite stark differences in funding, teacher quality, curriculum, and class sizes, the

prevailing view is that if students do not achieve, it is their own fault. If we are ever to get beyond the problem of the color line, we must confront and address these inequalities. (p. 1)

World War II: Expanded Opportunities for Women and Minorities

The chapter begins with an historical overview, starting in the World War II (WWI) era, a watershed moment for major changes in U.S. social policy. World War II ushered in new opportunities for women and minorities. With so many men serving in the military abroad and with such a need for skilled workers at home, women gained new jobs and skills as they took positions of responsibility that had been denied them in the past. And, even though the military was segregated, African Americans, Hispanics, Native Americans, and other minorities found new opportunities for advancement because of their military service. At the end of WWII, the GI Bill financed college education for returning veterans, individuals who otherwise would not have had opportunities for higher education or access to the careers that opened up to them as a result of their college degrees. The immediate impact of the GI Bill was different for white veterans than for minorities who experienced higher levels of discrimination and poverty when back on U.S. soil, making access to higher education harder to achieve. Nonetheless, as Hilary Herbold (1994–95) wrote,

> Clearly, the G.I. Bill was a crack in the wall of racism that had surrounded the American university system. It forced predominantly white colleges to allow a larger number of blacks to enroll, contributed to a more diverse curriculum at many HBCUs, and helped provide a foundation for the gradual growth of the black middle class. (p. 108)

Altogether, the GI Bill produced a new group of educated professionals and disrupted the old pre-WWII social orders. The educational structure in the U.S. was changed forever.

What Kind of Science Gets Taught and to Whom? 1940s to 1960s

From the 1940s to the 1960s, a major issue for many African American and other minority groups was access to schools that provided an equal opportunity to learn. Because of the Supreme Court's decision in *Plessy v. Ferguson* in 1896, U.S. schools in the post-WWII years remained separate, but they were not equal. Even when the "separate but equal" doctrine was declared unconstitutional in *Brown v. Board of Education* in 1954, many states and jurisdictions continued *de facto* segregation practices in schooling for many years. This usually took the form segregation in neighborhood housing, which then affected the ethnic/racial composition of schools. With the Civil Rights Act of 1964, *de jure* (by law) segregation was outlawed in all areas of American society (U.S. Government Guide, 2002), but *de facto* segregation remained.

Having schools that provided equal educational resources was particularly important in science. In order to have any opportunity to enter scientific and technical careers, students had to have access to high quality science and mathematics courses in middle and high school that would enable them to succeed in rigorous science courses in college. Without such educational experiences, students are unlikely to be successful in college science, technology, engineering, and mathematics (STEM) courses and careers. In addition, when STEM courses are taught at low levels, students leave high school not only unprepared to pursue science careers, but also without adequate knowledge, skills, and habits of mind needed as citizens and for many entry-level jobs.

Throughout this period and before, there was a long-standing debate about the type of science that ought to be provided to U.S. K–12 students (see DeBoer, 1991). The progressive camp saw school science as the socially-relevant, practical knowledge necessary to operate in the world as fully functioning adults. Traditionalists saw the sciences as disciplinary bodies of knowledge, each with its own unique history and aesthetic. They were more interested in rigor, discipline, and the logical organization of the disciplines than in social or personal relevance. The traditional view also often views school science as the conduit for preparing students to be scientists, mathematicians, and engineers. This dual role that science education has been asked to fulfill in U.S. education—to provide all students with an understanding of the world around them while at the same time opening up access and opportunities to students who are preparing for STEM careers—often has led to debates about equity versus excellence and differentiated programs, especially, how to educate academically talented students while at the same time providing a high quality science education for all. There has been a struggle to balance these common goods by enabling a deeper and richer understanding of science for science literacy while opening up access and opportunities to students who will someday need substantial science knowledge, skills, and habits of mind for STEM careers. The tension between the two approaches and the drive for equity pressures schools to provide *all* students with the quality experiences needed to achieve college and work readiness and the preparation necessary to pursue careers in advanced technical fields.

As an indicator of how well schools were doing during this time period in providing the access and encouragement needed to pursue scientific careers, in 1952, less than 1% of the students who received National Science Foundation (NSF) fellowships were female. From an equity perspective, there is no reasonable explanation for such a disparity without looking to issues of fairness in the treatment of women and how women were influenced as students when making educational and career decisions. (The National Science Foundation tracks such trends and up-to-date data, c. 2010, can be found at http://www.nsf.gov/statistics/women/)

Cold War Worries and Civil Rights: 1960s to 1990s

New Curriculum Materials and Tracking

The focus on the science pipeline and how to prepare students for careers in science became stronger in the Cold War years. This also was the time of intense civil rights activity. As the country moved through the civil rights era (1950s through 1970s), the science community was trying to produce more scientists to respond to a perceived threat from the Soviet Union. In the post-WWII years, the U.S. could no longer rely upon the supply of European science talent. This tension was voiced clearly in the policy document, *A Nation at Risk* (National Commission on Excellence in Education [NCEE], 1983). This document reflected a new set of concerns about the U.S. economic position with respect to other industrialized economies, particularly Germany, Japan, and Korea. (The influence of this document persists today, and is one of the most often cited documents in laying the groundwork for current science, mathematics, technology, and engineering education policies and concerns.)

One solution to the declining number of U.S. scientists was to offer advanced science courses to the most talented students in the K–12 system so that they would be better prepared for college science majors and careers. Other efforts to provide intellectually challenging opportunities for students identified as gifted and talented included

> the creation of separate schools for the gifted, honors classes, use of a two-track system so some could advance more rapidly, acceleration though the curriculum, supervised work experience outside of school, individualized projects, class projects, small-group projects, use of a gifted student as an assistant in class, math and science clubs, and contests and exhibits. (DeBoer, 1991, p. 137)

The rest of the students received a mere cursory introduction to science, with high school graduation requirements usually demanding a single science course, often general science or biology. Tracking was commonly practiced, and many students were streamed into vocational tracks where science was not seen as important. For those who set their sights on college and the professions, opportunities to learn high quality science were vastly different for women and men, for people of color and the white majority, and for the economically well-off compared to those who were not.

The science curriculum reforms of the late 1950s and 1960s were intended to raise the level of science instruction in schools. New comprehensive textbooks in high school biology, chemistry, and physics were produced by committees made up primarily of scientists and psychologists. The intellectual level of the new curriculum materials was high, and the accompanying laboratory activities were complex and challenging and required specialized materials and equipment. Substantial amounts of time were to be spent in science laboratories, doing guided inquiry science activities. Within about 25 years, however, the number of schools using these reform materials dwindled and most of these published curriculum materials were no longer used. Nevertheless, the practice of streaming some science students into the most challenging courses continued as tracking.

Civil Rights Goals Encounter National Needs

By the 1980s, the uneven participation of women, minorities, and persons with disabilities in scientific careers was an increasing concern. Empowerment movements were led by African Americans, women, persons with disabilities, Native Americans, Latinos, and others, and were spurred on by legal challenges. The Science and Engineering Equal Opportunities Act of 1980 (NSF, 1994) declared that:

> it is the policy of the United States to encourage men and women, equally, of all ethnic, racial, and economic backgrounds to acquire skills in science, engineering, and mathematics, to have equal opportunity in education, training and employment in scientific and engineering fields, and thereby to promote scientific and engineering literacy and the full use of human resources of the Nation in science and engineering. (p. 1)

The goal of the 1980 Act was to promote equality of access to quality science education and equality of opportunities to pursue science careers. The Act, which initially focused on gender, race, and economic background, was amended in 1985 to include persons with disabilities. As a result of this Act, in 1982 the National Science Foundation published, for the first time, a report on the status of various groups in science careers, *Women and Minorities in Science and Engineering: 1982.* This report provided statistics on the rates of participation of various groups of students (women and minorities), from elementary school through postdoctoral careers, and it tracked changes over time. Although this first NSF report did not include statistics on persons with disabilities, they were included in subsequent volumes, issued at 2-year intervals by NSF. All of the reports may be found on line at NSF's website (http://www.nsf.gov/statistics/wmpd/ or http://www.nsf.gov/statistics/women/).

School-level policies that created separate tracks for students taking science and mathematics increasingly were seen as a problem because they institutionalized unequal opportunity to learn. Low, medium, and high level science courses had radically different student expectations, resources, and quality of teachers. Students were not randomly distributed among tracks; students of color and students from low SES backgrounds predominated in low track classrooms. The NSF commissioned Jeanne Oakes with the Rand Corporation to examine these issues in depth. Two reports, *Lost Talent: The Underparticipation of Women, Minorities, and Disabled Persons in Science* (1990a), and *Multiplying Inequalities: The Effects of Race, Social Class and Tracking on Opportunities to Learn Mathematics and science* (1990b) made a convincing case that many students had reduced opportunities to learn and pursue STEM careers:

> the use of tracking and ability grouping in mathematics and science stems from the widespread belief that children's intellectual differences are so great that students with different perceived ability levels need to be taught in separate classes and that much of the curriculum, especially at secondary level, is not appropriate for many students ... (seeing) the coincidence of these differences with students' racial and socioeconomic status as distressing, but not a matter over which schools have much control ... (but this) ignores the overall ineffectiveness of such grouping practices in increasing achievement. (Oakes, 1990b, p. iv)

Grouping practices in the elementary and middle school grades affected children who had been clustered in "low-ability classes" for years on end. By the time these students reached high school, their science education experiences were strikingly different from their peers in high track classes, with markedly different expectations for achievement, access to resources, and chances of having competent science teachers. Consequently, these students entered high school ill-prepared to take the gatekeeper courses that would allow them to succeed in college, especially in college-level STEM courses. The net effect was that students from low SES backgrounds and students of color could leave high school with less of the knowledge and skills considered necessary to become scientifically literate or to pursue STEM careers at any level of entry into the workforce.

The tracking research pushed policymakers to confront the uneven distribution of resources for teaching and learning science within schools and exposed some of the shocking differences in distribution of resources among schools and school districts. Urban and rural school districts often could not provide the same opportunities to learn as suburban districts, such as teachers who had expertise in science, the number and quality of science courses offered (such as Advanced Placement courses), access to materials necessary to do science, and access to new technologies.

The NSF's 1994 volume of *Women, Minorities and Persons with Disabilities in Science and Engineering: 1994* focused on reasons for the under-representation of groups of students and the quality of the education infrastructure, and sought to develop an equity agenda:

> Reaching the Nation's goal for science and mathematics achievement will not occur unless all participate, including female and minority students and students with disabilities (Oakes, 1990[a]). In general, these students graduate from high school ill prepared for the technology-oriented employment and they are less likely than white, male students or students without disabilities to enter science and mathematics fields in post-secondary education.... This underrepresentation begins in elementary and secondary schools. (p .13)

Equity and the Economy: 1990s to Present

By the 1990s, the importance of opportunity to learn science was becoming obvious to stakeholders and policymakers. Students who had participated in gateway courses in science and mathematics in high school were more likely to achieve in college. As equity research was beginning to demonstrate that too many U.S. women and minority students were failing to enter the STEM pipeline and to offer reasons for this problem, there also was increasing dissatisfaction and alarm among policymakers and stakeholders about the state of science education in general. This frustration had been voiced earlier in the policy document, *A Nation at Risk* (NCEE, 1983), reflecting the worries about the U.S. economic position with respect to other industrialized economies. These economic pressures often also were felt at the state level, as states struggled to keep their economies afloat in changing times. The traditionally industrial states were hardest hit, but Southern states, too, felt the pressure to compete and to provide better economic opportunities and growth. Consequently, state governors came to demand improved education outcomes for their citizens. The Commission also was concerned about the equity versus excellence argument and a possible end to the push for higher standards. In the Commission's words:

> We do not believe that a public commitment to excellence and educational reform must be made at the expense of a strong public commitment to the equitable treatment of our diverse population. The twin goals of equity and

> high-quality schooling have profound and practical meaning for our economy and society, and we cannot permit one to yield to the other either in principle or in practice. To do so would deny young people their chance to learn and live according to their aspirations and abilities. It also would lead to a generalized accommodation to mediocrity in our society on the one hand or the creation of an undemocratic elitism on the other. Our goal must be to develop the talents of all to their fullest. Attaining that goal requires that we expect and assist all students to work to the limits of their capabilities. We should expect schools to have genuinely high standards rather than minimum ones, and parents to support and encourage their children to make the most of their talents and abilities. (NCEE, 1983, p. 13)

In 1994, the *Goals 2000: Educate America Act* (Goals 2000, 1994) was signed by President Clinton with wide support of the National Governor's Association. The most ambitious aim of this widely publicized effort was to have each state launch a concerted effort to make U.S. students first in the world in science and mathematics by the year 2000. The passage and support of the *Goals 2000* legislation demonstrated that state governments increasingly believed that the economy of each state was dependent on the quality of its workforce and that knowledge of science was important in developing a high quality workforce. This meant that ways to improve state-level education systems, particularly in the STEM areas, had to be found. The language of excellence and equity could be found throughout the language of the Act. Among the purposes of the act were:

- supporting new initiatives at the Federal, State, local, and school levels to provide equal educational opportunity for all students to meet high academic and occupational skill standards and to succeed in the world of employment and civic participation;
- creating a vision of excellence and equity that will guide all Federal education and related programs

Even though *Goals 2000* may have been more a rallying cry than significant public policy, it did result in annual reports of hard-to-ignore, state-by-state comparisons of key indicators for science achievement and other aspects of education. For science, the National Assessment of Educational Progress (NAEP) scores were used as the common metric. Despite the cheerleading approach of *Goals 2000* and substantial funding to initiate systemic changes in education at the state level, science NAEP scores did not change much during the 1990s, and achievement gaps were stubbornly intransigent (National Science Board [NSB], 2008). However, *Goals 2000* demonstrated that state governments were increasingly aware that the economy of each state was dependent on education to improve the quality of its workforce.

During the 1990s through the present, the U.S. participated in international comparisons of science and mathematics curriculum, instruction, and achievement such as the Third International Mathematics and Science Study (TIMSS), later renamed the Trends in Mathematics and Science Studies, as well as the more recent Program for International Student Assessment (PISA). These international comparisons consistently showed that U.S. student performance to be about average, but they also revealed that U.S. expenditures in K–12 education to be high in comparison to those of other countries (NSB, 2006, 2008). A recent analysis of the TIMSS and PISA data compared U.S. progress in science and mathematics with the Group of Eight (G8) countries (Miller, Malley, & Burns, 2009), that are highly developed economically. Japan tends to score well among this group while U.S. scores tend to be in the low to middle range, with increasing achievement gaps as U.S. students get older. Although such U.S. achievement statistics often are reported and are no longer surprising, economic comparisons among the G8 are eye-opening. The U.S. spends a higher percent of its Gross Domestic Product (6.7%) on education than the other countries, but its students have the lowest percentages of STEM-related degrees awarded. Parent occupational status is higher in the U.S. than in other G8 countries, but U.S. students from families from the bottom quartile of the occupational index were outperformed by students in this quartile in several countries. One of the more disturbing findings from an equity perspective is that low income children and children of immigrants in the U.S. are not achieving as well as their counterparts in other G8 countries. In other words, although U.S. education spending is high, the money is not benefitting poor students.

Such international comparisons confirmed what policymakers and stakeholders had feared, and they further illuminated equity issues. Disaggregated achievement data showed that in order to increase U.S. student achievement, state and school district educational agencies could no longer ignore the achievement gaps between various groups of students. Given the documentation of the large numbers of students underserved by STEM education and the relatively high cost of U.S. education, it appeared that resources were not well used, nor equitably distributed. The need to compete in a global economy meant that previously underserved groups would now be considered as important human resources to be developed.

THREE PERSPECTIVES ON EQUITY IN SCIENCE EDUCATION

This next section examines equity issues in science education from three different perspectives. The first perspective focuses on equity as *equality of inputs*, equity that demands equal access and equal opportunity to learn.

This has long been a recurrent theme and, despite civil rights legislation, remains a problem with shifting dimensions. The second perspective sees equity as *equality of outcomes*, with the goal of closing the achievement gaps among demographic subgroups of students (described by gender, ethnicity, SES, as well as by English language learner or special education status). The third view of equity is seen through the lens of an accountability system that mandates equality of outcomes without specifying exactly how states and local districts are to achieve that goal. This inevitably leads to trade-offs in how resources are spent and to competition among various constituents. From this perspective, it must be acknowledged that education funding is not unlimited. If money is used to serve the needs of some students, it has to come from programming for others. Equity policy decisions are, therefore, affected by pressures from different interest groups, economics, and societal realities.

Equality of Inputs

High Standards of Learning for All

A major goal of the standards movement, initiated in the early 1980s with the publication of *A Nation at Risk* (NCEE, 1983), was to raise standards for *all* students. As noted earlier, the report emphasized that the "goal must be to develop the talents of all to their fullest" (p. 13). Although the heart of the standards movement may have been the perceived need to improve schools by making them more accountable for student learning, the equity goal was also inherent in its rationale: one common set of standards for *all* children in *all* schools provides for equality of inputs—equal opportunity to learn—and this should close achievement gaps (Lauer et al., 2005). This view followed, in part, from the tracking research in the 1990s showing that not only were resources vastly unequal in their distribution to different groups of students, but so were underlying expectations for women, students of color, students of low SES, those with disabilities, and those learning to speak English (Lynch, 2000). High standards would ensure that all students would learn at a high level rather than holding one set of expectations for the privileged and another for students who were not well enough situated to demand and obtain a high quality education. As Robert Maynard Hutchins put it: "The best education for the best is the best education for all" (see DeBoer, 2006 for a history of the standards movement and the idea of achieving the best for all).

Changing Science Requirements for High School Graduation

As part of the effort to raise the bar for all students, states have increasingly been requiring students to complete more STEM coursework to graduate from high school. These policies were developed as part of the general effort to raise expectations for all students, to increase the number of students who would be eligible to take college-level science courses and possibly pursue science careers. In 2002, over half of the students intending to attend college were unprepared to take college-level science because they had not taken the requisite core high school courses (ACT, 2006 and Boznik, Ingels, & Daniels, 2007 as cited in NSB, 2008). Data also showed that White and Asian American students took the core courses at far higher rates than did African American or Hispanic students. A longitudinal study by ACT found that students who had taken a certain set of "core" college preparatory courses scored higher on the ACT than students who had not taken the core. Other studies showed a similar relationship between SAT scores and students who took high school chemistry or physics or upper level mathematics courses (NSB, 2008). Consequently, choosing to view such correlations as causal, policymakers began to mandate certain core courses for high school graduation, meaning that all students would have to complete the core courses, not just the college bound students. Figure 11.1 shows the steady increase in science course completion between 1990 and 2005 (NSB, 2008). Despite these changed requirements, the pattern of taking science courses still differs by ethnic-racial group, as well as gender, at least for the advanced-level courses. White and Asian American students take the core courses at higher rates than do African American or Hispanic students as seen in Figure 11.2. Perhaps even more telling is Figure 11.3, which shows the variation in course taking patterns by school poverty level. These data suggest there are fewer opportunities to complete upper level mathematics and science courses in high poverty schools (i.e., schools in which more than half of the students are eligible for federal meal assistance). In high-poverty schools, high level science courses are less likely to be offered, either because the school's facilities and teacher resources cannot support such courses or there is less demand by students. Because students of color attend high poverty schools at higher rates than White or Asian American students, a large part of this problem is economic and systemic. It is worth noting that demographic subgroups of students reported here are vastly more complex than broad categories like "Asian American" indicate; for instance, the demographic "Asian American students" constitutes many ethnic groupings, some of which have had few advantages and access to high quality schooling, cf. Asian Nation: Asian, http://www.asian-nation.org/index.shtml or Lynch, 2000.

Clearly, youth who attend high poverty schools do not have an equal opportunity to learn in STEM fields, and they are underprepared for higher education and for life.

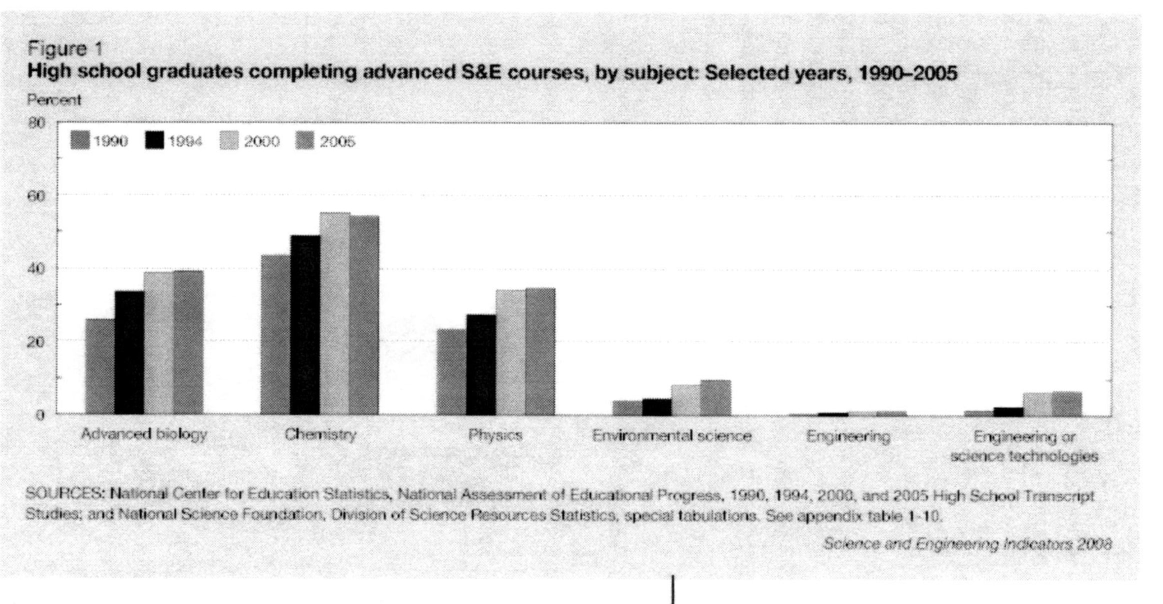

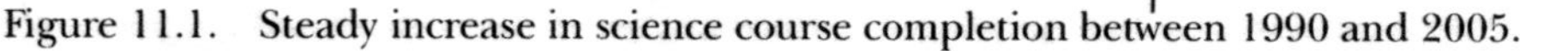

Figure 11.1. Steady increase in science course completion between 1990 and 2005.

As states increased graduation requirements in science, schools have been pressured to add resources for students who needed extra help in order to graduate. Such resources may include tutoring programs, summer programs, or new arrangements in how a course is delivered. In mathematics, for instance, adjustments have been made such as offering an Algebra 1 course that meets for two periods a day for a school year or is extended to a 2-year-long course. Some schools have increased the number and variety of science courses offered, rather than limiting students to the traditional biology-chemistry-physics sequence, which may be too difficult for all students to pass and that may be viewed as unappealing and too removed from everyday life. Consequently, there is an increasing variety of science courses offered to students to fulfill graduation requirements such as marine biology and environmental science, and applied courses such as pre-engineering, robotics, nano-technology, and biotechnology. Such course offerings vary by school or school system. This diversity in science course offerings means that a wider variety of students

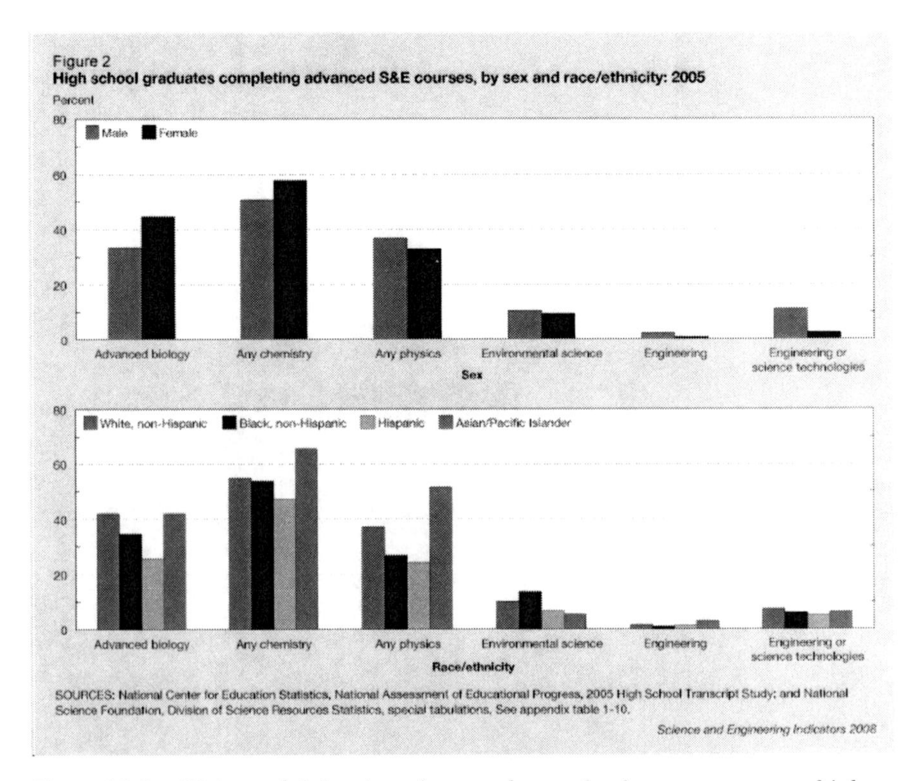

Figure 11.2. White and Asian American students take the core courses at higher rates than do African American or Hispanic students.

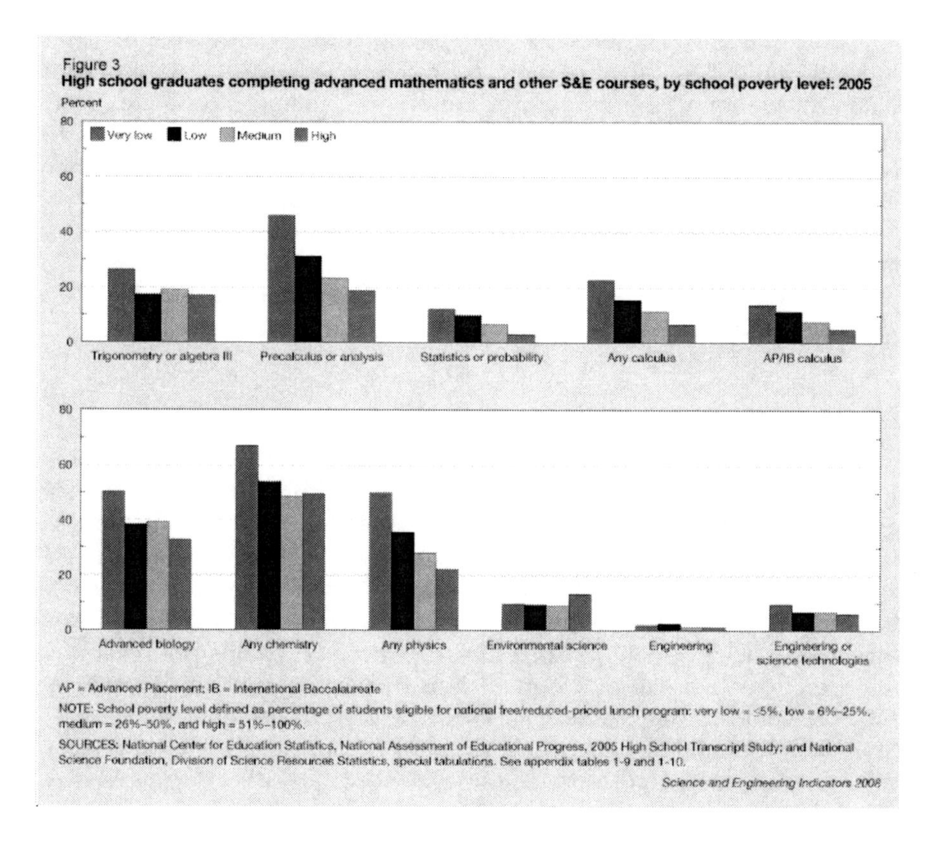

Figure 11.3. Variation in course taking patterns by school poverty level.

can be successful in science courses even if they are not taking the courses that are traditionally associated with college preparation. (See also Halverson, Feinstein, & Meshoulam, Chapter 13, this volume.)

Although Figure 11.1 shows a steady increase in course-taking for all science courses between 1990 and 2005, Figure 11.2 shows that for physics and engineering, there are troublesome gaps in course completion by gender and ethnicity/race. One response to this problem has been Project Lead The Way (PLTW), a pre-engineering program initiated in 1997 and promoted by business leaders such as the Business Roundtable and the Committee on Prospering in the Global Economy of the 21st Century (NSB, 2008; PLTW, 2009). The course has reached over 500,000 students in 3000 schools. Its goal is not only to increase the number of engineering graduates, but to do so by seeking participation of students proportionate to the gender and race/ethnicity distribution in the school. Although a 2004–05 evaluation of PLTW still showed gender and racial-ethnic differences in participation, with White men overrepresented in the program, the differences were less pronounced than those found in the distribution of college graduates with engineering degrees. To further increase the attractiveness of the program for women and minorities, PLTW also intends to introduce new biomedical courses. All of these special courses are embedded in the traditional, required program of studies in science and mathematics that are essential for those who want to pursue STEM degrees. The PLTW program students are guided to take the courses that will provide them with opportunities to succeed, and the equity implications are apparent.

Qualified Science Teachers

Another important way to ensure equality of inputs is to improve the quality of teaching for all students. The need for high quality science teachers is one area that has achieved remarkable consensus among policy makers as being critical to improving U.S. science education, and it has emerged as the single most important evidence-based factor for raising student achievement (NSB, 2008). For instance, results from National Assessment of Education Progress (NAEP) science assessments indicate the importance of science teacher characteristics to student achievement. Students who had teachers who have college degrees in science, were certified to teach science (for the eighth grade only), or who had more years of teaching experience were more likely to achieve higher scores on the science portion of NAEP (NSB, 2000). Such results, replicated in many studies, prompted the authors of the NCLB Act of 2001 to require that every classroom have a highly qualified science teacher. (See also Kahle & Woodruff, Chapter 3, this volume, for an analysis of research related to this policy area.)

Teachers who are well prepared in science, who understand how to make science ideas clear to students, and who have previous experience teaching are not evenly distributed across U.S. schools. Science teachers with undergraduate or graduate degrees in science, or who have teacher certification in science, are more likely to be found in schools where there are fewer students of color (NSB, 2000) and are far less likely to be found in schools where more than 50% of the students have families at the poverty level (Orfield & Lee, 2005). High poverty schools are characterized by high rates of student mobility, problems with disorder and discipline, higher dropout rates, lower homework completion rates, and watered down coursework with fewer course offerings. Teachers in such schools have much higher mobility rates, less education, are less likely to be licensed, and are less experienced. A 2004 U.S. Department of Education report showed that in schools where at least 75% of the students were low-income, there were three times as many uncertified or out-of-field teachers in science (Orfield & Lee, 2005).

High minority schools (i.e., schools in which the majority of students are African American or Hispanic) have a higher percentage of teachers who did not participate in practice teaching and who participated in alternative certification programs (NSB, 2008). Such schools also have a greater concentration of new science teachers, and research shows that more teacher experience contributes to higher student science outcomes (NSB, 2008). Science teachers in high minority schools also express less confidence that they are well-prepared to teach; and teachers who lack confidence are more likely to leave teaching, resulting in high teacher turnover rate and instability.

In summary, there is still a general pattern in high minority and high poverty schools with unequal access to qualified teachers. In order to change this inequitable distribution of qualified teachers, policies and incentives are needed to encourage more qualified teachers to teach in these rural or urban schools and to retain science teachers who know their discipline and are dedicated to the success of students in these more challenging settings. Some urban school districts rely heavily on alternative certification programs and other innovative measures to staff schools, such as Teach For America (2009), which attracts college graduates with strong backgrounds in the academic disciplines and a desire to teach in schools with high need. There also have been proposals to pay science teachers higher salaries to work in high poverty urban and rural schools. Some school districts regard their strongest teachers as resources that can affect change in new ways, and encourage them to take assignments in schools identified as failing. Such troubleshooter assignments come with additional compensation and a new view of the steps on the teacher career ladder. Research on the success of such experiments needs to be conducted so that the results can be used to develop the equity policy agenda in science education. Requiring that all students be taught by a qualified teacher, as NCLB does, is a very important first policy step toward that goal, but additional policies, regulations, programs, and research, as well as adequate funding and resources, are needed to ensure its success.

Curriculum and Instruction

There are no federal-level policies that directly determine what science curriculum and instruction will be. Rather, such policy decisions occur at the state and local levels. At the national level, curriculum and instruction as a tool for addressing issues of equal access to science education is influenced by the recommendations of national science and science education groups such as American Association for the Advancement of Science, National Science Teachers Association, and the National Research Council, but there is currently no way that the federal government can use the curriculum as part of an equity policy.

If policy for curriculum and instruction is to address the inequalities in access that have characterized science education in the United States, then curriculum developers should provide much stronger evidence that their materials are effective and beneficial to various subgroups of students. It is beyond the scope of this chapter to describe the research on how different forms of curriculum and instruction affect diverse student groups, and the reader is referred to Lee and Luykx (2006) for a discussion of that research. A review of the effects of new standards-based curriculum materials found that they generally have positive influences on student achievement, including students characterized as being "at risk." Weaker results were found for reform-based *instructional* strategies. Unfortunately, at risk students may experience less access to reform-oriented strategies than more advantaged students, despite some evidence that they can benefit from such strategies (Lauer et al., 2005).

The SCALE-uP is one research program that examined middle school standards-based curriculum materials, studying the responses of students of varied backgrounds (ethnicity, SES, or eligibility for English language for speakers of other languages [ESOL] or special education programs). Results of this study generally showed a trend indicating that if a standards-based unit increased student performance overall, it also was effective for the demographic subgroups of students listed above. Effects were often most dramatic for subgroups of students with worrisome achievement gaps. The most successful units were focused on a particular standard or science learning goal, were hands-on and student-centered, and employed guided inquiry to encourage student thinking. Less successful units were not as focused and were more teacher-centered (Lynch, Szesze, Pyke, & Kuipers, 2007; Lynch, Taymans, Watson, Ochsendorf, Pyke, & Szesze, 2007).

Facilities for Teaching Science

Science, more than any other academic subject, requires specialized classroom facilities and materials. Such facilities and materials are expensive to build and maintain. This means that high poverty schools often do not have the resources needed. Instructional technology also has become integral to doing science, and carries additional expenses for Internet access and maintenance, for hardware and software, and for the professional development of teachers who must be provided time to update their skills and knowledge. These facilities, materials, and technologies should not be viewed as luxuries to be enjoyed by the privileged few. They are essential to the kind of science education experts believe all students should be receiving. *Taking Science to School: Learning and Teaching Science in Grades K–8* (Duschl, Schweingruber, & Shouse, 2007), for example, summarizes how many in the field of science education currently view science learning and teaching. The report is clear on the need for students to engage in science activities and to talk and do science in order to learn science. In order to engage meaningfully in science, students need access to materials and labs, and this is an equity issue.

America's Lab Report: Investigations in High School Science (Singer, Hilton, & Schweingruber, 2005) demonstrates the inequalities in the facilities for teaching science, and includes a summary of a U.S. Government Accountability Office survey of 10,000 schools in 5,000 school districts, which found the following:

> The survey identified three trends. First, inadequate laboratory facilities varied by community type. The highest percentage of ill-equipped schools was in central cities, followed by urban fringes or large towns, and the smallest percentage of ill-equipped schools was in rural areas or small towns. Second, inadequate laboratory facilities varied by proportion of minority students, with less adequate laboratory facilities in schools with higher concentrations of minorities.... Third, inadequate laboratories were associated with the proportion of students approved for free or reduced-price lunch, with less adequate facilities in schools with higher concentrations of students eligible for reduced-price meals. (p. 178)

Other research in *America's Lab Report* indicates that rural schools generally have less money to spend on laboratory supplies than urban or suburban schools. Often schools with large numbers of non-Asian minority students spend less on laboratory supplies and equipment. Surveys conducted in Washington DC, Chicago, and New York City all indicated that laboratories were in disrepair and student use was seriously curtailed as a result. Rural and urban teachers spend more money out-of-pocket to conduct laboratories, and anecdotal reports indicate that the lack of decent laboratory facilities is one reason why some science teachers leave urban schools for better-equipped suburban assignments.

It is somewhat surprising that so little policy attention has been paid to the importance of science laboratories, equipment and supplies, and safety equipment, when considering science education equity policy (see Lynch, 2000). But, the policy implications are clear. If students learn science better by handling actual physical objects and by directly observing phenomena in the world—which the laboratory and various field activities provide (Minner, Levy, & Century, 2009)—and, if minority students have less access to such facilities, then their education is diminished. Because science laboratories are expensive to build and expensive to maintain, rather than trying to correct the inequalities that exist, some schools are increasingly looking to technological solutions as a less expensive and "more modern" substitute for hands-on science laboratory experiences. The next section looks at the impact of technology on science education.

Information Technology in Science Education

A major effort has been made to upgrade technology in K–12 U.S. science classrooms. Sometimes, this technology is used to provide students with simulations of the real world as a partial substitute for the real world experiences that they have in the laboratory or in the field. The technology also provides new opportunities for examining data models and computer-based simulations that scientists are developing to help explain natural phenomena. As with the laboratories themselves, however, there is a gap in the technology available to students who attend poor or predominantly minority schools. How this technology will be used in science classes and how it will be made available equally to all students is a significant policy issue in science education, whether hands-on laboratoriess are supplemented with instructional technology that enhances the recording and analysis of laboratory data or are replaced by virtual laboratories in whole or in part (NSB, 2006). Certainly the more technology-savvy students are better positioned to succeed in those new environments, something that depends upon access to technology at home and at school.

Another goal of getting technology into the schools is to provide students with opportunities for good jobs in the twenty-first century and to improve state and national economic competitiveness (see NSF Task Force on Cyberlearning, 2008 and Partnership for 21st Century Skills, 2009). Because this continues to occupy an important policy niche, the equity implications are important. Although gaps in the quality and number of computers, and access to them, have declined in the last five years for urban and other types of schools, students in high-poverty schools remain at a disadvantage. Internet use has been found to be related to race/ethnicity, family income, and parental education. The data in Figure 11.4 indicate that these technology gaps are far more prevalent for home use than school use. Students from affluent families are more likely to use computers at home to complete school work, so more computers in school won't be enough to solve the problem of inequitable access to essential educational inputs (NSB, 2006). Any policy initiatives that will affect equality of inputs would have to address access to technology outside of school (NSB, 2006) as well as in school.

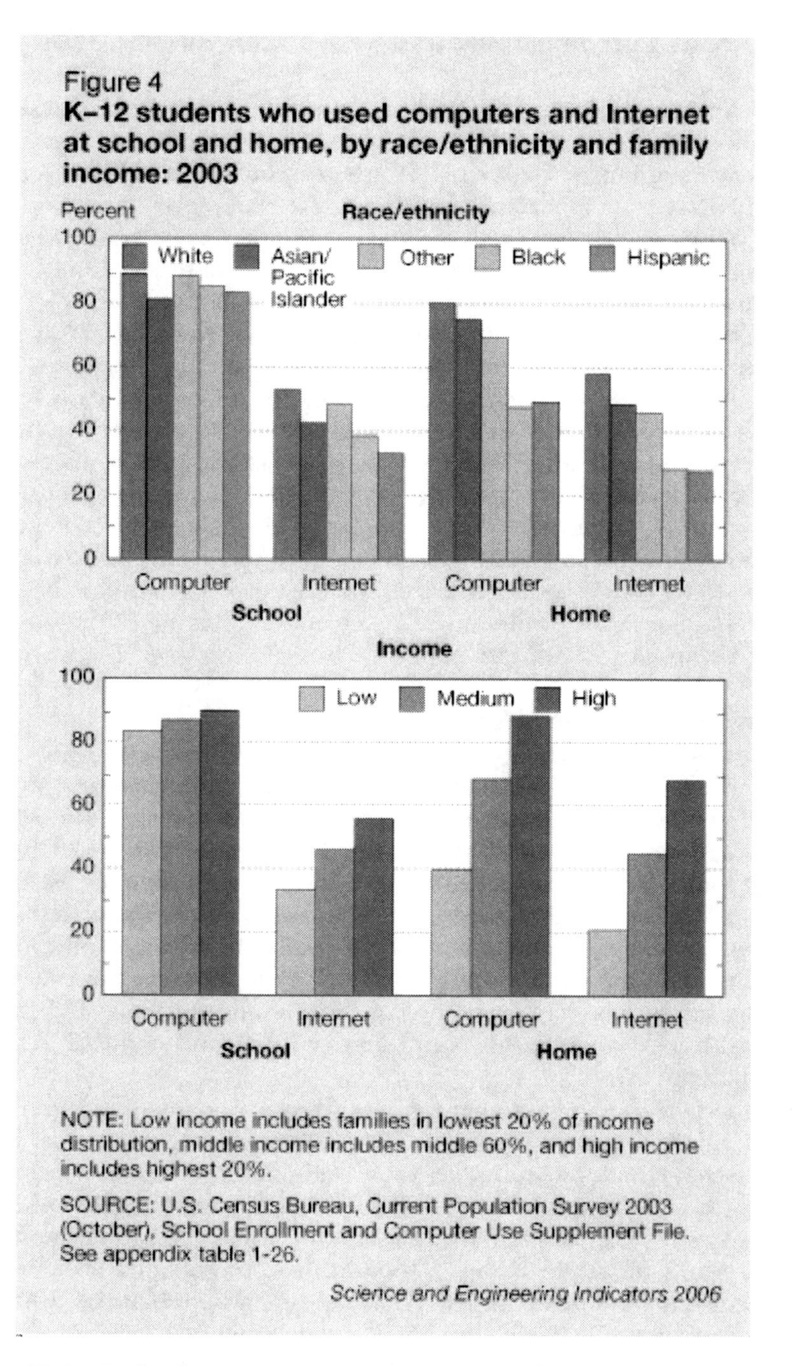

Figure 11.4. Technology gaps are far more prevalent for home use than school use.

Equality of Outcomes: Understanding Achievement Gaps

The previous section examined the importance of resources in science education and the lack of consistent access to science education resources for some subgroups of students. Efforts to redress inequalities in inputs have had, at best, mixed success. Given the differences in achievement and attainment that persist, policymakers also have concentrated on ways to close that achievement gap by focusing directly on the outcomes side of the equation. There is now a national consensus about the need for more equitable outcomes for all in science. A synthesis study of STEM policy documents conducted by the Education Commission for the States found six recently issued policy documents calling for improvements in science and mathematics, half of which called directly for an outreach to minorities or economically disadvantaged students (Zinth, 2006). Achievement gaps found in K–12 education spill over to affect the entire society as students graduate from high school unprepared for jobs or for college, and find college majors in STEM careers closed to them. It is nearly impossible for students to enter the science pipeline unless preparation begins in high school, or most would argue, much earlier, because of the opportunity structure in U.S. education.

Achievement Gaps in Science: Trends

Gaps in science outcomes often are measured by contrasting differences in achievement and participation by gender, ethnicity/race, and SES as well as disability status. In addition, many researchers take note of English Language Learner (ELL) status (see for example, Lynch, 2000). The NAEP fourth grade science assessment showed gaps between ELLs and non-ELLs of about 50 points in 1996 and 2000, with a significant decrease to about 30 points in 2005. A similar trend was found for students tested at the eighth grade. For 12th grade students, however, there were no significant changes in the last three administrations of NAEP science, with gaps of about 40 points between ELL and non-ELL students (NAEP, 2006).

In the United States, gender gaps on science assessment results are relatively small. For example, the 1999 NAEP long-term trend assessments showed that females scored lower than males on 2000 science assessments at the 4th and 8th grade levels, but there were no statistically significant differences by 12th grade (NSF, Division of Science Resources Statistics, 2002). The same study, however, showed there was about a 30 point gap between white and black students that stayed constant between 1990 and 1999. Similar results were seen for Hispanic/white gaps. However, from 1996 through 2005, fourth grade African American students were able to narrow the black-white achievement gap (NSB, 2008). There also was some small progress made in closing the gap between students eligible for federally subsidized lunches (a proxy for SES) and those who were not. On the whole, however, substantial achievement gaps in science remain stubbornly in place (NSB, 2008).

Origins of Achievement Gaps in Science

What are the origins of these achievement gaps? The Early Childhood Longitudinal Study (ECLS) followed the science and mathematics achievement of a large number of students who entered kindergarten in 1998 through the fifth grade (Princiotta, Flanigan, & Hausken, 2006; Rathbun, West, & Hausken, 2004, as cited in National Science Board, 2008). Gaps in mathematics learning by ethnicity/race already were apparent at the kindergarten level, and gaps in science were apparent for the first assessment given during third grade. Boys outscored girls slightly, White and Asian American students outscored African American students, and Hispanics outscored their African American peers. By fifth grade none of the gaps had narrowed, and the gaps between White students and African American and Hispanic students had increased.

Other factors measured in this study illuminate possible explanations for the results. For example, children whose mothers had more education outscored the children of less educated mothers, and children who lived below the poverty level scored lower than those who did not. By fifth grade, children from low SES backgrounds produced science achievement scores similar to their wealthier peers in *third* grade. In others words, by fifth grade, children from poor families were 2 years behind in science on this assessment. Such studies indicate that students from low SES homes or whose mothers are not well-educated, have skill gaps when they enter school and that these gaps tend to widen as they grow older. Longitudinal studies such as these show that achievement gaps among students of different ethnic/racial groups are related to social, cultural, and economic factors arising early in students' lives.

To understand achievement gaps in science, it is increasingly clear that what best describes a child's chances in school is social class and geography rather than race/ethnicity (Kahlenberg, 2001; Orfield & Lee, 2005). The 2008 *Kids Count* (Annie B. Casey Foundation, 2008) reported the U.S. youth poverty rate at 18%, an extraordinarily high rate of poverty compared to other developed countries. In 2004, 33% of Black children, 28% of Hispanic children, and 14% of White children were living in poverty. This translates to 3.6 million Black children, 3.9 million Hispanic children, and 7.8 million White children in poverty in 2004.

Research has shown that poverty has replaced race as the most pervasive index of social disadvantage and school failure (Burbridge, 1991; Hodgkinson, 1995, as cited in Lynch, 2000). Although African Americans, on average, have the highest poverty rates and show the largest achievement gaps, many Black children are doing well in school. This is more likely when their parents are affluent and they attend low poverty schools. A middle class Black student in a middle class school stands a better chance of success than any poor child attending a school where there is a high concentration of poor children. Further, a middle class student who attends a low SES school is at a greater disadvantage than a poor student who attends a middle class school. But, Black students at low SES schools are more disadvantaged than White students attending the same schools (Kahlenberg, 1995a and 1995b, as cited in Lynch, 2000). So, when considering ways to address the achievement gap, it is important to understand that it is not only the social class of the individual student that is related to school success, but also the collective poverty level of the school.

Disturbingly, for the last 20 years, the U.S. has been experiencing school re-segregation for African American children and increasing segregation for Latinos (Kahlenberg, 2001; Orfield & Lee, 2005). The Civil Rights Project (originally housed at Harvard University and now at UCLA) follows these trends and notes the confluence of factors that make poor, segregated schools difficult places to learn:

> U.S. schools are now 41% non-White and the great majority of the non-White students attend schools which now show substantial segregation. Levels of segregation for Black and Latino students have been steadily increasing since the 1980s.... Achievement scores are strongly linked to school racial composition and so is the presence of highly qualified and experienced teachers. The nation's shockingly high dropout problem is squarely concentrated in heavily minority high schools in big cities. The high level of poverty among children, together with many housing policies and practices which exclude poor people from most communities, mean that students in inner city schools face isolation not only from the white community but also from middle class schools. Minority children are far more likely than whites to grow up in persistent poverty. (Orfield & Lee, 2005, p. 5)

Reporting achievement gaps in science by racial/ethnic demographic categories obfuscates the problems of high poverty schools and the disproportionate number of children who attend them: more than 60% of Black and Latino students attend high poverty schools (> 50% poor), compared to 30% of Asians and 18% of Whites (Orfield & Lee, 2005). In extreme poverty schools (90 to 100% poor) in the Northeast and Midwest, Blacks comprise more than half of the students, while in the South the number increases to 62%. Latino students in the West make up 76% of the student body in these extreme poverty schools (Orfield & Lee, 2005). Moreover, the poorest performing high poverty schools tend to be located in urban areas. These trends are crucial to understanding K–12 science education policy and equity issues. The problem—poverty and housing segregation—is clear, but solutions are not. Richard Kahlenberg's (2001) proposal is to create public policy that requires "desegregation" of public schools, not by race but by SES so that high poverty schools are essentially eliminated. Housing policies, however, are challenging to make and require both political will and community sensitivity and resolve.

Mandating Equality of Outcomes: Raising the Stakes With High School Exit Exams

The accountability movement that began in the 1980s resulted in a state-level trend to increase the number and kind of STEM courses required for high school graduation (discussed earlier in this chapter) in order to enforce a sort of "equality of inputs." Mandating a set of STEM core courses was seen as a means to increase the number of students having the requisite experiences for STEM college majors or to better prepare them for jobs that require STEM knowledge and skills. The next step was to create a policy lever to raise the quality of these courses by requiring students to pass high school exit exams in science (and other subjects) as part of graduation requirements. This could be viewed as a means to "enforce" equality of outcomes. In other words, if all students had to pass these tests, the quality of instruction would have to be at least good enough for the students to learn the science needed to pass. According to work done by the Center for Education Policy (CEP), which has followed these trends carefully, high school exit examinations are now taken by 68% of the nation's public high school students, who attend schools in the 23 states with such policies. By 2012, approximately 74% of the nation's public high school students will be affected, as three more states are expected to add such policies. Moreover, because there is a greater concentration of students of color in states adopting the exams, a full 84% of students of color will be required to pass such exams to graduate from high school by 2012 (Zabala, Minnici, McMurrer, & Briggs, 2008).

Science exit exams will be required in 19 of the 25 states by 2012 (Kober et al., 2006). Some states require students to take a general comprehensive assessment and others require students to take a number of state-developed end-of-course examinations tied directly to the courses taken (see Bybee, Chapter 8, this volume). The state of Virginia, for example, requires end-of-course exams in biology, chemistry, and earth science (Zabala et al., 2008). Among states requiring students to pass science tests to graduate from high school, initial pass rates ranged from 68% in Georgia to 95% in Tennessee. Because states withhold diplomas when students don't pass the science assessments, the fairness of the exit exams has been challenged. However, supreme courts of the states of California and Arizona have upheld these high school graduation requirements. One of the most contentious issues is whether to withhold diplomas from students with disabilities or for ELLs, and whether to provide these students alternative exams or other routes to graduation. Of the states requiring exit exams, all have alternative routes to graduation for students with disabilities, but only three offer such routes to ELLs. States with greater student diversity face special challenges (Zabala et al., 2008).

On the positive side of this development, it is possible that exit exams may be producing high school graduates who actually are learning more science, are better prepared for work or college, and have not been shut out of the STEM career pipeline, which is the intent of such policies. Encouragingly, results from the more ubiquitous mathematics high school exit exams show that as the number of white students who pass the exams increases, the number of African American students who pass also increases. The same holds true for other students of color. In other words, the exit exam policy and the programs being put in place to ensure that students can pass the exams appear to be affecting all subgroups of students positively. This also includes ELLs and students with disabilities. Furthermore, states with higher overall pass rates seem to also have decreased their achievement gaps. For instance, states with the highest pass rates report gaps as small as 10 percentage points, while states with the lowest pass rates report gaps as large as 25 to 40 percentage points (Zabala & Minicci, 2008).

But raising the bar in science for high school graduation also may have negative consequences for students who were already struggling to stay in school and graduate. Higher standards for graduation may contribute to the drop-out rate, especially for students of color and those from low SES families. What happens to the students who fail? Some research has suggested that increasing requirements for graduation has increased dropout rates, especially for minority students, while others show no differences, suggesting that other policies (e.g., retention of students in grade or tougher courses) may be more strongly associated with dropping out (Chudowsky & Gaylor, 2003).

Increasing high school graduation requirements is a policy heavily freighted with equity implications that could work either for or against the science success of subgroups of students. If exit exams impose substantial new requirements on students, it is reasonable to expect that without additional support or resources that enable students to meet the additional requirements, exit exams will lead to increased drop out rates. If the students who fail or drop out because of the higher graduation requirements are disproportionately minority students, then this has serious equity implications, given the importance of a high school diploma for employment. This may be a case where the effort to increase the science knowledge of all students is in conflict with the desire for all students to complete high school and find productive and satisfying work. Untangling these relationships through research still needs to be done.

Equity as Trade-Offs: The Consequence of Legislating Equality of Outcomes

A third perspective of equity is seen through the lens of an accountability system that must balance limited resources to achieve the goal of equitable outcomes. Schools have always had to decide how to deploy and balance resources to help children succeed, and this inevitably requires trade-offs. School leaders, buffeted by competing interest groups, determine which students get the best science teachers, laboratories, materials, and access to resources, in what Loveless (1995) calls the politics of aggregation (see also Green, 1980, 1983). The politics of aggregation involves the competing goals of different social institutions. In education, the voices of the prominent and powerful can influence the distribution of limited resources and affect the opportunity structure within the school (Lynch, 2000). McDonald (2009) believes that because schools are critical democratic institutions, it is crucial to study how they are governed, who participates, how resources are allocated, and who benefits from the resources, all having long-term consequences beyond the individual school.

At the time of this writing, the No Child Left Behind (NCLB) Act of 2001 is the preeminent federal policy on K–12 public education (Stecher, et al., 2008). The NCLB Act does not legislate equality of inputs, but leaves it to schools, school districts, and states to decide how to use the funds provided to help students learn and meet school accountability targets. To accomplish this, data must be disaggregated by categories such as gender, ethnicity/race, and social class, and schools must show adequate yearly progress (AYP) for all demographic subgroups. So, NCLB forces the equity issue by requiring schools to actually accomplish specified outcomes. This brings into sharp focus the question of how each school is to allocate resources to accomplish NCLB goals, and epitomizes the view of equity as leading to trade-offs in resource allocation, as creative interventions are developed to help students learn and demonstrate improved performance. Besides focusing on higher outcomes in mathematics and reading, NCLB has affected the equitable teaching and learning of science. Because of this, we examine in some detail the complexities and implications of this federal education legislation.

Federal Education Legislation (NCLB)

The NCLB Act grew out of the standards movement of the 1990s and the Improving America's Schools Act of 1994, which required that states adopt standards in reading and mathematics and administer aligned assessments in order to receive Title I federal funds for low-income students (Lauer et al., 2005). The NCLB and the Improving America's Schools Acts are reauthorizations of the original Elementary and Secondary Education Act of 1965, a significant part of President Johnson's War on Poverty that provided funding to "educationally disadvantaged" children, most often from poor families. The NCLB Act was initiated through the efforts of President George W. Bush and a nonpartisan coalition from Congress, state governors, and major leaders in business and industry. One noteworthy goal of this legislation is equity for all students. President Bush promised that NCLB and the expansion of high stakes testing to high schools would end the "soft bigotry of low expectations." Inherent in this view of equity was something new, not only equality of existing outcome measures, but raising the bar so that all students would be required to meet specified standards.

The NCLB Act does not mandate specific standards, assessments, or targets for annual yearly progress (AYP), leaving it to each state to make such decisions. It does require states to have these accountability tools in place, however. Schools and school districts decide how to use the funds provided (seen by many as hopelessly inadequate) to help students learn and meet school AYP targets. As noted earlier, education standards are used to equalize opportunities for all students, regardless of where they live or their SES, by spelling out and making public what all students should learn and providing additional funding to schools that need more resources (The Education Trust, 2009). Although NCLB's limitations and problems are well-documented (cf. Ryan & Shepard, 2008), there is no consensus on the overall value of NCLB. Some look at the law's effects and see progress through the accountability measures, while others see NCLB's effects as weakening the educational system. However, passionate supporters and opponents of the law cross political boundaries (McDonald, 2009).

NCLB: Science Education Left Behind?

So far, the effects of NCLB on science education have been mixed, and they are only beginning to emerge. This is due largely to the fact that science was not scheduled for required student assessment until 2007, and then at only three times in a student's K–12 education, at least once during Grades 3–5, Grades 6–9, and Grades 10–12. Although NCLB mandates reporting AYP for reading and mathematics, the same is not true for science. For science, NCLB only requires that each state must administer assessments and report the results. Unlike reading and mathematics, however, where schools and school districts in each state must meet goals for AYP, science education is handled in a different way. Each state has the option of using its science assessment scores as an indicator for calculating whether it has made AYP according to the current version of NCLB, or it can choose from among other indicators of success. At the present time no states report using their science assessments for AYP (Penfield & Lee, 2010).

Prior to the introduction of science in NCLB, it was widely believed that the emphasis on mathematics and reading in the early years of NCLB took time away from science. However, Lee and Houseal (2003, as cited in Griffith & Scharmann, 2008) reported that even prior to NCLB, there were factors that resulted in diminished time spent on elementary science education. The factors included the added costs of supplies, material, and equipment, as well as issues having to do with classroom management, working with diverse learners, responding to individual differences, and support from colleagues, administrators, and the community at large. They also cited problems with elementary school science teachers' content preparation, self-confidence levels, anxiety, attitude, and professional identity.

Much of the concern generated by NCLB and its effects on science education has focused on the fact that science education has been pushed to the back burner due to the emphasis on reading and mathematics. Undoubtedly, one of the most negative effects of NCLB on science education to date is the reduction of class time allotted for science in elementary schools and in some middle schools (Shaver, Cuevas, Lee, & Avalos, 2007; Stecher et al., 2008). (See also DeLucchi & Malone, Chapter 12, this volume.) The Center on Education Policy (2006, as cited in Griffith & Scharmann, 2008) found that 71% of school districts surveyed reported reducing elementary school instructional time for science (Griffith & Scharmann, 2008; Vasquez, Teferi, & Schicht, 2003). Thus, the subjects that are tested annually and determine whether a school is making annual yearly progress, that is, reading and mathematics, are the subjects that have had the greatest concentration of effort (McMurrer, 2007). The neglect of science education could affect student science learning for years to come. It may have a particularly negative impact on low-income students who, compared to their more affluent peers, probably would benefit more from science taught in elementary school because they are less likely to be exposed to science from other venues (Lynch, 2000). Low income children also are more likely to be found in schools that struggle to meet NCLB's adequate yearly progress standard. As a consequence, they may receive even more instruction in reading and mathematics to compensate for low performance, but with the consequence of even less academic time for science. This may have far-reaching equity consequences. If science is important preparation for both college and careers, and if a strong foundation at the elementary school level helps students be successful later, then NCLB may have inadvertently thwarted progress in U.S. STEM education.

The science education community worries about NCLB, and their concerns can roughly be divided into three camps with much overlap among them. The first camp is pragmatic. It sees that children are having fewer chances to learn science and that science is increasingly left out of the elementary and middle school curricula. In order for science to "count" in the era of accountability, it must be assessed. Thus, this camp is cautiously willing to step up to the challenge of wrestling science education into the accountability framework, knowing full well that if the science assessments are weak, then science education can be set back. For example, if high stakes science assessments test only factual information, this could lead to even more instruction that focuses on the memorization of science facts. But, given that the teaching of science has become an increasingly low priority under NCLB, the pragmatic camp wants to get the policy map set in the right direction (see Banilower, Heck, & Weiss, 2007; Marx & Harris, 2006; and Penfield & Lee, 2010). This camp is cautious and aware, yet open to using federal education legislation as a positive force for change in science education. Science education may be able to learn from the mathematics education community, which has, over time, managed to focus aspects of accountability on deep changes in the way that mathematics is taught and learned (Ferrini-Mundy, personal communication, June 1, 2009).

A second camp of concerned science educators and others see NCLB as an equity lever that lifts students who have traditionally been bypassed in schools onto the stage of concern. Under NCLB it is no longer possible for schools and school districts to ignore the performance of students of color, ELLs, or students with disabilities in science because science must be assessed at least three times during K–12 education, and more often in some states. With the advent of public accountability in science, student performance cannot be swept under the rug as it often has been in the past. The Council for Exceptional Children; the National Council of La Raza, an advocacy group focused on reducing poverty and discrimination and increasing opportunities for Hispanics; and the Education Trust, which works to close the achievement gaps that separate low income students and students of color from other students, all backed NCLB, but generally with concerns about inadequate funding and the nature of the assessments. The National Association for the Advancement of Colored People (NAACP) did not support NCLB as it was originally written, but the Connecticut state chapter of NAACP (Inquiring News, 2008) has used various provisions of the law to mount legal challenges to reduce education inequities within that state. Moreover, NAACP, along with 150 other education, civil rights, religious, children's, disability, and civic organizations, has signed on to a statement of commitment to No Child Left Behind Act's objectives of strong academic achievement of all children and of closing the achievement gap. The statement acknowledges the federal government's critical role in the use of an accountability system to ensure that all children, including children of color, from low-income families, with disabilities, and who are developing English proficiency, are prepared to be successful, participating members of our democracy (Fairtest, 2009). Although these groups are interested in all areas of the curriculum, not just science, others believe that accountability policies can increase the chances of academic success for marginalized students in science (Tate, 2001). For example, Vasquez, Teferi, and Schicht (2003) are persuaded by evaluation research from the St. Louis Public Schools, a school system where about 80% of the students are African American and are eligible for federal meal assistance. The state of Missouri has administered science assessments since 1998, before they were mandated by NCLB. The St. Louis science assessment data show students making steady gains in science from 1998 to 2002, closing Black-White achievement gaps, as well as approaching the state average for science in third grade. Vasquez et al. find the equity rationale for NCLB compelling.

A third camp of science education researchers worries that the accountability system, as it is currently constructed under NCLB and interpreted by states, will actually weaken science education (cf. Wood, Lawrenz, Huffman, & Schultz, 2006). For instance, Southerland, Smith, Sowell, and Kittleson (2007) invoke the frequently made distinction between first and second order changes in education to explain their position.

> First-order change requires small alterations of or additions to existing practices (e.g., changes in texts, number of students in a classroom, length of day, equipment), basically any attempt to increase the effectiveness and efficiency of current schooling practices. In contrast, second-order change is meant to alter the fundamental patterns of schooling; these changes are much more radical and transformative because they challenge the structures and rules that constitute traditional schooling practices. Second-order changes "challenge the cultural traditions of schools" (Romberg & Price, 1983, p. 159) and require fundamental changes in both teacher thinking and classroom practice. Thus, they are inherently more difficult to implement and sustain. (p. 46)

These authors see NCLB, as it is interpreted by the states, to simply elicit first-order changes because it does not articulate clear goals for learners or explicitly describe any particular form of instruction or assessment. They believe NCLB is not capable of instigating second-order changes. However, they do recognize that to ignore the scope of what is occurring at the state and national levels is not likely to be productive, and they urge engagement with some of the accountability issues that surrounded NCLB's implementation in science education in 2007.

Science Education and NCLB: Preliminary Research Findings

At the time of this writing, it is difficult to ascertain the effects of NCLB on science achievement or on the science achievement gap because most states only recently began mandatory, standards-based science assessment in 2007. The effects of NCLB on reading and mathematics scores, however, show achievement gains since the passage of NCLB in 2001, as measured by state level tests and by NAEP (Kober, Chudowsky, & Chudowsky, 2008). Moreover, achievement gaps have somewhat narrowed, especially for African American and low income students. This effect, however, is more dramatically seen on state tests than on the NAEP scores, a national measure that has been in place for decades. Reliable data for students with disabilities and for ELLs are lacking because state testing policy for these groups has fluctuated a great deal over the years, making test score trends difficult to interpret. Similarly, the population of Hispanic students has grown sufficiently within states to cause similar problems with data interpretation (Kober, Chudowsky, & Chudowsky, 2008). But, overall, the results are encouraging and the accountability model may bode well for science as well. Perhaps NCLB is a policy lever that can have a positive impact on student learning in science and can be an effective way of equalizing the outcomes of science learning by a more fair and equitable use of inputs.

The Rand Corporation (Stecher et al., 2008) conducted an NSF-sponsored study of the effects of NCLB on mathematics and science in three states (California, Georgia, and Pennsylvania) over 3 years, 2003 through 2006. (At the time of the study, science assessments were not used to calculate AYP in any of these states, but were nonetheless mandated, administered, and the results reported.) The study's goal was to examine the strategies used by states, districts, and schools to implement standards-based accountability under NCLB and how these strategies are associated with classroom practices and student achievement in mathematics and science in elementary and middle schools. School superintendents, principals, and teachers were asked to respond to their perceptions of the implementation of NCLB in their schools. Although there were differences between how each state responded to NCLB, there were remarkable similarities as well.

This study, one of the most comprehensive studies of NCLB focusing on science education, found that all three states had constructed the necessary infrastructure to support standards-based accountability, standards, curriculum frameworks, assessments, and support and technical assistance for schools. Most educators understood the standards-based reform's theory of action: set clear goals, develop measures, and establish consequences to encourage educators to achieve the goals. The alignment among the levels of the accountability structure was a major focus (standards, assessment, and curriculum all should be closely aligned). However, across states, there were important differences in the content of the academic standards, the difficulty of the performance standards set by each state, and the kinds of technical assistance available to schools. Teachers who enacted these initiatives were somewhat more positive about them over the 3-year period of the study but expressed reservations about how the reforms might narrow the curriculum, the ability of some students to pass mandated tests, and teacher morale.

Summarizing across all three states, this study found that, because science is tested less frequently than mathematics and reading and does not count toward AYP, there has been less effort to implement standards-based accountability in science. This is reflected by the fact that fewer schools report aligning their science programs with state standards, allocating more time to science, or providing more professional development in science. The study concludes that subjects that do not "count" as much do not command resources or attention. The experiences of these three states are likely not much different from the other 47 states, and the implications seem clear for science education. Either science education must come to "count" more in accountability systems, or science is likely to be left behind.

The accountability movement is not likely to lose impetus as a policy lever during the Obama administration. Indeed, statements by President Obama and Secretary of Education Arne Duncan show that although NCLB may be renamed and re-organized, the call for standards-based accountability is likely to increase, rather than diminish. In his speech to the National Academies in the spring of 2009, President Obama said:

> That is why I am announcing today that states making strong commitments and progress in math and science education will be eligible to compete later this fall for additional funds under the Secretary of Education's $5 billion Race to the Top program ... I am challenging states to dramatically improve achievement in math and science by raising standards, modernizing science labs, upgrading curriculum, and forging partnerships to improve the use of science and technology in our classrooms. And I am challenging states to enhance teacher preparation and training, and to attract new and qualified math and science teachers to better engage students and reinvigorate these subjects in our schools. (The White House, Office of the Press Secretary, 2009)

Post NCLB: Mobilization of Varied Interests to Improve STEM Education

As a result of this dissatisfaction with the inability of the U.S. education system to produce human capital for STEM-related careers and frustrated by the inability of the current system to remedy inequities in student performance, a new aggregate of social forces is becoming increasingly influential. Governmental, business, and other institutions associated with economic issues within states and nationally are increasingly vocal. Kaestle (2007, in McDonald, 2009) found 10 categories of education policy organizations ranging from the traditional education establishment to think-tanks, foundations and for-profit firms. They advocate for conditions of schooling that would allow more children to be successful in science. They have been powerful forces for change, and a new politics has arisen that does not follow traditional partisan or ideological lines (McDonald, 2009). Although the motivation of these new groups may be as much about the economy as about equity, they influence STEM education policy in ways that may serve subgroups of students too readily ignored by social forces in the past. These organizations and coalitions call for improved standards-based reform in K–12 science.

The National Science Teachers Association is leading an effort called *Science Anchors* (NSTA, 2009) that intends to bring greater focus, clarity, and coherence to science education. *Science Anchors* would draw from current K–12 science education standards, but would be more streamlined and focused. Science content would be organized around a smaller number of big ideas rooted in the major fields of science that develop over the K–12 span. The purpose of *Science Anchors* is to bring more consistency to science standards across states (NSTA). This effort has been guided by an Interagency Steering Committee composed of representatives from NSTA, the National Research Council (NRC), Project 2061 of the American Association for the Advancement of Science (AAAS), and Achieve, Inc.

Another influential group that includes the National Governors Association, the Council of Chief State School Officers (CCSSO), and Achieve, Inc. has issued a report called *Benchmarking for Success: Ensuring U.S. Students Receive a World-Class Education* (NGA, CCSSO, Achieve, 2008). This report is clearly focused on the need to establish international mathematics and science benchmarks that would allow states to make international comparisons. Such benchmarks would allow states to focus and stimulate state-level education systems to produce students with improved STEM achievement. The rationale is unabashedly economic. However, it draws these connections to equity:

> State leaders also should tackle "the equity imperative" by creating strategies for closing the achievement gap between students from different racial and socioeconomic backgrounds.... Reducing inequality in education is not only socially just, it's essential for ensuring that the United States retain a competitive edge. Research shows that education systems in the United States tend to give disadvantaged and low-achieving students a watered down curriculum and place them in larger classes taught by less qualified teachers—exactly opposite of the educational practices of high performing countries. (p. 6)

Perhaps the most ambitious effort to place science education at the center of U.S. education policy has been mounted by the Carnegie Corporation of New York and the Institute for Advanced Study (2009). This report, *The Opportunity Equation: Transforming Mathematics and Science Education for Citizenship and the Global Economy*, has had endorsements from major policy, business, and advocacy organizations including the National Governors Association, the National Science Teachers Association, the National Education Association, La Raza, and the National Association for Research on Science Teaching. Secretary of Education Arne Duncan endorsed this report's recommendations on behalf of President Obama and the U.S. Department of Education. The report challenges the nation to mobilize for coordinated action to:

- Establish common standards for the nation in mathematics and science—standards that are fewer, clearer, and higher—along with high-quality assessments
- Improve math and science teaching—and our methods for recruiting and preparing teachers and for managing the nation's teaching talent
- Redesign schools and systems to deliver excellent, equitable math and science learning

The kind of reform imagined in this report goes well beyond prior reports, and it calls for sweeping changes in the way that science is taught in U.S. schools. It does not stop there, but calls for major changes in the way U.S. schools are organized and run, so that science and mathematics can have a central role in a new literacy that prepares all students for life in the twenty-first century. The report addresses equity squarely, and sees high quality science education as a means for students to invest in themselves and in their futures. This requires doing away with dreary science and mathematics classrooms, finding teachers who can bring science into the lives of students, not relegating students to poor schools or low tracks within schools that produce diminished opportunities to learn, and taking bold new steps to create models of STEM schools that provide students with a student-centered, personal, and challenging education and provides them the means to succeed in work and in post-high school studies.

Given the wide support backing the Opportunity Equation report, new leadership in Washington that is committed to school reform in ways that go beyond improved test scores, and an abundance of federal stimulus money that is being distributed to school systems, this reform may have a chance of succeeding in changing the course of U.S. science education.

IMPLICATIONS FOR THE FUTURE OF EQUITY AND SCIENCE EDUCATION POLICY

This chapter illustrates the efforts that have been made to achieve equitable science education over the last 70 years, as well as documenting how far the nation has yet to go to achieve equity. Because of the links that are being increasingly made between high quality science education and the economy, however, equity in science education is increasingly a national, state, and local priority. In President Obama's speech to the Hispanic Chamber of Commerce he argued that economic progress and educational achievement have always gone hand-in-hand in America: "It's the most American of ideas, that with the right education, a child of any race, any faith, any station, can overcome whatever barriers that stand in their way and fulfill their god-given potential" (in McDonald, 2009, p. 420). The standards-based accountability movement and NCLB, for all their flaws, have fomented a sea-change in how students are viewed in schools. No longer is it possible to relegate students of color, ELLs, or students with disabilities to the poorest teachers and provide them with the least resources. These students exist and they now "count." For every school system that has merely gamed the accountability system, there are others that have set new goals and higher sights. But, because of the emphasis on economic issues and the needs of business communities, it is also important for policymakers and educators to be wary of an onset of regressive educational measures such as a tracking system designed for the children of the economically well-off that relegates other peoples' children to academic experiences that shunts them away from the STEM pipeline.

Instead, *Taking Science to School* (Duschl, Schweingruber, & Shouse, 2007) suggests children can learn science ideas at earlier ages than previously thought possible, and some schools are experimenting with these findings by offering very young children opportunities to learn STEM material, while other children in the same school must wait to have this opportunity to learn. How are the early STEM learners selected? Will this trend play out once again in science education according to ethnicity or SES, raising equity issues, as it has in mathematics where increasingly early access to Algebra 1 has become an issue complicated by social class and developmental concerns?

An equally worrisome issue is the increase in school re-segregation due to local/municipal housing policies that concentrate poverty in some neighborhoods and lead to inequity of inputs and outputs in community schools. The science taught and learned in underperforming schools may be a symptom of this larger equity issue. Such housing/school policies are likely to be extremely resistant to change, so more research needs to be done on its effects, especially in documenting the results for communities that have made progress by intentionally integrating housing along lines of income.

Although a positive aspect for equality of inputs are recent efforts to improve high school graduation requirements by providing more and better science course offerings, there should be better accounting of how the drop-out rate (equality of outcomes) is affected by these new regulations. Creative solutions should be developed to keep children in school, while providing them with a more meaningful science education. Experimental schools focused on STEM education are an exciting new development that may provide new ideas about how to accomplish this.

If science education is not to forever steam slowly on the education backburner, then student achievement in science must be assessed and progress evaluated. Crucial to improving science education is a system that creates science assessments of the highest quality possible that are consistent with the research about learning science. The current accountability system launched through NCLB falls short of the original vision of education reform; standards should be the basis for instructional change that incorporates constructivist, student-centered views of learning. This would require deep changes in pedagogy and in school structures, as well some increase in financial capacity. However, McDonald (2009) explains how, ironically, standards-based accountability may have resulted in a more fragmented and low capacity system of education, unable to make the large changes needed in STEM education because of the tendency to centralize control at the federal and state levels without building capacity. It seems that the 50 state educational systems and the federal system will continue to replicate inefficient and low-level changes to each state system, without gaining the traction needed to make substantive improvements.

Finally, this chapter was written to stimulate discussion about equity among the science education research and education policymaking communities. There is remarkably little science education policy research being conducted, compared to research in reading and mathematics.

Moreover, when science education policies are made, they are often made without much input of the science education research community. Circularly, this is because there little of the type of science education research done that can directly inform day-to-day K–12 STEM policy decisions. There needs to be a conscious expansion of research collaborations to inform STEM education policy that can affect the lives of students in ways that are far-reaching and equitable.

ACKNOWLEDGEMENTS

I am grateful to six anonymous reviewers for their thoughtful and careful reviews of this chapter. Every comment carefully considered and many changes to the manuscript were made as a result of the reviews. Reviewers wanted more nuance, depth, analysis of ideology, or more tables and figures. I responded whenever feasible, but there is much to discuss and limited pages. One author with one chapter on equity and science education policy cannot possibly capture the myriad points of view, expertise, and sensibilities of the reviewers or of other researchers/scholars in the field. The only solution is for others to think about these issues deeply, and to discuss, conduct research, and write about them. There is much important work that needs to be done. In addition, I thank George DeBoer, the editor of this book, for his patience, editing skills, thoughtful comments, and for providing me the incentive to update my earlier work by exploring equity issues in science education policy since the passage of No Child Left Behind.

AUTHOR'S NOTE

This work was made possible through Independent Research and Development agreement of the National Science Foundation, where I have spent the 2008–09 and 2009–2010 academic years working as a program officer in the NSF Education and Human Resources Directorate in the Division for Research on Learning in Formal and Informal Settings. The opinions expressed here are my own, and not those of the National Science Foundation.

APPENDIX

Science Education Policy and Civil Rights Landmarks[1]

AND

Policies, Documents, and Events That Shaped K–12 Science Education[2]

1944: The GI Bill pays for GIs returning from WWII to attend college and increases college enrollment to accommodate their large numbers. This makes a college education available to veterans from varying socioeconomic status and ethnicities who might not have found college accessible, prior to the War.

1945: *Science—The Endless Frontier* is written by Vannevar Bush in response to a charge by President Roosevelt to apply lessons learned during the War in science and engineering to peacetime. This provides a rationale for science fellowships, based on academic promise to reduce reliance on European science and for improvements in science education that fails to awaken interest in science or provide adequate instruction for men and *women* who want to attend college.

1947: *Science and Public Policy* is delivered to President Harry Truman. Its goal is to sort out relations between research in the federal government, industrial, and academic sectors, and to improve the condition of science teaching at all levels, including K–12, including the preparation of students to become scientists.

1950: The National Science Foundation is created to promote the progress of science and to advance the national health, prosperity, and welfare. Although the U.S. is in midst of the Korean war, NSF is designed to be separate from defense-related research, and to re-emphasize research and education in the sciences with both industry and academe with goals to fund basic research and the development of science talent.

1952: NSF provides fellowships to study science at the graduate level for the first time. Of the 535 fellows, 32 are women.

1954: In *Brown v. Board of Education,* a decision widely regarded as having sparked the modern civil rights era, the Supreme Court rules that deliberate public school segregation is illegal, effectively overturning "separate but equal" doctrine of *Plessy v. Ferguson.* Chief Justice Earl Warren, writing for a unanimous Court, notes that to segregate children by race "generates a feeling of inferiority as to their status in the community that may affect their hearts and minds in a way unlikely ever to be undone."

1957: The Soviet Union successfully launches the Sputnik satellite. The United States government sees this as a threat, and challenges the U.S education system to produce better students in science and mathematics. The NSF supports development of reformed curriculum materials in science and mathematics with the goal to expand the pool of science talent, producing more scientists. Over 60% of school districts were estimated to use the new materials, but ten years later the number was halved.

1964*: The Civil Rights Act of 1964 (*Pub.L. 88-352, 78 Stat. 241, July 2, 1964) was a landmark piece of legislation in the United States that extended voting rights and outlawed racial segregation in schools, at the workplace and by facilities that served the general public ("public accommodations")

1962: The *United Farm Workers Union* , under the leadership of *Cesar Chavez*, organizes to win bargaining power for Mexican Americans.

1966: *National Organization for Women* (NOW) is founded to fight politically for full equality between the sexes.

1968: March 1, The National Advisory Commission on Civil Disorders, popularly known as the Kerner Commission after chairman Otto Kerner, Governor of Illinois, issues its report warning that the nation is moving toward two separate societies—one Black and poor, the other affluent and White. The commission, appointed by President Johnson following the 1967 disorders in Detroit and other communities, calls for major anti-poverty efforts and strengthened civil rights enforcement to eliminate the causes of the disorders.

1968: The Supreme Court, in *Green v. County School Board of New Kent County* (Virginia), rules that "actual desegregation" of schools in the South is required, effectively ruling out so-called school "freedom of choice" plans and requiring affirmative action to achieve integrated schools.

1971: The Supreme Court, in *Swann v. Charlotte-Mecklenburg Board of Education*, upholds busing as a legitimate and sometimes necessary tool to achieve desegregation and integration. But the Court does not rule on segregation in public schools in northern states where it is not imposed by statute.

1972: Title IX of the Education Amendments of 1972, now known as the *Patsy T. Mink Equal Opportunity in Education Act* in honor of its principal author, but more commonly known simply as Title IX, is a United States law enacted on June 23, 1972 that states: "No person in the United States shall on the basis of sex, be denied the benefits of, or be subjected to discrimination under any education program or activity receiving Federal financial assistance."

1973: Congress passes Section 504 of the *Vocation Rehabilitation Act* barring discrimination against disabled people with the use of federal funds.

1974: The Supreme Court rules that public schools must teach English to foreign language-speaking students (*Lau v. Nichols*). The case involves the San Francisco school system, which does not provide any instruction in English to some 1,800 Chinese-speaking pupils. The court holds that, under the Civil Rights Act of 1964, districts receiving federal funds must provide either a bilingual or English as a second language program whenever students of a non-English speaking minority are enrolled in significant numbers.

1977: MILLIKEN II (a court ruling) set the pattern for a number of court-ordered and voluntary plans that followed in Missouri, Ohio, Indiana and Arkansas, among other places. The concept of requiring states to provide resources for improving education as a remedy to segregation was further expanded in the Supreme Court's 1990 decision, Missouri v. Jenkins, which permitted courts to order school authorities to increase spending on education remedies even when voters rejected referenda raising taxes.

1978: The Supreme Court, in the *Regents of the University of California v. Bakke* case, upholds the principle of affirmative action but rejects fixed racial quotas as unconstitutional. The case involves Alan Bakke, denied a slot at the University of California medical school at Davis. Bakke claims he is a victim of reverse discrimination because a minority student, with lower test scores, is admitted instead on affirmative action ground.

1980: The Science and Engineering Equal Opportunity Act (1980) declared that the policy of the United States is to encourage men and women and persons of various racial, ethnic and socioeconomic groups to learn science and engineering and to find employment in these fields.

1982: Supreme Court rules in *Plyer v. Doe* that children of illegal immigrants have a right to free public schooling. Poverty reached its highest level—14%—since 1967. African American poverty rate is 34.2%; Latino rate is 26.2.

1983: A Nation at Risk was written and lamented a rising tide of educational mediocrity. This report recommended increasing graduation requirements, increasing instructional time, school year time, the development and implementation of rigorous national standards, higher standards for teachers, and the implementation of highs school exit exams in core subjects.

1983: Educating Americans for the Twenty-First Century in which NSF sets an ambitious goal to provide high standards of excellence for all student whatever their race, gender, or economic or immigration status, or whatever language is spoken at home.

1985: Jeanne Oakes publishes *Keeping Track*, a highly influential book that pointed out the dangers of tracking, and how it disadvantages poor and ethnically diverse students. Although this was not a "policy document," this and similar research on tracking raised the issue and resulted in court challenges of tracking.

1989: The National Council for Teachers of Mathematics publishes the first mathematics standards to be used by states to create state-level standards. Science for All Americans published by the American Association for the Advancement of Science which advocated for science literacy for all students and described the science content and process skills necessary to be scientifically literate.

1990: Congress passes -and President H. W. Bush signs the landmark *Americans With Disabilities Act*, banning job discrimination against people with disabilities and requiring buildings, businesses and public transportation to be accessible. Most provisions take effect in 1992–93.

1991: National Education Goals Report: Building a Nation of Learners was produced as the result of the National Governors Summit in 1989, backed by President H. W. Bush. Goal 4 was: By the year 2000, U.S. students will be first in the world in science and mathematics achievement (National Education Goals Panel, 1991). Interest in systemic reform took hold which suggested the reform of educational policy structures (such as assessment), curriculum materials, and instruction by teachers. There was an acknowledged need for higher standards for all students and improved student achievement through the purposeful alignment of policies within a state or school district.

1993: Benchmarks for Science Literacy, the first national attempt to define science literacy for all students published by AAAS.

1994: *Science in the National Interest* (Clinton & Gore, 1994) was published on the Internet and called for improvements in science education and raising the scientific literacy of all Americans. The document includes health, economic prosperity, environmental responsibility, quality of life, contributions to culture, urging improved communication between scientists and the people. A goal was to maintain U.S. science preeminence internationally by educating children who can compete economically in global economy.

1994: In *Adarand,* the Supreme Court ruled in a 5–4 vote for the first time that all federal laws creating racial classifications, regardless of an intention to burden or benefit minorities, when challenged, must be tested by the same stringent standard. Federal set aside and affirmative action programs benefiting minorities then are subject to strict scrutiny and must be narrowly tailored.

1996: The National Research Council published the *National Science Education Standards* in which equity and excellence are treated as equally important goals.

1999: Third International Mathematics and Science Studies show that U.S. students are not performing as well as expected in international comparisons, with scores in low or medium ranges at for various science and mathematics subjects tested at three grade levels.

2000: *Goals 2000: Educate America Act* provided some funding for large scale reform focusing on alignment and policy, with goals for all children start school ready to learn; a 90% graduation rate; measured competencies in science and mathematics in Grades 4, 8, and 12; for adult literacy in mathematics and science and for the United States to become first in the world in achievement in math and science.

c. 2000: Schools growing more diverse, with the doubling of the population of Hispanic children in public schools, and with increasing concentrations of African Americans and Hispanics in city schools. Schools are presented with more English Language Learners than ever before. Family incomes dropped in the 1990s and poverty increased with more black and Hispanic students likely to live in poverty.

2000: The National Commission on Mathematics and Science Teaching for the 21st Century publishes *Before It's Too Late,* often referred to as the Glenn Report. The report charges that the preparation that U.S. students receive in mathematics and science "unacceptable" (National Commission on Mathematics and Science Teaching for the 21st Century, 2000, p. 7) and that (1) American students must improve their performance in mathematics and science if they are to succeed in today's world and if the United States is to stay competitive in an integrated global economy and that (2) the most direct route to improving mathematics and science achievement for all students is better mathematics and science teaching.

2002: President George W. Bush signs the *No Child Left Behind Act* into law. This alters the way that federal government approaches educating elementary and secondary students in math and reading.

2003: Trends in International Mathematics and Science Studies (TIMSS) show that performance of U.S. students slightly improved in science and mathematics, but still in medium score range for most subjects and grade levels.

2005: *Rising Above the Gathering Storm,* the National Academies' report that the U.S. would lose its competitive edge if strong measures were not take to improve STEM education, among other things.

2007: *America Competes Act,* written by a bipartisan, bicameral group of lawmakers, authorizes $33.6 billion dollars over fiscal years 2008-2010 for science, technology, engineering and math (STEM) education programs across the federal government.

2007: The *No Child Left Behind Act* (NCLB) made content standards a required part of federal and state accountability systems and requires that in the 2007–08 school year, schools must administer annual tests in science achievement at least once in Grades 3–5, 6–9, and 10–12.

REFERENCES

Annie B. Casey Foundation. (2008). *2008 KIDS COUNT Data Book: State Profiles of Child Well-being*. Retrieved July 19, 2009, from http://www.aecf.org/KnowledgeCenter/Publications.aspx?pubguid={AD7773E9-4971-4A1C-B227-D04AB9D79065}http://www.aecf.org/

Apple, M. W. (1995). Taking power seriously: New directions in mathematics education and beyond. In W. G. Secada, E. Fennema, & L. B. Adajian (Eds.), *New directions for equity in mathematics education* (pp. 329–248). Cambridge, England: Cambridge University Press.

Banilower, E. R., Heck, D. J., & Weiss, I. R. (2007). Can professional development make the vision of the standards a reality? The impact of the national science foundation's local systemic change through teacher enhancement initiative. *Journal of Research of Science Teaching, 45*, 375–395.

Brown v. Board of Education, 347 U.S. 483 (1954) 347 U.S. 483B. Retrieved August 16, 2010, from http://caselaw.lp.findlaw.com/scripts/getcase.pl?court=us&vol=347&invol=483

Carnegie Corporation of New York and the Instititute for Advanced Study. (2009). *The opportunity equation: Transforming mathematics and science education for citizenship and the global economy.* Retrieved on July 19, 2009, from http://www.opportunityequation.org/

Chudowsky, N., & Gaylor, K. (2003, March). *Effects of high school exit exams on dropout rates*. Washington DC: Center for Education Policy.

Clinton, W. J., & Gore, A., Jr. (1994, August). *Science in the national interest.* Washington, DC: Executive Office of the President. Retrieved October 16, 2010, from http://eric.ed.gov/PDFS/ED373994.pdf

Darling-Hammond, L. (1998, spring). *Unequal opportunity: Race and education.* Washington DC: The Brookings Institute.

DeBoer, G. E. (1991). *History of ideas in science education: Implications for practice*. New York, NY: Teachers College Press.

DeBoer, G. E. (2006). History of the science standards movement in the United States. In D. W. Sunal & E. L. Wright (Eds.), *Research in science education: . The impact of state and national standards on K–12 science teaching* (Vol. 2, pp. 7–49). Greenwich, CT: Information Age Publishing.

Duschl, R. A., Schweingruber, H. A., & Shouse, A. W. (2007). *Taking science to school: Teaching and learning science in grades K–8.* Washington DC: National Academy Press.

Education Trust. (2009). General *information on No Child Left Behind*. Retrieved February 10, 2009, from http://www2.edtrust.org/EdTrust/ESEA/ESEA+General.htm.

Expanding the Circle Resources. (2010). *Brief history of American Indian education.* Retrieved January 10, 2010 from http://ici.umn.edu/etc/resources/briefhistory.htm.

Fairtest. (2009). *Joint Organizational Statement on No Child Left Behind (NCLB) Act*. Retrieved July 21, 2009, from http://www.fairtest.org/joint%20statement%20civil%20rights%20grps%2010-21-04.html

Goals 2000. (1998, April 30). *Goals 2000: Reforming Education to Improve Student Achievement.* Retrieved September 12, 2009, from http://www.ed.gov/pubs/G2KReforming/g2exec.html

Goals 2000. (1994, March, 31). *Goals 2000: Educate America Act* (P.L. 103-227). Retrieved October 16, 2010, from http://www.ncrel.org/sdrs/areas/issues/envrnmnt/stw/sw0goals.htm

Green, T. F. (1980). *Predicting the behavior of the educational system.* Syracuse, NY: Syracuse University Press.

Green, T.F. (1983). Excellence, equity and equality. In L. S. Shulman & G. Sykes (Eds.), *Handbook of teaching and policy* (pp. 318–341). New York, NY: Longman.

Griffith, G., & Scharmann, L. (2008). *Initial impacts of No Child Left Behind on elementary science education.* Retrieved on June 15, 2008, from http://findarticles.com/p/articles/mi_hb6515/is_3_20/ai_n29459089.

Herbold, H. (Winter, 1994–95). Never a level playing field: Blacks and the GI Bill. The *Journal of Blacks in Higher Education, 6,* 104–108.

Inquiring News. (2008, April 29). *CT conference NAACP wins NCLB law suit.* Retrieved July 21, 2009, from http://www.inqnews.com/Article.php?id=831

Kahle, J.B. (1996). *Thinking about equity in a different way.* Washington, DC: American Association for the Advancement of Science.

Kahlenberg, R. D. (2001). *Socioeconomic school integration.* Washington DC: The Century Foundation.

Kober, N., Chudowsky, N., & Chudowsky, V. (2008, June). *Has student achievement increased since 2002? State test score trends through 2006–07.* Washington, DC: Center on Education Policy.

Kober, N., Zabala, D., Chudowsky, N., Chudowdsky, V., Gayler, K., & McMurrer, J. (2006, August). *State high school exit exams: A challenging year.* Washington, DC: Center on Education Policy.

Lauer, P.A., Snow, D., Martin-Glenn, M., Van Buehler, R.J., Stoutmeyer, K., & Snow-Renner, R. (2005, August). *The Influence of standards on K–12 teaching and student learning: A research synthesis.* MCREL: Regional Educational Laboratory Contract ED-01-CO-0006. Retrieved June 17, 2009, from http://www.mcrel.org/pdf/synthesis/5052_RSInfluenceofStandards.pdf

Leadership Conference on Civil Rights. (2009). *Civil Rights chronology.* Retrieved on February 10, 2009, from http://www.civilrights.org/resources/civilrights101/chronology.html

Lee, O., & Luykx, A. (2006). *Science education and student diversity: Synthesis and research agenda.* New York, NY: Cambridge University Press.

Loveless, T. (1995). *Parents, professionals, and the politics of tracking policy.* Cambridge, MA: Harvard University, Kennedy School of Government. (ERIC Document Reproduction Service No. ED 390121

Lynch, S. J. (2000). *Equity and science education reform.* Mahwah, NJ: Erlbaum.

Lynch, S., Szesze, M., Pyke, C., & Kuipers, J. (2007). Scaling-up highly rated middle science curriculum units for diverse student populations: Features that affect collaborative research, and vice versa. In B. Schneider (Ed.) *Scale-up in Practice* (Vol. II). Lanham, MD: Rowman and Littlefield.

Lynch, S., Taymans, J. Watson, W., Ochsendorf, R., Pyke, C., & Szesze, M. (2007). Effectiveness of a highly-rated science curriculum unit for students with disabilities in general education classrooms. *Exceptional Children, 73*(2), 202–223.

Marx, R. W., & Harris, C. J. (2006). No Child Left Behind and Science Education: Opportunities, Challenges, and Risks. *The Elementary School Journal, 106*, 467–478.

McDonald, L.M. (2009). Repositioning politics in education's circle of knowledge. *Educational Researcher, 38*, 417–427.

McMurrer, J. (2007, July 24). *Choices, changes and challenges: Curriculum and instruction in the NCLB era*. Washington, DC: Center on Education Policy.

Miller, D. C., Malley, L. B., & Burns, S.D. (2009-039). *Comparative indicators of education in the United States and other G-8 countries: 2009*. Washington, DC: National Center for Educational Statistics, Institute of Education Sciences, U.S. Department of Education.

Minner, D. D., Levy, A. J., & Century, J. (2009). Inquiry-based science instruction— What is it and does it matter? Results from a research synthesis years 1984 to 2002. *Journal of Research in Science Teaching*. Retrieved February 27, 2010 from www3.interscience.wiley.com/journal/123205106/abstract

NGA, CCSSO, Achieve. (2008). *Benchmarking for success: Ensuring U.S. students receive a world-class education*. Washington DC: National Governors' Association. Retrieved June 13, from http://www.nga.org/Files/pdf/0812BENCHMARKING.PDF

NSF Task Force on Cyberlearning. (2008). *Fostering learning in the networked world: The Cyberlearning opportunity and challenge*. Arlington, VA: The National Science Foundation.

National Assessment of Educational Progress. (2006, May). *The Nation's Report Card* Retrieved on September 3, 2009, from http://nationsreportcard.gov/science_2005/s0115.asp?printver

National Commission on Excellence in Education. (1983). *A Nation at Risk: The Imperative for educational reform*. Washington, DC: Author.

National Commission on Science and Mathematics for the 21st Century. (2000, Jan. 23). Before it's too late. Retrieved on October 16, 2010, from http://www2.ed.gov/inits/Math/glenn/toolate-execsum.html

National Education Goals Panel. (1991). *National Education Goals Report: Building a Nation of Learners*. Washington, DC: Author

National Science Board. (2000). *Science and engineering indicators—2000*. Arlington, VA: National Science Foundation.

National Science Board. (2006). Science *and engineering indicators, 2008*. Two volumes. Arlington, VA: National Science Foundation (Volume 1, NSF 008-01).

National Science Board. (2008). *Science and engineering indicators, 2008*. Two volumes. Arlington, VA: National Science Foundation (Volume 1, NSF 008-01).

National Science Foundation. (1992). *Women and minorities in science and engineering*. Arlington, VA: Author.

National Science Foundation, Division of Science Resources Statistics. (2002). *Women, minorities and persons with disabilities in science and engineering: 2002*. Arlington, VA: Author.

National Science Teachers Association. (2009, July 9). *Science Anchors—A Vision for Clear, Coherent, and Manageable Science Standards*. Retrieved July 9, 2009, from http://scienceanchors.nsta.org/?lid=tnav

No Child Left Behind Act of 2001, Pub. L. No. 107-110, 115 Stat. 1425. (2002). Retrieved August 22, 2005, from http://www.ed.gov/legislation/ESEA02

Oakes, J. (1990a). *Lost talent: The Underparticipation of women, minorities, and disabled persons in science*. Santa Monica, CA: The Rand Corporation.

Oakes, J. (1990b). *Multiplying inequalities: The Effects of race, social class and tracking on opportunities to learn mathematics and science*. Santa Monica, CA: The Rand Corporation.

Orfield, G., & Lee, C. (2005). *Why segregation matters: Poverty and educational inequality*. Cambridge MA: Harvard University, The Civil Rights Project.

Partnership for 21st Century Skills. (2009). Retrieved on June 17, 2009, from http://www.21stcenturyskills.org/

Penfield, R. D., & Lee, O. (2010). Test-based accountability: Potential benefits and pitfalls of science assessment with student diversity. *Journal of Research in Science Teaching, 47*, 6–24.

Project Lead The Way. (2009) Retrieved September 3, 2009 from http://www.pltw.org/

Plessy v. Ferguson, 163 U.S. 537 (1896). Retrieved on August 16, 2010, from http://caselaw.lp.findlaw.com/scripts/getcase.pl?court=US&vol=163&invol=537

Rawls, J. (1971). *A theory of justice*. Cambridge, MA: Belknap Press.

Ryan, K. E., & Shepard, L. A. (2008). *The future of test-based educational accountability*. New York: Routledge.

Secada, W. G. (1994). Equity in restructured schools. *NCSMSE Research Review, 3*(3), 11–13.

Shaver, A., Cuevas, P., Lee, O., & Avalos, M. (2007). Teachers' perceptions of policy influences on science instruction with culturally and linguistically diverse elementary students. *Journal of Research in Science Teaching, 44*, 725–746.

Singer, S. R., Hilton, M. L., Schweingruber, H. A. (2005). *America's lab report: Investigations in high school science*. Washington DC: National Academies Press.

Southerland, S. A. , Smith, L. K., Sowell, S. P., & Kittleson, J. M. (2007). Resisting Unlearning: Understanding Science Education's Response to the United States' National Accountability Movement. *Review of Research in Education, 31*, 45–77.

Stecher, B. M., Epstein, S., Hamilton, L. S., Marsh. J., Robyn, A., McCombs, J. et al. (2008). *Pain and gain: Implementing No Child Left Behind in three states, 2004–06*. Santa Monica, CA: The Rand Corporation.

Tate, W. F. (2001). Science Education as a Civil Right: Urban Schools and opportunity to learn considerations. *Journal for Research in Science Teaching, 38*, 1015–1028.

Teach For America. (2009). Retrieved September 3, 2009, from http://www.teachforamerica.org/

U.S. Government Guide. (2002). *The Oxford Guide to the United States Government*. J. J. Patrick, R. M. Pious, & D. M. Ritchie (Eds.), Retrieved January 11, 2010 from http://www.answers.com/topic/de-facto-and-de-jure-segregation

Vasquez, J. A., Teferi, M., & Schicht, W. W. (2003). Science in the city: Consistently improved achievement in elementary school science results from careful planning and stakeholder inclusion. *Science Educator, 12*(1), 16–22.

White House Office of the Press Secretary. (2009). Speech of President Obama to the National Academy of Science, April 29, 2009. Retrieved July 18, 2009, from http://www8.nationalacademies.org/onpinews/newsitem.aspx?RecordID=20090427

Wood, N. B., Lawrenz, F., Huffman, D., & Schultz, M. (2006). Viewing the school environment through multiple lenses: In search of school-level variables tied to student achievement. *Journal of Research in Science Teaching, 43*, 237–254.

Zabala, D. & Minnici, A. (2008, February). *High school exit exams: Patterns in gaps and pass rates*. Washington, DC: Center on Education Policy.

Zabala, D., Minnici, A., McMurrer, J., & Briggs, L. (2008, August). *State high school exit exams: Moving toward end-of-course exams*. Washington, DC: Center on Education Policy.

Zinth, K. (2006, March). *A Synthesis of recommendations for improving U.S. science and mathematics education: Highlights science and mathematics*. Denver, CO: Education Committee for the States.

CHAPTER 12

THE EFFECT OF EDUCATIONAL POLICY ON CURRICULUM DEVELOPMENT

A Perspective From the Lawrence Hall of Science

Linda De Lucchi and Larry Malone

INTRODUCTION

Science curriculum developers are design engineers, weaving together ideas about the natural world, guided by research from the cognitive sciences, the experiences of teachers, and what is known about the complexities of school systems. The work is contoured by government and school policies and humanized by the realities of actual classroom practice. The goal of this enterprise is the production of tools that can be used by teachers and students to make teaching effective, and learning rich and engaging for all students.

The Role of Public Policy in K–12 Science Education, pp. 355–394

When design and development result in a viable commercial product, developers collaborate with publishers to bring the product to the public. Because developers tend to be deeply invested in the curriculum they create, their energies often extend beyond marketing to supporting actual implementation of the materials. Successful curriculum materials developers often create long-term curriculum development *projects,* which encompass the ongoing work of the developers as they design, disseminate, support, and continuously improve the materials. Since the early days of the Science Curriculum Improvement Study project (SCIS) in the 1960s, the Lawrence Hall of Science, housed at the University of California at Berkeley, has been home to a community of curriculum developers where the ongoing project has always been, and continues to be, the locus of activity and the creative engine for the development of new ideas in curriculum materials development (Fuller, 2002; Thier, 2001). This chapter is written from the perspective of that nearly 50-year history. As such, it does not claim to represent the thinking of the many other noncommercial materials developers who specialize in small niche markets, let alone the much vaster enterprise of commercial textbook publishers. It is a story about our experience at the Lawrence Hall of Science and how federal and state policies in science education have affected our work.

In this chapter we discuss the interplay between education policy at the national, state, and local levels and the development, dissemination, and implementation of elementary school science instructional materials. First, we describe how federal accountability policy in the No Child Left Behind Act (NCLB) has influenced state and local policies and reshaped the science education landscape. We share our experience with issues related to (1) state science standards and the burden they place on the design and development of research-based instructional materials intended for a national audience; (2) the disparate policies that influence the procedures that educators use to select instructional materials; and (3) the policies that affect the implementation of science instructional materials in the era of NCLB.

Although we acknowledge that there are many positive features of NCLB and that the legislation was well intended, our discussion emphasizes the ways that the legislation has had a negative impact on elementary science education. We are also critical of the ways in which state policies regarding textbook adoption have made it difficult for developers to introduce innovative materials into schools on a large-scale basis.

In the second part of the chapter, to give context to these issues, we present the Lawrence Hall of Science (LHS) story as a case history, describing how one institution, dedicated to quality science education for all students through curriculum development and support, has survived countless policy transitions through the years by being both agile and proactive. Embedded in the LHS history is the story of how one successful elementary science curriculum, Full Option Science System (FOSS), originated and how it has been repeatedly reinvented, partly in response to policy changes, and partly in anticipation of policy initiatives taking shape just over the horizon. In the final section, we discuss some of the trends that we now observe in policy and practice which curriculum materials developers should be aware of if they are going to continue to support the science education enterprise.

WHAT SCIENCE WILL WE TEACH OUR CHILDREN?

The Role of Content Standards

The science content taught to our children is determined mostly by state standards. Since the release of the Project 2061 *Benchmarks for Science Literacy* (American Association for the Advancement of Science [AAAS], 1993) and the *National Science Education Standards* (NSES) (National Research Council [NRC], 1996), all 50 states have developed science content standards. Many states used *NSES* and/or *Benchmarks* to guide the development of their science standards, although a number did not, wanting instead unique state standards documents. California went so far as to insist that no mention of *NSES* or *Benchmarks* appear in any of the instructional materials submitted for state consideration. As a result, state science standards vary widely in their structure, content, quality, and approach.

More recently, there has been a transition in the format of the state standards from the grade-banded standards proposed by *Benchmarks* and *NSES* (e.g., K–4, 5–8, 9–12) toward specific grade-level standards, driven at least in part by the influence of the NCLB legislation. The NCLB Act requires testing in reading and mathematics each year from grades three through eight, and even though the legislation requires that science testing be done only once during the elementary grades, once in middle school, and once in high school, many states have developed grade-level expectations and tests in science to match what they have done in reading and mathematics. As a result, the idea that science standards are statements of the knowledge and skills expected at the end of a grade band has been virtually abandoned by states. The intent of the national content standards was to provide a consensus of what students should know, the level of specificity and complexity at which they should know it, and the time in their academic careers by which they should be able to demonstrate that they know it. *Benchmarks* and *NSES* were intended to provide guidance for the development of curriculum, instruction, and assessment while at the same time offering the flexibility to achieve those learning goals in a variety of ways. At the state level, that broad and flexible approach to standard setting has been replaced by much greater specificity concerning what students should know and the particular grade where it should be taught. (See Osborne, Chapter 2, this volume, for a discussion of a shift in the opposite direction that has taken place recently in the United Kingdom.)

This increased specificity (which also varies from state to state) has made it difficult for curriculum developers to design and produce instructional materials to address the diverse science education needs of the country. In the next section, we discuss some specific ways that the standards movement, as it has been interpreted and implemented by the individual states, has impacted the design of instructional materials.

Content Burden

Perhaps the biggest problem with state science standards from the perspective of an elementary science curriculum developer is the excessively large number of content standards currently assigned by the states to each grade level. This creates unrealistic expectations for teaching and learning even in those exceptional elementary schools where science is taught every day. Too many standards at each grade level results in the production of instructional materials that promote superficial treatment of concepts and shallow coverage rather than deep conceptual understanding.

Results from the Third International Mathematics and Science Study (TIMSS) 1999 Video Study indicated that students from countries that score higher on international science tests at grade eight have more opportunities to develop strong scientific concepts, establish links between concepts, and build coherent content storylines than do their peers in American classrooms (Roth et al., 2006). In the TIMSS study, successful science students came from classrooms where there was a focus on one main learning goal (one big science idea) at a time, where the focus question was clearly stated and revisited, where the activities and content representations were matched to the learning goals, and where students' thinking about science ideas was activated before, during, and after the activity. Understandably, matching those kinds of learning outcomes is difficult to achieve when policymakers dictate that so much needs to be covered.

Two prominent science educators and curriculum developers, Myron Atkin and Paul Black, offered this warning on content burden.

> A key issue in designing the science curriculum, and a major threat to the quality of any orientation toward science, is the pressure for coverage. Whatever the mix of approaches to the subject, the biggest error would be to try

> to address too many discrete topics, either within a discipline or in examining how they relate to one another in addressing practical issues. Whether science is taught on a disciplinary or cross-disciplinary basis, the greatest threat to quality is superficiality. Unfortunately, the threat is real. Most textbooks and curricula put the emphasis on the number of topics that are included rather than the soundness of how each one is treated. Examinations often exacerbate the problem. Many teachers, through predilection or pressure, skim the surface of the subject. (Atkin & Black, 2003, p. 79)

Research and best-practices in science education suggest that instructional materials developers should design in-depth learning experiences that result in high quality student outcomes. But, in-depth learning experiences take additional classroom time, making it impossible to cover all the material specified in state standards in the way that it should be covered. The goals of depth and breadth constrain each other given the realities of limited instructional time. If you want to focus on depth, a certain amount of breadth has to be sacrificed; and if you want to focus on breadth, a certain amount of depth has to be sacrificed.

Fragmentation

A second problem is that, in an effort to specify grade-level expectations, state policymakers often just pull apart broader concepts and assign the parts to different grades, and then have students revisit them at several different grades in the name of a spiraling curriculum. A true spiraling curriculum would make use of what we know about conceptual learning progressions and would have students revisit big ideas at more complex and deeper conceptual levels periodically through the grades. One of the guiding principles of effective curriculum design is that instructional materials need to tell a coherent story. The *Atlas of Science Literacy Volumes 1 and 2* (AAAS, 2001, 2007) describes the vertical articulation of concepts through the grade bands and the horizontal connections between related content themes. Some states are using this resource to create the coherence that is called for, but others simply pull apart the science content and distribute it incoherently throughout the grades, without paying attention to what is needed to create conceptual flow. The result is that concepts are taught as disconnected fragments with no effort to achieve real conceptual depth at a particular grade level or across multiple grade levels. This represents a clear failure to understand what it means to revisit a content area to take students to a deeper and more sophisticated understanding of the topic as they move through the grades.

The challenge for a curriculum developer is to create a coherent instructional sequence that embraces a logical conceptual flow, is embedded in an engaging story line, and provides context for understanding the big ideas of science. This challenge cannot be met when educational policies lead to the parsing out of discrete and disconnected bits of information at different grade levels. To make the point about fragmentation, let us look at several examples from the standards document of one large and influential state to see what sometimes happens when policymakers attempt to organize standards in a grade-by-grade format. Here is an example from the Earth and Space Science standards for Grades 3, 4, and 5 in the standards document in that one state:

> Content Standard
>
> *The student knows there are recognizable patterns in the natural world and among Sun, Earth and Moon system.* (Texas Essential Knowledge and Skills, Science. [TEKS], 2009, p. 27)
>
> Grade 3: Students are expected to:
>
> Construct models that demonstrate the relationship of the Sun, Earth, and Moon system including orbit and position (TEKS, 2009, p. 25)
>
> Grade 4: Students are expected to:
>
> Collect and analyze data to identify sequences and predict patterns of change in shadows, tides, seasons, and the observable appearance of the Moon over time. (TEKS, 2009, p. 27)
>
> Grade 5: Students are expected to:
>
> Demonstrate that Earth rotates on its axis once approximately every 24 hours causing the day/night cycle and the apparent movement of the Sun across the sky. (TEKS, 2009, p. 30)

Clearly, there is a big idea in this content standard: the relationships in the Sun/Earth/Moon system include position, motion, appearance, and time; and the position and motion of the Sun, Earth, and Moon account for observable celestial phenomena, including day and night, seasons, and Moon phases. Classroom experience suggests that these variables and phenomena should be taught together in a carefully developed sequence of activities. In a curriculum that presents features of the Earth/Sun subsystem in a coherent way, for example, students might begin by observing the day/night cycle over a 24-hour period. They could record the position of the Sun in the sky during the daylight hours and analyze the data to identify a pattern. They could compare their findings to those of other students in other locations to see if day/night is the same in different parts of the Northern Hemisphere and in the Southern Hemisphere. Students then could construct, and work with, models to discover the relationship of the Earth and Sun and to confirm the kinds of motion that could produce day and night. Key questions addressed would include, "What is the reason that we see the Sun move across the sky during the day?" "Where is the Sun at night?" "What causes day and night?" After a thorough investigation of the Earth/Sun system, students would be prepared to incorporate a third element into their thinking, the Moon. But, instead, what are presented in these grade-level statements are disconnected fragments that are almost impossible for a curriculum developer to weave into a coherent story.

At whatever grade or grades this idea of the relationship between the Sun, Earth, and Moon is taught, and regardless of the specific activities students engage in, each time it is taught it must be taught with enough richness and complexity that it has meaning for the students at that time. Ideas cannot be broken into fragments and distributed over each of several years unless the learning experiences provide meaning each time the particular idea or set of ideas is taught. In addition, a judgment must be made regarding *how* to engage students in this inquiry, which might include outdoor observations, data collection and analysis, comparison with data from other observers, and model building to explain complex phenomena. It also is important to know *when* students will be able to engage in the critical inferential thinking required to integrate these celestial relationships into useful explanatory models.

These are not easy decisions to make, but based on experiences that instructional materials developers have had with these and similar ideas, they can provide insights into effective ways to organize instruction to create conceptual coherence, and that is critically important for state policymakers to be aware of and pay attention to. Unfortunately, in their enthusiasm for prescribing a rigorous set of standards for elementary school students, those who write state standards often set expectations that exceed the effective cognitive capabilities of the students they intend to serve. There is a tendency for state policymakers to push the content down into lower and lower grades in an effort to give the appearance of being rigorous or having "world class" standards, but that approach often does a disservice to students who are not ready for the cognitive demand presented by the subject matter.

A second problem with moving from standards developed for grade bands to grade-level expectations is that there often is unproductive repetition from grade to grade, with no increase in the sophistication of the presentation. This is an inefficient approach wasting valuable instructional time. In the example below, certain words are italicized to help make the point.

Content Standard:
The student knows that organisms undergo similar life processes and have structures that help them survive within their environments. (TEKS, Science, 2009, p. 28)

> Grade 3: Students are expected to:
>
> Explore how *structures and functions* of plants and animals allow them to survive in a particular environment;
>
> Explore that some characteristics of organisms are *inherited*, such as the number of limbs on an animal or flower color, and recognize some behaviors are *learned* in response to living in a certain environment such as animals using tools to get food. (TEKS, Science, 2009, p. 25)
>
> Grade 4: Students are expected to:
>
> Explore how *adaptations* enable organisms to survive in their environment, such as comparing birds' beaks and leaves on plants;
>
> Demonstrate that some likenesses between parents and offspring are *inherited*, passing from generation to generation, such as eye color in humans or shapes of leaves in plants. Other likenesses are *learned*, such as table manners or reading a book, and seals balancing balls on their noses. (TEKS, Science, 2009, p. 28)
>
> Grade 5: Students are expected to:
>
> Compare the *structures and functions* of different species that help them live and survive, such as hooves on prairie animals or webbed feet in aquatic animals;
>
> Differentiate between *inherited* traits of plants and animals, such as spines on a cactus or shape of beak, and *learned* behaviors, such as an animal learning tricks or a child riding a bicycle. (TEKS, Science, 2009, p. 31)

In this example, there is very little difference between the standards listed at these three grade levels. "Structure and function" appear in Grades 3 and 5 with the word "adaptation" introduced as a substitute in Grade 4. "Inherited" and "learned" are repeated in all three grades. This is an example of redundancy, not spiraling. Either of these core ideas (i.e., "structures and functions of organisms are related to their survival," and "some traits of plants and animals are learned and others are inherited") could be developed to the depth suggested by the statements in just one grade level. An appropriately spiraled curriculum strand might develop the ideas of structure and function in an elementary grade and then revisit these ideas at middle school in the context of adaptation, that is, the idea that a structure or behavior has survival value because of specific environmental pressures.

We have provided these examples to illustrate some of the ways that state science education policy, in the form of content standards, creates difficulties for instructional materials developers. But, it does not have to be that way. We have already mentioned there are content standards that are well written and that provide guidance for materials developers.

Benchmarks and *NSES* represent good science, include topics that are appropriately correlated with child development, present ideas that can build on each other throughout an academic career, and have the flexibility to accommodate local and regional specificity to the curriculum designed for children. If national standards like these were to guide standards development, curriculum designers would have the direction they need to design coherent instructional materials. A laudable set of science standards at the state level comes from the state of Washington, the Washington State K–12 Science Standards (2008). The Washington State standards formed the central structure for *Science Anchors*, a project spearheaded by the National Science Teachers Association (NSTA) in 2009, offering a three-dimensional structure or Abacus Model (see Figure 12.1), with science content topics displayed across the x-axis, grade bands displayed on the y-axis, and dimensions of the science enterprise (scientific inquiry, technology and design, nature of science, and habits of mind) on the z-axis (NSTA, 2009). This working draft model of science standards would give curriculum designers structure and flexibility simultaneously, providing opportunities to explore creative, progressive science instructional materials that address twenty-first century goals for STEM education. The Washington state standards suggest a way forward, with the caveats that there is still far too much content to be covered in the time available, and the cognitive expectations for students are sometimes developmentally inappropriate, both of which act in opposition to the overarching goals of the state's content standards.

Another effort to reshape state standards and encourage greater conceptual coherence was recently undertaken by the National Research Council (NRC). The NRC commissioned a committee of experts, chaired by Mark. R. Wilson of the University of California at Berkeley, to make recommendations for the design and implementation of large-scale, standardized science assessments. Although the report is focused on assessment, it also points to the importance of having a clear idea of what the target learning outcomes are and how learning develops over time to guide the development of sound curriculum and instruction as well as assessment. In their report, *Systems for State Science Assessment* (Wilson & Bertenthal, 2005), the authors call for coherence between what is taught and what is tested. The report acknowledges that content standards serve as the basis for developing and selecting instructional materials, so those standards should be reasonable in scope and built around a conceptual framework that reflects sound models of student learning.

> State standards should be organized and elaborated in ways that clearly specify what students need to know and be able to do and how their knowledge and skills will develop over time with instruction. Learning progres-

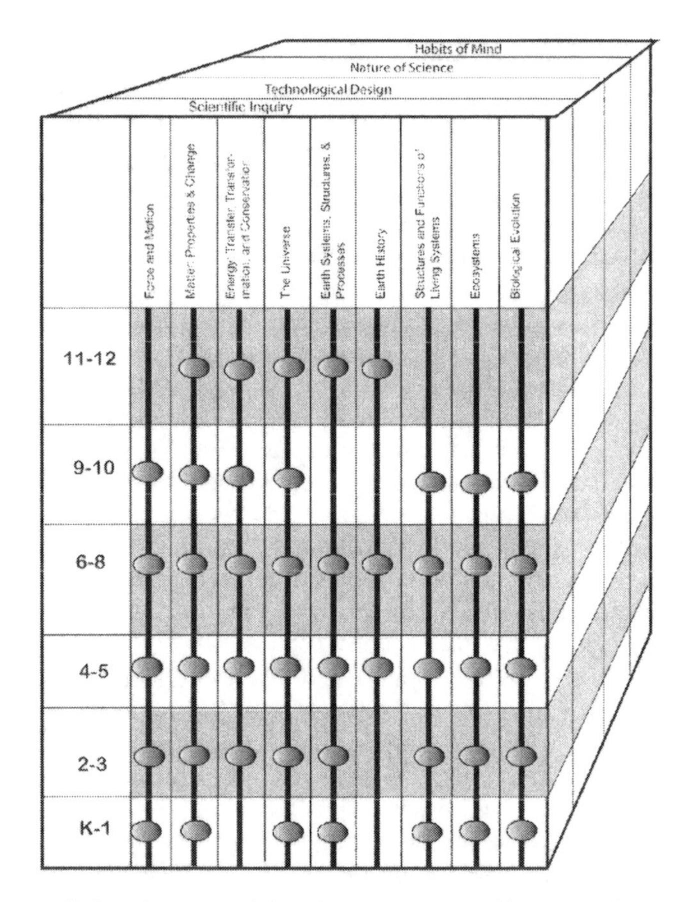

Figure 12.1. "The Abacus Model, as it may appear online, as an interactive graphic. Each bead represents a science unit for a grade band. Each string of beads represents a Big Idea. Science units are to be taught just once during the specified grade band, so they can be presented in-depth. Crosscutting concepts and abilities, represented by the third dimension, are to be taught explicitly, to illustrate commonalities among the sciences" (working draft for feedback, not a final document, used with permission, NSTA Science Anchors, 2009).

sions and learning performances are two strategies that states can use in organizing and elaborating their standards to guide curriculum, instruction, and assessment. Learning progressions are descriptions of successively more sophisticated ways of thinking about an idea that follow one another as students learn: they lay out in words and examples what it means to move toward more expert understanding. Learning progressions should be developed around the organizing principles of science.... A potentially positive outcome of reorganization in state standards from discrete topics to big ideas is a shift from breadth of coverage to depth of coverage around a relatively small set of foundational principles and concepts. Those principles and concepts should be the targets of instruction so that they can be progressively refined, elaborated, and extended over time. Creating learning performances is a strategy for elaborating on content standards by specifying what students should be able to do if they have achieved a standard. (Wilson & Bertenthal, 2005, p. ES2)

The conflict between a policy that leads to coverage of highly specified grade-level state standards, on the one hand, and the body of research supporting systematic conceptual engagement on the other hand, poses a dilemma for curriculum developers. As much as developers would like to build instructional materials based on the best knowledge they have of student learning, they also have to pay attention to what state standards say must be covered. Otherwise, their materials will be ignored. Commercial developers, in particular, are driven by the need to make a profit. Smaller niche developers, often funded by federal dollars, can be more innovative and exploratory, but there too, in the end, if schools refuse to buy the product, the materials are of no use to students. At its extreme, this can be cast as a business model versus a research model of materials development. The business model is driven primarily by market forces. To maximize the usefulness of the materials to the largest audience, the developer designs a program that covers all of the grade-level standards and meets all of the required technical regulations of a wide range of states. The goal of the development process is to produce a competitive product, one that wins as many adoptions as possible. This approach can be illustrated by the reaction taken by many commercial publishers immediately following the introduction of science content standards in the mid- 1990s. According to an SRI report at that time:

Publishers admit that texts continue to be laden with superfluous details because "Even though people ask for 'less is more,' when they go to make their decision, they want everything.... From a business point of view, we can't make the decision to cut content. Every state looks at content differently ... to cut content would be financial suicide." (SRI International, 1996, p. 18)

In contrast to the business model, where market forces loom large, the research model of curriculum development is guided primarily by research from the fields of science, science education, cognitive science, assessment, and professional development, and by observation and testing in typical classrooms; and it is influenced by the culture, values, and policy structures of the institution that supports the developers. The goal of the research model of development is to design instructional materials that provide the most efficient instructional tools for teachers and the most effective learning experiences for all students, even though the development process may take several years and may involve numerous revisions along the way. But, as already noted, even the research-based, niche developers cannot totally ignore costs, market forces, and whether schools will want to purchase the product.

What Determines Which Curriculum Gets Adopted?

To this point we have discussed issues that affect instructional materials design and development, largely in relation to how state science content standards are written and organized. In this section we discuss how developers get their curriculum materials into the hands of teachers and students. The approach they take depends to a large extent on the adoption policies on the consumer side of the equation. Some districts focus on standards compliance and have explicit policies to ensure alignment with state standards. Other districts have policies that are more related to instructional philosophy and have a more relaxed interpretation of standards compliance. We go into considerable depth on this topic so that the reader can appreciate how important the adoption process is and, especially, how important it is that teachers and administrators have sound principles to guide their instructional materials selections.

Science instructional materials are renewed periodically in pubic schools across the country. The renewal process starts at the state level. Some states simply announce that money will be available for science materials adoption, and schools use the state funds to make their purchases. Other states leave the process up to the local educational agencies to decide how to use their budget, maybe science this year, maybe next year. Twenty-one states (see Figure 12.2) conduct a formal adoption process, with additional states developing lists of "approved" instructional materials, a step just shy of a large-scale state adoption process. The 21 adoption states mount a formal, centrally coordinated process, during which instructional materials are vetted and approved. The state adoption process occurs on average every seven years, and it is driven by the state standards and/or science framework (Zinth, 2005). Money is then made available to schools to purchase materials from the state approved list. We begin with a discussion of the formal state adoption process and then discuss the variety of ways that materials are selected at the district and school levels.

State Instructional Materials Approval Process

According to Finn and Ravitch (2004), the textbook adoption process had its origin in the aftermath of the Civil War,

> when most publishers had their headquarters in the North. Embittered ex-Confederates distrusted Yankee publishers and wanted Dixie schoolchildren to have their own textbooks—so southern states established textbook adoption processes to make sure anti-Confederate books stayed out of their schools. (Finn & Ravitch, 2004, p. 6)

California is one of the 21 adoption states. The California State Board of Education (CSBE) approves the academic content standards as well as the various curriculum frameworks, which serve to interpret how the implementation of standards might look for science, English-language arts, mathematics, and history-social science classrooms to meet the requirements of No Child Left Behind and the state education code. The Curriculum Commission serves as an advisory body to the State Board of Education and is responsible for overseeing the development of the standards, frameworks, and instructional materials criteria, all of which drive the adoption processes (see Figure 12.3). The adoption process involves three concurrent steps: (1) legal compliance, that is, review of the social content in the materials (e.g., discriminatory material, appropriate age,

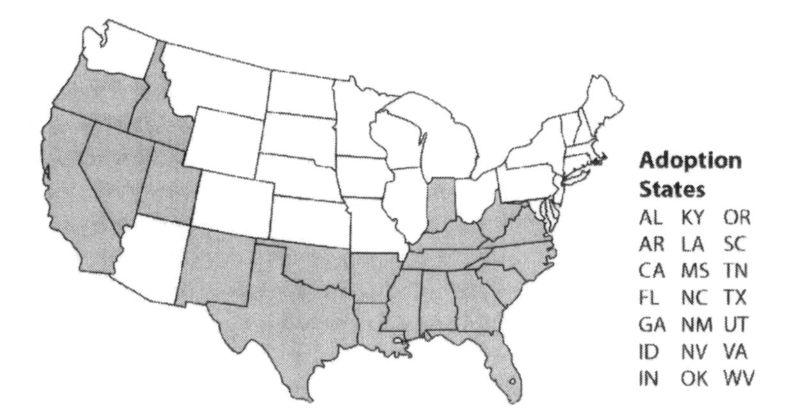

Figure 12.2. Twenty-one states have a formal adoption process.

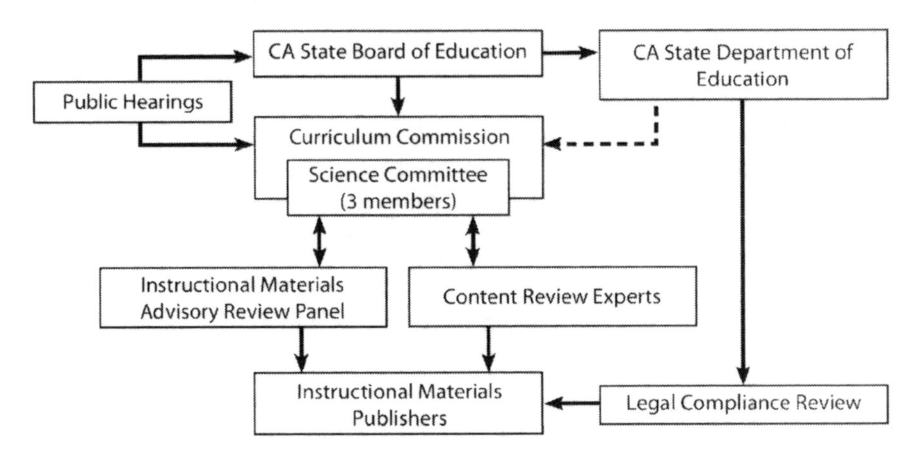

Figure 12.3. Example of a state-wide adoption process.

minority, and gender representation, safety, brand recognition, environmental awareness, etc.), (2) public review and comment, and (3) educational content review by members of the Instructional Materials Advisory Review Panel and Content Review Experts (California Department of Education, February 2006). At this point a final list of approved materials is developed.

Textbook adoptions or instructional materials adoptions? Before describing the variety of approaches used in making adoption decisions at the state level, it is important to raise a subtle policy issue that affects the ability of instructional materials developers to get their materials adopted in the various states. When you review the education code for each state, you see the word "textbooks" more often than "instructional materials." Intended or not, the choice of terminology often is making a policy statement. Traditionally, a textbook has been thought of as a presentation of the facts and principles of a subject, and a student is expected to learn by reading about those facts and principles and studying representational exhibits of them. The textbooks may include an occasional activity that the teacher or student can use to demonstrate what the student has read, but the driver for instruction is the written text. "Instructional materials," on the other hand, tend to be more broadly defined. They may include written text for the student to read, but reading is just one part of a student's engagement with the science ideas and concepts, not the main focus. Student engagement may be driven by experiences with objects, organisms, and systems through observations, investigations, and experiments; opportunities to gather, organize, and analyze data; and collaboration with peers to generate knowledge and solve problems. These experiences are organized carefully and sequenced by the developer so that efficient teaching and effective learning can occur. The learning experiences involve more than just a "presentation" of ideas. Well-crafted instructional materials provide support to teachers so they can find out and take into consideration the prior knowledge of their students and the systematic, progressive development of student ideas from novice to advanced levels.

The word "textbook" in a state's education code may be an artifact from an earlier time when textbooks defined curriculum, but regardless of why the term is still used, its' use can (and does) raise issues about compliance with a state's laws governing educational practice. The terminology sometimes presents a legal stumbling block for developers when their materials do not conform to the idea that a textbook is a hardcover book that describes the science facts and principles the students should know. There is no consistent use of terminology across states, and materials developers must adjust to the requirement and definitions of each individual state.

One of the policy battles that science instructional materials developers have engaged in over the past decade is to change that language in state education regulations so that state adoptions will be more inclusive, allowing educators to consider a broader selection of teaching resources (interactive web-based materials and project-based materials, for example). This implies more than just a change in the physical form that approved materials will be allowed to take. It also suggests a redefinition of what it means to provide each child with the opportunity to learn. This is a case where state policy is situated in the meaning and use of a particular word or phrase. How that word is interpreted can have a tremendous impact on the kinds of learning materials students will have available to them.

The process and politics of state adoption. The first step for publishers in the state adoption process is to send a letter of intent to submit materials for review, a process that takes place about 5 months later. The submission policy is different in each state, but it always involves delivering multiple sets of the materials being offered for consideration and systematic review by a panel. This is the stage in the adoption process during which policy and politics can become conflated. In the best of environments, the process is open and transparent, and leaders in the science education community (teachers, administrators, researchers, curriculum developers, scientists, and parents) are able to participate in thoughtful and productive ways. But, there is also the possibility of the process being manipulated or subverted by special interest groups (e.g., intelligent design advocates) or individuals with power and strongly held ideas (ideologues) on the state board or review committee.

One example of how a few individuals can influence the state adoption process and determine what science materials get adopted occurred in California in late 2003 in preparation for the 2006 science adoption. Science standards for K–12, which contain a concise description of what to teach at specific grade levels, were adopted by the State Board of Education in 2000, and the state science framework, which outlines the *implementation* of those standards by providing the scientific background and the classroom context, was completed in 2003. But the Criteria for Evaluation of Instructional Materials document for the 2006 adoption had not yet been finalized as of December 2003. This Criteria document, critical to the adoption process, is written by members of the science subject matter committee of the Curriculum Commission (three members in this case). The Criteria document then goes out for public review and comment, is revised, and then is presented to the State Board for approval in the spring. Once approved, the Criteria document guides the entire science adoption process over the next 30 months.

The Criteria document, written by the science committee, was posted on the California Department of Education website for public review in December, 2003, several days before winter break. The deadline for comments back to the committee was in early January, 2004. The timing of the release and the deadline for feedback so soon afterward seemed designed to ensure that the Criteria would not be challenged. Several science education leadership groups, however, were following the process closely and worked quickly to mobilize different sectors of the research, education, business, and political community to formulate a consistent and strong message dealing with fundamental problems with the Criteria. The California Science Teachers Association (CSTA) was instrumental in facilitating this ad hoc working group—The Science Education Coalition—and was able to meet directly with the executive director of the State Board of Education to discuss the major issues with the Criteria and to propose alternative language. A small group of Coalition members carried the message, and over the next few months, the Criteria document was changed. The most significant change to the Criteria was in the language that limited hands-on activities to "no more than 20 to 25 percent of total instructional time in K–8 science instruction." The "no more than 20 to 25 percent" clause was changed to "at least 20 to 25 percent." The new version of the pertinent paragraph in the Criteria now reads:

> A table of evidence in the teacher edition demonstrating that the California Science Standards can be comprehensively taught from the submitted materials, which includes hands-on activities composing at least 20 to 25 percent

of the science instructional program. Hands-on activities must be cohesive, connected, and build on each other to lead students to a comprehensive understanding of the California Science Standards. (California Department of Education, 2004, p. 4)

This was a dramatic policy change that took place during a high-stakes adoption process, and it had enormous consequences for materials developers. It also had the potential to have significant impact on the instructional methods practiced by teachers and on the kind of engagement children in the state of California would have with science ideas. The original 20 to 25% language was written by three science committee members who genuinely thought that the most equitable way to deliver science information to students was with a textbook. They believed that teachers could not effectively conduct active learning experiences for children, and they thought that hands-on activities should be minimized. When their position was challenged by professional science educators, administrators of major school districts, science-related corporate leadership, and teachers, in good faith they revised the language to embrace active science learning. Working through a strong professional society (such as CSTA) with knowledge of state politics and access to policymakers was critical to making this policy change. And, without the combined voices representing education, business, research, and science, the message would not have been heard.

The effect of state adoption policies on market competition. To have a chance of having one's instructional materials adopted requires a large investment of resources with no assurance of reaping rewards at the end. Small publishers or companies with innovative materials that focus on a singe subject area or grade level have little chance of competing in adoption states where the materials must address all of the grade-level standards and do so for a range of grades, such as K–5, 6–8, or K–8. According to Finn and Ravitch (2004), this has given the large commercial textbook companies significant dominance in the textbook market.

> Textbook adoption created a textbook cartel controlled by just a few companies. Requiring publishers to post performance bonds, stock outmoded book depositories, and produce huge numbers of free samples have all raised the costs of producing textbooks. This has frozen smaller, innovative textbook [instructional materials] companies out of the adoption process and put control of the $4.3 billion textbook market in the hands of just four multi-national publishers. (p. ii)

The result of state policies to require formal state-level textbook adoptions has been to narrow the successful competitors to a small number of large commercial textbook companies. In nonadoption states, there is more likelihood of success for small developers, but there, too, large commercial publishers tend to have more success because they typically show coverage of all of the state standards in a single book and they spend more money to promote their product.

Science Curriculum Selection at the District or School Level

Instructional materials that successfully emerge from the state-level materials review process have cleared the first hurdle on their way to classrooms. But, approval by a state board of education simply provides the publisher the *opportunity* to market the materials at the district or school level. In both adoption and nonadoption states, the publisher must now engage with a local decision maker who has authority to invest his or her state allotment of funds in curriculum materials. This thrusts the publisher into the local policy arena where policy is less well articulated than it is at the state level. The local policy that guides the purchase of materials usually depends on the relationship that the individual schools have to the school district, and it can take many forms. For example, a district might have an overriding policy of site-based decision making. Whether by legislation or tradition, if decision making is site based, the responsibility for the curriculum selection is delegated to the school site. In other cases, the district policy may be to mount a thorough district-wide review process to systematically evaluate all of the approved materials and decide which instructional materials will be implemented throughout the district.

Districts that have a systematic district-level process of review usually engage in predictable kinds of activities. The process is typically coordinated by a superintendent-level curriculum director or designee. The coordinator forms an adoption committee which, depending on the size of the district, could include principals, science specialists, and teachers as well as representatives from the community, local institutions of higher education, and specialists from intermediate educational service agencies, such as a county office of education or a regional collaborative. Using a formal set of evaluation criteria, a committee would typically propose two or more programs to be piloted by a small number of classroom teachers over a period of from 2 to 6 months. Then, based on reports from the piloting teachers and evaluation data regarding the technical and pedagogical merit of the competing programs, the committee would decide which program will be implemented. The committee's decision is then acted upon, materials are purchased, and the adopted curriculum is implemented uniformly in all the schools in the district.

County or regional offices of education often provide "Science Adoption Tool Kits" to assist in this review process. The tool kits can be simple, consisting of little more than an inventory or checklist of available resources that the program provides for teachers and students, or they can be comprehensive, such as the Analyzing Instructional Materials (AIM) process, developed by WestEd and BSCS, which asks questions about science content (standards alignment, accuracy, concept development, sequencing, context), student activity (engaging prior knowledge, metacognition, abilities to do inquiry, understandings about inquiry, accessibility), assessment (quality, multiple measure, use of assessments, accessibility), and work teachers do (instructional model, effective teaching strategies for content, teacher support). To prepare teachers to use this more in-depth evaluation process calls for strong district leadership.

The particular evaluation process used in the selection process is critically important when decisions are made at the site level. This is where the greatest variation in local policy occurs. At one extreme, a school might organize an evaluation committee and enact a scaled-down version of the district process described above. At its best, the school leadership would discuss selection criteria with the members of the evaluation committee to help them choose the materials that present the most accurate and developmentally-appropriate content and that are the most pedagogically sound. At the other end of the evaluation spectrum, program selection is often simply a matter of personal preference, or what some might call a purely "democratic" process, which reduces decision-making to an all-teacher vote. Several state-approved science programs are put out on display so that the whole school staff can inspect them. Then, all too often, what happens is that based on a cursory review, a publisher's sales pitch, and minimal discussion, each teacher votes for his or her favorite. At this extreme, teachers are given no guidance on what criteria to use to make an informed selection. Without guidance, an all-teacher vote represents a lack of leadership in science education. This approach to instructional materials adoption virtually guarantees that the district will reject an active-learning, inquiry-based program that might be unfamiliar to teachers and appear more challenging in favor of a traditional textbook that appears familiar and safe. Without guidance, many teachers will base their decisions on their own comfort level and not on what is the best material to engage students and facilitate learning.

National Dissemination and Implementation Centers

In recognition of the problems inherent in current materials selection practices, in 2001 the National Science Foundation (NSF) provided funding for four national dissemination and implementation centers—Education Development Center (EDC), National Science Resources Center (NSRC), Biological Sciences Curriculum Study (BSCS), and the Center for the Enhancement of Science and Math Education (CESAME)—to provide guidance in the selection process and to promote the use of innovative curriculum materials for elementary and middle-school science instruction. The goal also was to help school staff make more informed decisions when selecting instructional materials and to provide assistance in implementing them (Berns & Sandler, 2009). Each Center developed tools and strategies that clearly described a process that included "analysis of science content, student learning activities, teaching activities, teacher content information, and assessment strategies" (Taylor, 2009, p. 25). The Centers provided coaches and consultants to help school and district leaders integrate the selection process into their community culture and to help participants give thoughtful consideration to issues around curriculum design and rigor, including content, pedagogy, equity, and developmental appropriateness. The Centers "hoped to show schools and districts how to think more seriously about curriculum that engages students, deepens science knowledge, and advances the profession of science education" (p. 33). This process, recommended by the Centers, was a

> far cry from the frequent status quo around the country, in which superficial impressions or preference for familiar books or publishers bested other selection criteria. Instead the Centers advocated for deep discussion among a wide variety of participants about expectations, teaching methods, standards, and achievement data ... as well as a systematic, evidence-based examination of curriculum materials. (p. 34)

The work of the Centers had a significant impact on the relatively small number of school districts that were involved in the project and only for a short period of time. For those individuals who participated as members of a school or district team at a leadership institute, the experience often was transformative. But, to sustain and institutionalize an innovation like this requires ongoing administrative and leadership support at the school and district level and the infusion of outside resources and facilitation. (For a discussion of the importance of school-level leadership in science education reform, see Halverson, Feinstein, & Meshoula, Chapter 13, this volume.) While there is still much work for these Science Implementation Centers to do, there is very little funding available for such long-term technical support projects.

What Determines How Instructional Materials are Implemented?

We have now looked at issues of design and development, and approaches to selection and adoption. After instructional materials have been selected by a school district or individual school, they now must be implemented by the teaching staff in such a way that they will in fact support student learning in science. For successful implementation to occur, several components are needed. First, enough time must be allocated in the school day for the teacher to carry out the learning activities that the materials suggest. Second, the teachers must understand and appreciate the intent of the materials developer. This is particularly true in the case of the smaller scale, more innovative materials, where the developer usually has followed specific design principles consistent with a particular theory of teaching and learning. Third, the teachers must have support throughout the early stages of implementation so that the materials are implemented as intended. This extra support to achieve "fidelity of implementation" is particularly critical when introducing materials that use approaches that are unfamiliar to teachers. (See Halverson, Feinstein, & Meshoulam, Chapter 13, this volume, for a further discussion of the support teachers need in the early stages of program implementation.)

Professional Development Malaise

The decline in elementary science learning imposed by the meager time devoted to science instruction is further exacerbated by the lack of resources to assist science teachers in their work. Schools are amply populated with reading resource teachers, grade-level reading coaches, reading aides, and ancillary reading programs, as well as professional development resources directed toward reading issues. Many fewer resources are available to support active-learning science curricula, which are complex in design and demanding in pedagogical content. In short, elementary teachers need opportunities to learn how to teach inquiry science, time to learn the content and instructional design, time to try out the new materials and methods in their classes, time to debrief with experts and colleagues, and time to grow into the curriculum.

To counter this neglect of elementary science teaching, science curriculum developers have found they need to direct more of their creative energy toward developing alternative professional development models for teachers who have fewer traditional ways to access training. Many of these newer vehicles are in the form of digital courses. Providing digital professional development products, either self-contained or web-based, gives teachers opportunities to access information that they would not otherwise have. Another way in which instructional materials developers can support teachers is to design materials that are more educative than they have been. Finally, research has shown that professional development is most effective when it is linked to or embedded in the curriculum materials themselves. The National Science Foundation-funded State Systemic Initiatives, Urban Systemic Initiatives, and Local Systemic Initiatives have provided useful information regarding successful professional development strategies. But, for any of these to be successful, there must be leadership at the school level.

> For curriculum reform to be successful, there must be a close and continuing relationship between those who create and distribute the new curriculum and those who use it. Long-term implementation strategies must be designed to take into account the need for this continuing support, and developers must devise a way to respond appropriately to what is learned in the implementation process.... [F]or the process of innovation to be successful, it must be institutionalized within the school system so that when the initiating administrator-innovator moves on to other responsibilities, as he or she often does, the new program will continue to receive on-going support. School systems are by nature conservative institutions, and no innovation will last long unless it has strong administrative backing and unless provisions are made for continuing assistance at every stage of the implementation process. (Dow, 1999, p. 263)

Our own experience as instructional materials developers tells us that the most effective professional development happens at the school site on a regular basis in the company of grade-level colleagues and focuses on student work. Professional learning communities (PLCs), facilitated by a district leadership educator or a school-site mentor, have great potential for improving science teaching and learning. As curriculum developers, we should develop tools and make them available online to support and encourage learning communities. A library of student work samples or exemplars, classroom video clips to generate teacher discussion, and next-step classroom strategies for specific investigations are the kinds of resources that we can provide to support learning communities. When it becomes policy that teaching is a public, not a private, enterprise, and that risk-free environments are as important for teacher learning as student learning, then the PLC can become a part of the school culture.

Time Deprivation

Successful implementation of innovative instructional materials requires that there is enough time in the school schedule for science to be taught well. National educational policy speaks volumes about what is valued for the intellectual and cultural development of our youth. Education policy is communicated in proclamations, laws, funding opportunities, assessment activities, and reward structures. Unfortunately, recent national policy has made it more difficult for quality science instruction to reach teachers and students. Between 2000 and 2009, science often was devalued, restricted, and even distorted at the highest levels of the federal government. This general disregard for science extended to science education, and was felt in two significant ways: funding for all aspects of science education was reduced, and a greater emphasis was placed on reading and mathematics.

This is most clearly communicated in the No Child Left Behind legislation of 2001. The law mandated progressively higher reading and mathematics scores on annual high-stakes state tests for all children. With mandated high-stakes testing in mathematics and reading, national reading and mathematics policy was translated into state and local policy. The predictable response across the country was to increase the number of hours devoted to reading and mathematics instruction. The effect on elementary classroom practice was to shift hours in the school day from content education (science and social studies) to skill education (reading and mathematics). The institutional panic to produce incrementally more proficient readers drew time away from early elementary science. As a consequence, high-quality, research-based science curricula were placed on the shelf. The attention of elementary school administrators and, therefore, their teachers had been pulled away from science; their energies were spent on activities that were more highly valued by NCLB policymakers.

A study by Dorph, Goldstein, Lee, Lepori, Scheider, and Venkatesan (2007), using surveys and interviews, examined the status of elementary school science in nine San Francisco Bay Area counties. The study revealed a considerable reduction in the amount of time spent on science.

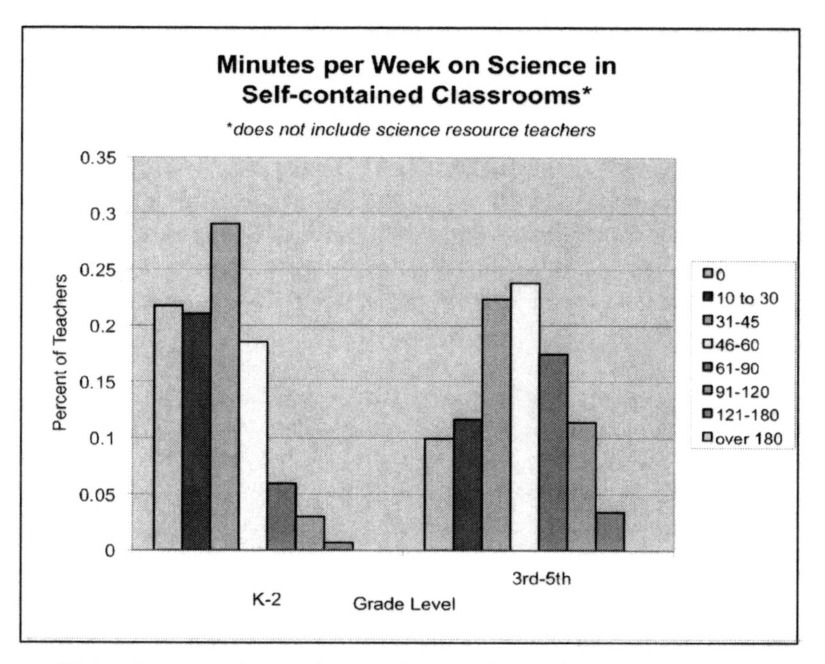

Figure 12.4. Survey and interview results examining the status of elementary school science in nine San Francisco Bay Area counties (used with permission, Dorph et al., 2007)

> Eighty percent (80%) of K–5th grade multiple-subject teachers who are responsible for teaching science in their classrooms reported spending 60 minutes or less per week on science, with 16% of teachers spending no time at all on science. The Figure [see Figure 12.4] displays related teacher survey results by grade band in greater detail.... This estimate is considerably lower than the 125 minutes per week reported as a result of a national survey conducted in 2000.... Echoing a recent national study, nine district representatives who responded to our survey and participated in interviews reported that a diminishing amount of time has been spent on science since the enactment of *No Child Left Behind (NCLB)*. Those districts with schools in *Program Improvement* status, due to their prior language arts and mathematics test results, report little to no time for science at all. In a few exceptional circumstances (special programs, community priorities), individual schools go against that trend, focusing adequate time and attention on science instruction. (Dorph et al., 2007, p. 1)

One of the strategies that elementary science educators have used to adapt to the increase in emphasis on reading and decreasing emphasis on science is to argue for the value of science reading in developing broad linguistic skills, and science materials developers have built strategies to enhance reading skills into their instructional materials. What is interesting about this is that it is an approach that is being tried elsewhere as well. (See, e.g., Sarmant, Saltiel, & Léna, Chapter 6, this volume, for a discussion of a similar debate taking place in France.)

The importance of reading and writing in the elementary science curriculum had already been articulated in a number of documents and studies prior to NCLB. A 2001 collaborative publication from the Mid-continent Research for Education and Learning (McREL), the Council of State Supervisors of Science, and The National Network of Eisenhower Regional Consortia and Clearinghouse, summarized research and best practices in science and reading in this way:

> Many of the process skills needed for science inquiry are similar to reading skills ... predicting, inferring, communicating, comparing and contrasting, and recognizing cause and effect relationships. In language as well as science learning, students analyze, interpret and communicate ideas.... Students' comprehension of text improves when they have had hands-on experiences with a science concept. Prior knowledge, which is developed and enhanced through science inquiries, is the strongest predictor of student ability to make inferences from text. In a four-year study of elementary students who participated in an active inquiry-based science program, a direct correlation was shown between higher language test scores and years of participation in the science program. (Krueger & Sutton, Eds., 2001, p. 52)

The study referred to above was the Valle Imperial Project in Science (VIPS), a four-year comparison of student achievement data from 1995–1999 conducted with NSF funding (Klentschy, Garrison, & Amaral, 2001). An outgrowth of this project and other elementary science professional development projects in the 1990s was the implementation of the student science notebooks as an integral part of the science program. Linda Gregg, former Coordinator of Math and Science for Clark County School District in Las Vegas, described the journey of educators in these words in the foreword to *Science Notebooks: Writing about Inquiry*:

> The charm of the students' work and their pride in ownership of the knowledge recorded in their notebooks motivated teachers to move into unknown territory. The teachers worked together figuring out how to create time and authentic reasons for students to record the stories of their investigations and how to best have students use their notebooks to share their findings. Over time, more and more teachers and administrators began to consider notebooks an important part of their science programs. Teachers and administrators were beginning to share stories about student writing that communicated increased understanding of the science content they were studying and the process of inquiry they were using. Teachers began to describe how students were referring to their notebooks when sharing their findings and questioning their peers. (Campbell & Fulton, 2003, p. xiv)

In response to these studies, curriculum developers shifted their creative energy into new products designed to insinuate science back into the core curriculum and to improve the chances that their materials would actually be used after they were adopted. Reading materials and strategies to use them effectively in science became increasingly important to science materials developers. It was thought that if there was an emphasis on reading in the science materials, school leaders might be willing to allocate more time to science teaching and science teachers might be willing to enhance the hands-on activities with literacy activities. For example, science notebook strategies that had been developed by teachers and shown to be successful in the classroom were incorporated into active-learning investigations, giving all teachers one more opportunity to integrate science into the mandated language arts time. When these language arts skills are exercised for the purpose of learning science, then the science instructional time for a science-centered school could be increased to 150 minutes per week for K–2 students and 300 minutes per week for Grades 3–6 students.

At two National Science Teachers Association (NSTA) conferences in 2004 and 2005, the National Science Foundation supported a number of special meetings to focus on the science and literacy connection, with the goal of providing practitioners and policymakers with scientifically-based research on the effective integration of literacy in the pre-K–8 science curriculum. This was done in part because of the "current emphasis on literacy and mathematics in the NCLB legislation with the resulting lack of focus on science instruction" (Douglas, Klentschy, Worth, & Binder, 2006, pp. xii). One of the presentations at the conference (and in the resulting NSTA book based on the conference) described the Seeds of Science/Roots of Reading Project, a joint effort of the Graduate School of Education and the Lawrence Hall of Science at the University of California at Berkeley. The goal of the project is to transform existing inquiry-based units from the Great Explorations in Math and Science (GEMS) curriculum series into materials that help students make sense of the physical world through firsthand experiences while addressing foundational dimensions of reading, writing, and language (Cervetti, Pearson, Bravo, & Barber, 2006).

How science and English language arts will intersect and share time in the elementary grades in the future is a significant policy decision that will have an impact on what students learn and how they learn it. Any reauthorization of federal legislation for elementary and secondary education will need to take this into account.

A CASE HISTORY—SCIENCE EDUCATION POLICY AND FOSS

The authors of this chapter are members of an instructional materials development group at the Lawrence Hall of Science (LHS), the public science center of the University of California at Berkeley, and an innovative leader in the field of science and mathematics education since 1968. The authors are co-directors of the development team that created and sustains the Full Option Science System (FOSS) project (Regents University of California, 2005–2009), a major curriculum project of the Center for Curriculum Development and Implementation at LHS. Looking back to its roots, FOSS has been in continuous development, revision, and expansion since 1976. During that time, the evolution of FOSS has been largely supported, but sometimes hindered, by the ebb and flow of American science education policy. What follows is a brief review of key historic events and decisions, the policies they generated, and the impact they had on the development of FOSS. This more than 30-year case history presents a story of how science education policy and curriculum development inevitably intersect over the long term.

Curriculum Reform of the 1950s and 60s

In 1957, the earth-orbiting satellite Sputnik was launched by the Soviet Union. This signature technological event created anxiety and fear among American citizens and political leaders because our Cold War enemy had beaten us into space, and it created a firestorm of recrimination against the American educational system for its failure to produce the technical expertise that the Soviet educational system apparently had produced. In 1958, President Eisenhower signed the National Defense Education Act into law, the most comprehensive education reform bill to date in the United States. The act funded the National Science Foundation's nascent education division, which provided support for innovation in elementary and secondary science curriculum. For the first time, school and curriculum improvement became a focus of national concern, primarily through the efforts of the National Science Foundation. From the beginning of the curriculum reform movement, scientists and educators at the University of California at Berkeley became involved in science curriculum development, and in 1968 the Lawrence Hall of Science (LHS) was dedicated, and immediately became the home of the legacy curriculum project, the Science Curriculum Improvement Study (SCIS) project. Dr. Robert Karplus, a theoretical physicist at Berkeley, and his staff created a foundation for curriculum development that has had lasting influence on generations of educators.

Education for All Handicapped Children Act of 1975 and the Beginnings of FOSS

The FOSS program had its origins in two projects that were designed to provide access to students with disabilities. In 1975, Public Law 94–142, the Education for All Handicapped Children Act (now IDEA, Individuals with Disabilities Education Act) was enacted by Congress to support states and localities in protecting the rights and meeting the educational needs of individuals with disabilities. The Science Activities for the Visually Impaired (SAVI) project was funded by the U.S. Office of Education in 1976, and the Science Enrichment for Learners with Physical Handicaps (SELPH) project was funded by the Office of Education in 1979. The development of these special education materials at the Lawrence Hall of Science coincided with the new mandate to educate students with disabilities in "the least restrictive environment." Public Law 94–142 was popularly referred to as the mainstreaming law. The mainstreaming policy boosted the visibility of the SAVI/SELPH program because SAVI/SELPH provided effective strategies for serving all students in science classes.

A Nation at Risk and the Beginning of the Standards Movement

In 1983, The National Commission on Excellence in Education, a special commission created by the U.S. Department of Education during the Reagan administration to study the status of American education, released its report *A Nation at Risk,* which reported that:

> the educational foundations of our society are presently being eroded by a rising tide of mediocrity that threatens our very future as a National and a people.... If an unfriendly foreign power had attempted to impose on America the mediocre education performance that exists today, we might well have viewed it as an act of war. (National Commission on Excellence in Education, 1983, p. 1)

These strident words became a rallying cry and, again, there was a national focus on science education. The policy strategy the Commission proposed was for states and local schools to establish higher standards in five basic academic areas including science, mathematics, social studies, English, and computer science, and public accountability through national (but not federal) testing. The states and local governments were to be given flexibility to achieve those higher standards in the best way possible.

In 1986, the National Science Foundation became the standard bearer for a renewed policy of national curriculum reform. The NSF established the TRIAD program to encourage the development of several exemplary elementary science curricula. The program was called TRIAD because it had to have three partners: a research institution involving scientists and science educators, one or more school districts, and a publisher. According to an NSF press release in May 1988, the NSF Triad program was designed to

> join scientists and science educators with textbook publishers and school systems to improve content, provide more hands-on activities for students, and devote more time to the teaching of science in elementary schools ... The NSF program is unique in involving publishers, who are developing the school marketplace; committing staff and financial resources; participating with scientists, science educators, and schools in the development of the

program; selling it to the schools; and working with teachers to assure proper use of the published materials. (National Science Foundation, 1988)

The co-directors of the SAVI/SELPH program had been contemplating revising SAVI/SELPH to be a science education curriculum for all students, not just a program for students with visual, physical, and learning disabilities. The new curriculum reform policy provided an opportunity to reinvent SAVI/SELPH as FOSS, the Full Option Science System, "full option" because the modular format would provide implementation flexibility and allow districts to customize the program to fit their district learning goals. The FOSS program for Grades 3–6 was funded in 1989, followed by FOSS K–2 in 1992, and finally FOSS middle school in 1996 under a separate NSF solicitation.

In 1989, six years after *A Nation at Risk*, the American Association for the Advancement of Science (AAAS) published *Science for All Americans*, describing in greater detail than had ever been done before the knowledge, skills, and habits of mind that constitute adult literacy in science, mathematics, and technology. With this publication, the conversation about science for *all* students came to the forefront. In 1993, AAAS released *Benchmarks for Science Literacy*. This seminal document described for the first time a carefully thought out vision of a student's progression of understanding of science throughout an academic career by stating what students should know by the end of each of four grade bands—K–2, 3–5, 6–8, and 9–12—to achieve the goal of science literacy for all that had been laid out in *Science for All Americans*. Three years later, in 1996, after four years in development, the National Research Council published the *National Science Education Standards (NSES)*. The *NSES* established academic expectations for students in three grade-level bands—K–4, 5–8 and 9–12. The science content in *Benchmarks* and *NSES* was consistent with the model for science instruction represented in the FOSS program materials. The *NSES* also emphasized the importance of scientific inquiry, not only as a pedagogical strategy but also as a content area equivalent to Life, Earth, and Physical Science. Seeing the high level of agreement between *Benchmarks* and *NSES*, and envisioning the national policy that seemed certain to follow, the FOSS project revised FOSS to align clearly with the *NSES* content, assessment, and professional development standards. This provided another policy boost for FOSS because the nation was now paying attention to curriculum materials that were aligned with national standards.

In 1997, NSF funded four Centers for Dissemination and Implementation of science instructional materials, signaling a change in emphasis within the foundation. The NSF had decided that it was time to encourage the implementation of those curricula that were standards-aligned and research-based, scientifically accurate, developmentally appropriate, and pedagogically sound. So, FOSS became a beneficiary of this NSF policy. Because FOSS met NSF's dissemination criteria, the curriculum was disseminated through the Education Development Center (EDC), the National Science Resources Center (NSRC), and the Center for Excellence in Science and Math Education (CESAME), particularly in urban settings. (See Earle, Chapter 5, this volume, for a more detailed discussion of NSF's role in science education policy development.)

In January, 2002, President George W. Bush signed the No Child Left Behind Act of 2001 into law. Overnight, there was another new federal education policy. Schools would now be held accountable for students' adequate yearly progress (AYP) on state tests in reading and mathematics. Science testing would become a requirement in 2007–08, but schools would not be held accountable for their students' performance in science under the adequate yearly progress provisions. This dramatic policy shift toward annual testing of mathematics and reading in elementary school immediately impacted the teaching of science. Time for science was pushed aside by anxious and pressured administrators looking for immediate solutions to improve reading scores to meet accountability demands. The FOSS program implementation suffered, particularly in Grades K–3, where science instruction was replaced by additional instruction in reading.

In 2003, consistent with the accountability provisions of the standards movement, NSF adjusted its priorities to focus a larger portion of its declining resources on research on formative assessment, and the FOSS project was awarded funding to develop the Assessing Science Knowledge system (ASK). In an environment of heightened accountability, not only for schools but also for educational researchers and curriculum developers, the assessment system added a strong new dimension to the FOSS curriculum. The FOSS program would now be able to answer questions about its' effectiveness and provide tools for improving science instruction through formative assessment embedded in the materials.

All of the policies described above have had an impact on the evolution of FOSS. As the details of the story reveal, most of the policies have had positive effects on the direction the FOSS materials have taken. In just about all cases, the FOSS developers were able to embrace the changes in federal policy while at the same time maintaining the integrity of their research-based program. More recently, however, the emphasis on mathematics and reading at the expense of science, brought on by No Child Left Behind has had a significant negative impact on science teaching in elementary schools, and that policy has been difficult for the FOSS developers to respond to. As noted earlier, one strategy that has been used by elementary science materials developers to keep science alive in the schools has been to place a greater emphasis on the literacy aspects of science teaching, in an effort to convince school-level policymakers that science can contribute positively to improved reading skills. This has been done in the FOSS materials as well. In addition, FOSS continues to focus on enhanced professional development to ensure proper implementation of the materials when they are adopted by schools.

ANTICIPATING POLICY AND PREPARING FOR FUTURE POLICY DIRECTIONS

Now, additional policy initiatives are being discussed and, if implemented, will again have an impact on curriculum development. Since the release of *Benchmarks* in 1993, *NSES* in 1996, and the passage of the No Child Left Behind Act of 2001, the enterprise of K–8 science curriculum development has become increasingly demanding. Even though the national standards documents have provided needed direction to an educational system that had little content focus, the factors discussed earlier in the chapter—inconsistent and often incoherent application of science standards at the state level and a largely punitive accountability policy at the federal level—have inhibited systematic development of a sound approach to American science education, especially at the elementary level. There has been critical time lost in the development of a scientifically literate citizenry at a time when thoughtful engagement with difficult issues, such as finding alternative sources of energy and dealing with global climate change, is critical to the future of the global community. Yet, looking forward, there is reason for some optimism. With the change in government in 2009 comes the hope for a renewed recognition of the central role of science in society and the urgency of systematic, high-quality science education. In this next part of the chapter, we ask the questions, "How can instructional materials developers prepare for the changes that are likely to come?" "Is it possible to anticipate the policy decisions that will inevitably emerge as the country changes its attitude toward the importance of science in society and in our schools?" and "How can developers impact policy changes in order to provide educators in each state with a more a coherent and stable set of standards that benefit teachers and all students?"

Our mission at the Lawrence Hall of Science is to design, produce, implement, and support the highest quality science curriculum materials in the country. We know that materials development does not take place in isolation from policy mandates at the local, state, and national levels. So part of our commitment to successful development and implementation involves anticipating where elementary science education policy will go next, and being prepared to help move the field forward. Quality curriculum materials take time to develop, so it is imperative that developers keep a finger on the pulse of new trends and future policies so when the time comes, the instructional materials will be available to meet the educational needs. Examples of areas where instructional materials can support new and anticipated policy initiatives are described in the next several sections.

A Strong Formative and Summative Assessment System Seamlessly Integrated Into the Curriculum

Given the widespread belief that public accountability in education is a good thing (see DeBoer, Chapter 10, this volume), combined with the recognition that current systems of accountability being used under the federal legislation are creating problems for educators, we predict that new systems of assessment will continue to evolve. We also anticipate that new progress-based approaches for assessment will take into account students' classroom performance as well as their scores on statewide standardized tests.

> Assessment, seen as feedback between teachers and learners, is central to the business of learning. The explicit recognition of this principle is now spurring many activities for school improvement.... Yet one obstacle to its further development is the pressure exerted by high stakes tests.... Tests that were more sensitive to the task of exploring students' understanding would help to reduce such discordance, but externally designed tests can never meet all the requirements of valid assessment. Far more is needed, specifically ways to communicate the teacher's own knowledge of her students. She works with them day after day, month after month. She has the information available to no one else. (Atkin & Black, 2003, p. 126)

For FOSS, the assessment system known as ASK (Assessing Science Knowledge) utilizes embedded formative assessment materials and strategies, allowing teachers to monitor student learning continuously. Before instruction begins, again at critical junctures throughout the curriculum, and at the conclusion of instruction, teachers use valid and reliable assessment instruments to evaluate student knowledge of core conceptual understandings (Long, Malone, & De Lucchi, 2008).

To support this move toward more and better use of assessment, it is critical that curriculum developers provide teachers with useful strategies to gather evidence of student learning and ways to act on this information. One strategy, the reflective assessment technique, involves teachers' *anticipating* student outcomes by focusing on only one or two key concepts at a time in an instructional unit, *reviewing* student work with a focus on those specific concepts immediately after instruction, *reflecting* on the evidence of student understanding of those concepts, and *adjusting* their instruction by planning "next steps" to help students clarify their understanding. This approach has been shown to improve student performances on end-of-module assessments (Kennedy, Long, & Camins, 2009). Similar strategies are needed that involve having a clear idea of what students are expected to know, assessments that provide evidence of what students do know and the alternative ideas they have, and a plan for modifying instruction based on the results of those assessment results. Materials developers in science can contribute to the trend in this direction by incorporating these strategies into their materials.

An Outdoor Extension of the Classroom Curriculum

An environmental education mandate is building. Districts are initiating environmental education magnet schools, and language for environmental education policy has been drafted at the federal level (No Child Left Inside Act of 2009) and at state levels. The Senate version of the reauthorization of NCLB, which included language for environmental education, was introduced by Senator Jack Reed (D-RI), and the House version by Representative John Sarbanes (D-MD) on Earth Day, 2009. If the legislation were to pass, it would be the first piece of environmental education legislation to be passed by Congress in more than 25 years. In some states that policy will mandate that new curriculum materials in all subjects must integrate a set of environmental standards into the subject matter instructional practice.

We believe that there is a growing recognition among policymakers of this eroding relationship between America's youth and natural systems (Louv, 2006) and that curriculum materials developers can support this move toward environmental education. They can do this by creating materials that include activities that extend the classroom into the schoolyard and nearby environments as part of standard instructional practice. At LHS, our outdoor pedagogies will be informed by curriculum materials that were originally developed during the 70s—the Outdoor Biology Instructional Strategies (OBIS) project (NSF funding, 1972–1975), as well as extension activities developed by the Boston Schoolyard Initiative project. In addition, new environmentally oriented modules and courses will be developed to provide more in-depth experiences with ecology and natural history topics.

A Diversified Collection of Digital Products to Modernize and Augment the Classroom Curriculum

The promise of the digital American classroom has been delayed for a number of reasons, including awaiting the development of a coherent, uniform policy for digital instructional materials, and a commitment to the implementation of that policy. We anticipate that future instructional materials will utilize Web 2.0 interactivity, video streaming, user networking, and student assessment data processing, and that these digital resources can improve science education if used well. Materials developers can support the movement in this direction by incorporating the new technologies into their curriculum materials. Examples of science education projects making use of the new technologies that are currently in development at LHS include (1) a web-based digital science notebook using Universal Design for Learning methods and strategies; (2) web-based e-books of student reading materials; (3) a digital teacher guide; (4) interactive white board resources; and (5) a smart tutor. The digital notebook is designed to support students with high-incidence disabilities (learning disabilities and executive function deficits) to help them reach parity with their non-disabled peers. The smart tutor uses an artificial intelligence interface with voice recognition. Students who need additional opportunities to engage with the science concepts presented in an investigation can visit the tutor. With spoken language as the medium, student and computer have a conversation about the student's understanding of the science topic and, based on what the student tells the tutor, the student is directed to conduct simulated experiments, observe animations, and study visual images to get additional instruction on the target topic. The smart tutor will also support the classroom teacher in differentiating instruction to serve all students.

A Second Educative Voice in the Teacher Guide

We expect that in the future, teacher guides will include more research-based justifications for recommended activities and practices. Traditional lesson plans show teachers what to teach and offer suggestions about how to teach it. We anticipate that in the future, this "what to teach" discussion will be augmented with a second educative voice talking to the teacher about the rationale for the design of the lessons, *why* the instructional sequence is presented the way it is. In the materials we are developing, this educative voice appears in a sidebar of the teacher guide and explains fine points of pedagogy, research evidence for particular pedagogies, strategies for managing student engagement, all in the context of the recommended activities. The rationale appears right beside the description of instruction. Consistent with the work on educative curriculum described by Schneider and Krajcik (2002), the educative curriculum design turns the science lesson into a learning experience for both the teacher and the student simultaneously

A More Coherent Scope and Sequence for the Science Program

We also expect that the seminal work by AAAS Project 2061 and presented as a series of strand maps in the *Atlas of Science Literacy Volumes 1* and *2* (AAAS, 2001, 2007), which show the hierarchical relationships among concepts that contribute to an understanding of progressively more sophisticated ideas in science, will continue. This work and the more recent and ongoing work on learning progressions (Duschl, Schweingruber, & Shouse, 2007) should inform a revised conceptual structure for the elementary science curriculum. We believe that it is important for materials developers to make use of the work that has been done in this area so that the concepts developed in the disciplinary strands of the curriculum materials will be designed to elucidate these learning progressions. Thus, when a student encounters a conceptual area for a second or third time in her elementary science career, she will be engaging those ideas at successively deeper, more complex levels of understanding and not just repeating what was taught earlier. The policy of designing a curriculum that is internally coherent (with good conceptual flow) within a grade-level module should be part of a larger schema in which instructional sequences are articulated throughout Grades K–8, providing additional consistency and power to the science learning experience for students. We described the failure to do this as one of the key problems in the way that many states present their content standards. We believe that ensuring the coherence of curriculum materials is one of the most important ways that they can be improved in the future. (For an extended discussion of the importance of curricular coherence, see Kali, Linn, and Roseman, 2008.)

We believe that each of these current developments and policy trends in science education can be supported by materials developers and used to improve the teaching of science for all students. By being aware of them, developers ensure that the materials they develop will remain relevant and the prospects for student learning will be greatly enhanced.

SUMMARY

As we have shown in this chapter, the curriculum developer's job is significantly impacted by federal, state, and local educational policies. Some policies have benefited the curriculum development enterprise by infusing money and a new vision into the system. During those peaks, large strides were made in the technical excellence of instructional materials. Other policies have impeded advances in science education, especially at the elementary level, making the work of curriculum developers more difficult.

The history of the Lawrence Hall of Science, an institution dedicated to excellence in science curriculum development and teacher professional development, stands as an example of how policy can benefit or hinder the curriculum developer's work. Curriculum developers at the Lawrence Hall of Science have also tried to remain steadfast and diligent in anticipating and taking advantage of policy transitions in areas such as educational technology and environmental education. Over the years, curriculum development has benefited from funding opportunities generated by science education renewal. The Sputnik crisis and publication of *A Nation at Risk* inserted a surge of activity into science materials development. During the NCLB decade, however, national policy issues concerning language arts and mathematics drained resources away from science teaching.

The last decade saw progressively more stringent policies related to public accountability in education because of No Child Left Behind. The legacy of NCLB resides in the form that state science standards have taken, the impact the standards have had on the states' instructional materials adoption processes, and the reduction of professional development opportunities in science as resources have been shifted toward reading and mathematics. The specific state and local policies associated with standards, materials adoption, and professional development have complicated the science instructional materials development process, diverting energy and resources away from its primary mission to produce efficient, effective curriculum resources. The energy is squandered on state and local policy issues related to technical compliance, submission regulations and negotiations, access to consumers, and a host of other peripheral details. Many developers are dispirited by the requirement for absolute adherence to grade-level standards that vary state-by-state, which eliminate virtually all opportunity for creative exploration of promising approaches to the development of materials for adoption on a wide scale basis.

In order to regain momentum for elementary science education and make it possible for curriculum materials developers to create the instructional materials that will support student learning, there are two paths that policymakers might take. If grade-level expectations are going to continue into the future, then it would be better if all states, or at least a significantly large number of states, would share common grade-level expectations. That way materials developers would have a large enough potential audience for the materials they develop. To do this would mean revisiting the idea of national science standards. Two decades ago we began the difficult work of identifying what a shared vision of science education would look like. The AAAS *Benchmarks*, the NRC's *National Science Education Standards*, and the NSTA *Science Anchors* have all contributed to the specification of the science that is the most important for students to know. With a revised description of the science education that every student should receive that would be acceptable to a large number of states, coordinated with a national assessment system and a suite of incentives, we may be able to finally develop a coherent approach to science education for all Americans. Common core standards have been developed in the fields of mathematics and language arts and the process leading to common core science standards has begun (Carnegie Corporation of New York, Institute for Advanced Study, Commission on Mathematics and Science Education, 2009). In summer 2010, the National Research Council's Committee on Conceptual Framework for New Science Education Standards released the preliminary public draft of A Framework for Science Education (NRC, 2010). It is anticipated that this framework will guide the development of common core standards anticipated for release in 2012. This means, of course, that states would have to compromise their long standing right to determine the science learning goals of children in their individual states, so moving in that direction may be daunting. But to move forward with science excellence we need to solve the problem of an inconsistent, disconnected approach to American elementary science education that exists today.

A second approach would be to describe what students are expected to know at a more general level and then let states and local school districts fill in the details themselves. In that way, materials developers would have the flexibility of developing a wide range of materials, all of which would fit into the broader set of goals that had been described. This is the direction that the United Kingdom is now moving after a period of time when the National Curriculum specified in greater detail what students would be held accountable for (see Osborne, Chapter 2, this volume). Even if a common core is developed in science and a significant number of states buy into that common core, materials developers would still want that core to be described at a level that allows them flexibility so that they can explore a variety of innovative approaches to materials development.

The most important thing for curriculum materials development is that the core mission of preparing future generations of citizens to think critically about the natural world and be productive problem solvers be well defined, but defined with enough flexibility so that a variety of new approaches can be created. This would be the environment in which materials developers could best contribute to the science learning of all students.

REFERENCES

American Association for the Advancement of Science (1989). *Science for all Americans.* Washington, DC: Author.

American Association for the Advancement of Science (1993). *Benchmarks for science literacy.* Washington, DC: Author.

American Association for the Advancement of Science (2001). *Atlas of science literacy, Volume 1.* Washington, DC: Author.

American Association for the Advancement of Science (2007). *Atlas of science literacy, volume 2.* Washington, DC: Author.

Atkin, J. M., & Black, P. (2003). *Inside science education reform: A history of curricular and policy change.* New York, NY: Teachers College Press.

Berns, B. B., & Sandler, J. O. (Eds.) (2009). *Making science curriculum matter: Wisdom for the reform road ahead.* Thousand Oaks, CA: Corwin Press and EDC.

California Department of Education. (2004). *Criteria for evaluating instructional materials in science, kindergarten through grade eight.* Sacramento, CA: California Dept. of Education. Retrieved from www.cde.ca.gov/ci/sc/cf/documents/scicriteria04.pdf

California Department of Education. (February 2006). *Instructional materials in California: An overview of standards, curriculum frameworks, instructional materials adoptions, and funding.* Sacramento: CA Dept. of Education. Retrieved from http://www.cde.ca.gov/ci/cr/cf/imagen.asp

Campbell, B., & Fulton, Lori (2003). *Science notebooks, writing about inquiry.* Portsmouth, NH: Heinemann.

Carnegie Corporation of New York, Institute for Advanced Study, Commission on Mathematics and Science Education (2009). *The opportunity equation: Transforming mathematics and science education for citizenship and the global economy.* New York, NY: Author.

Cervetti, G. N., Pearson, P. D., Bravo, M. A., & Barber, J. (2006). Reading and writing in the service of inquiry-based science. In R. Douglas, M., Klentschy, K Worth, & W. Binder (Eds.), *Linking science & literacy in the K–8 classroom* (pp. 221–244). Arlington, VA: National Science Teachers Association.

Dorph, R., Goldstein, D., Lee, S., Lepori, K., Schneider, S., & Venkatesan, S. (2007). *The status of science education in the bay area: Research brief.* Lawrence Hall of Science, University of California, Berkeley, CA.

Douglas, R., Klentschy, M., Worth, K., & Binder, W. (Eds.) (2006). *Linking science & literacy in the K–8 classroom.* Arlington, VA: National Science Teachers Association.

Dow, P. B. (1999). *Schoolhouse politics: Lessons from the sputnik era.* Cambridge, MA: Harvard University Press.

Duschl, R. A., Schweingruber, H. A., & Shouse, A. W. (Eds.) (2007). *Taking science to school: Learning and teaching science in Grades K–8.* Committee on Science Learning, Kindergarten Through Eighth Grade. Washington, DC: The National Academies Press.

Finn, C E., & Ravitch, D. (2004). *The mad, mad world of textbook adoption.* Washington, DC: Thomas B. Fordham Institute. Retrieved from http://www.edexcellence.net/detail/news.cfm?news_id=335

Fuller, R. G. (Ed.) (2002). *A love of discovery: science education—the second career of Robert Karplus.* New York, NY: Kluwer Academic/Plenum.

Kali, Y., Linn, M. C., & Roseman, J. E. (Eds.). (2008). *Designing coherent curriculum: Implications for curriculum, instruction, and policy.* New York, NY: Teachers College Press.

Kennedy, C., Long, K., & Caminos, A. (2009). The reflective assessment technique: A new way of evaluating in-class student work. *Science and Children, 47*(4), 50–53.

Klentschy, M., Garrison L., & Amaral., O.M. (2001). Imperial Project in Science (VIPS) four-year comparison of student achievement data 1995–1999. National Science Foundation Grant #ESI-9731274.

Krueger, A., & Sutton, J. (2001). *Ed thoughts, what we know about science teaching and learning.* Aurora, CO: McREL.

Long, K., Malone, L., & De Lucchi, L. (2008). Assessing science knowledge: seeing more through the formative assessment lens. In J. Coffey, R. Douglas, & C. Stearns (Eds.), *Assessing Science learning: Perspectives from research and practice* (pp. 167–190). Arlington, VA: NSTA Press.

Louv, R. (2006). *Last child in the woods: Saving our children from nature-deficit disorde*r. Chapel Hill, NC: Alonquin Books.

National Commission on Excellence in Education. (1983). *A Nation at Risk: The imperative for educational reform.* Washington, DC: U.S. Department of Education.

National Research Council. (1996). *National Science Education Standards.* Washington, DC: The National Academy Press.

National Research Council. (2010). *A framework for science education, preliminary public draft.* Washington, DC: Committee on Conceptual Framework for New Science Education Standards, Board on Science Education.

National Science Foundation. (1988). NSF forges partnerships to develop new elementary school science materials. *NSF News PR, 88-24.*

National Science Teachers Association. (2009) *Science anchors, a proposed model framework.* Retrieved from www.nsta.org

Regents University of California (Lawrence Hall of Science, Berkeley) (2005–2009). *FOSS—Full Option Science System Program.* Nashua, NH: Delta Education.

Roth, K. J., Druker, S. L., Garnier, H. E., Lemmens, M., Chen, C., Kawanaka, T., et. al. (2006). *Teaching science in five countries: results from the TIMSS 1999 video study* (NCES 2006-011). Washington, DC: National Center for Education Statistics.

Schneider, R. M., & Krajcik, J. (2002). Supporting science teacher learning: the role of educative curriculum materials. *Journal of Science Teacher Education*, *13*(3), 221–245.

SRI International. (1996). *Evaluation of the American Association for the Advancement of Science's Project 2061, Volume I: Technical Report.* Menlo Park, CA: Author.

Taylor, J. A. (2009). Selecting curriculum materials: A critical step in science program design. In. B. B. Berns & J. O. Sandler (Eds.), *Making science curriculum matter: wisdom for the reform road ahead* (pp. 23–34). Thousand Oaks, CA: Corwin Press and EDC.

Texas Essential Knowledge and Skills (TEKS) Science. (August 2009). Texas Administrative Code (TAC), Title 19, Part II, Chapter 112 A. Retrieved from http://ritter.tea.state.tx.us/rules/tac/chapter112/index.html

Thier, H. D. (2001). *Developing inquiry-based science materials: A guide for educators.* New York, NY: Teacher College Press.

Washington state K–12 science standards, revised (December 14, 2008). Retrieved from http://www.k12.wa.us/Science/Standards.aspx

Wilson, M., & Bertenthal, M. (Eds.). (2005). *Systems for state science assessment. Report of the committee on test design for K–12 science achievement.* Washington, DC: National Academy Press.

Zinth, K. (January, 2005). *State Notes: State Textbook Adoption. Denver, CO: Education Commission of the States.* Retrieved from http://www.ecs.org/html/educationIssues/ECSStateNotes_2005.asp

PART III

POLICY IMPLEMENTATION

Part III includes a single chapter on policy implementation. Richard Halverson, Noah Feinstein, and David Meshoulam discuss the important role that school-level leaders play in policy implementation. They note in particular how the theory of action that typically guides school reform efforts focuses on standards, curriculum resources, and out-of-school professional development without adequately considering whether or not the reform is appropriate for the local context, how it could fit in, and then creating an environment for it to take hold and be sustained over the long term.

CHAPTER 13

SCHOOL LEADERSHIP FOR SCIENCE EDUCATION

Richard Halverson, Noah R. Feinstein, and David Meshoulam

INTRODUCTION: LEADERSHIP FOR SCIENCE EDUCATION AND THE PARADOX OF PLENTY

Kindergarten through Grade 12 (K–12) American schools sit amidst an extraordinary variety of resources for science education. Contemporary science educators have access to a range of networks offering curricular innovation and professional development opportunities that can enrich their practice and spark their own exploration of new scientific and technological fields. Science education and outreach receive a comparatively high degree of attention from federal funders such as the National Science Foundation (NSF) and the National Institutes of Health (NIH) who require investigators to specify education and outreach activities (American Association for the Advancement of Science [AAAS], 2007). The National Science Foundation alone commits over $800 million annually to science outreach, curriculum and professional development, and program evaluation activities (AAAS). Governmental and nongovernmental organizations have committed time and resources to training and recruitment issues that

The Role of Public Policy in K–12 Science Education, pp. 397–430

are central to science education reform, resulting in an array of new pathways into science teaching, as well as new access to science education resources for underrepresented groups of students. From the outside, it would seem inevitable that this wealth of science learning materials and professional development opportunities would make American science education a shining example of innovation and effective practice.

On closer examination, however, this "garden of plenty" looks very different. Kindergarten–12 schools have long been regarded by reformers as places that hamper or distort the implementation of innovative science curricula. Researchers and policymakers have identified numerous barriers to the widespread adoption of innovative practices, including the lack of alignment between state and local standards and innovative curriculum materials, the "mile wide and inch deep" nature of the standards that require teachers to focus on breadth rather than depth, and a chronic shortage of qualified teachers with the expertise to implement new approaches (NSF, 2006). These issues are compounded by pressures from outside the classroom. At the elementary school level, the testing mandates of the No Child Left Behind Act (NCLB) have until very recently emphasized reading and mathematics at the expense of science, making the teaching of science a lower priority in elementary schools (Center on Education Policy, 2007). At the high school level, anxious parents expect schools to reinforce the traditional science course sequences as a reliable pathway to college admission, which means that more innovative educational approaches look less appealing to them (Oakes, 2005). Science teachers, who may already struggle with inadequate subject-matter preparation, must cope with the combined demands of curriculum coverage, conservative community expectations, and high-stakes testing (Settlage & Meadows, 2002). These pressures create a chilly climate for local innovation and experimentation.

Although the push for instructional reform typically focuses on the classroom, teachers have little control over the out-of-school constraints on classroom practice. It falls to district and school leaders such as principals and curriculum coordinators to manage these constraints and make space for science education reform. It is these local school leaders who play a central role in establishing the conditions for improvement in science teaching and learning (Leithwood, Seashore, Anderson, & Wahlstrom, 2004; Spillane, Halverson, & Diamond, 2004). In this chapter, we take a distributed perspective on school leadership that focuses on the tasks rather than the official roles or the organizational conditions of leadership (Spillane, 2006). From a distributed leadership perspective, the key challenge is to determine how a variety of K–12 formal or informal school leaders, such as principals, instructional coaches, lead teachers, department heads, and district curriculum leaders, engage in the tasks that improve conditions for student learning (Spillane, Halverson, & Diamond, 2004). While teachers focus on classroom issues, leaders can take a school-wide, "meta-classroom" perspective, promoting classroom reform by carrying out organization-level tasks such as acquiring and allocating resources, monitoring instruction, establishing partnerships within and across schools, and legitimizing preferred reform strategies.

We begin with the premise that without the involvement of school leadership, the likelihood of meaningful, enduring change is small. Our goal is to explain why and how it is important to support school leaders in establishing the policy and practice conditions for science education reform. We do this by drawing a sharp contrast between the theory of action that has guided many previous science education reform efforts on the one hand, and theories of action grounded in local leadership practices on the other. A theory of action is the network of assumptions, strategies, goals and resources that guide behavior (Argyris & Schön, 1974). In the first section of the chapter, we describe the characteristics of the theory of action that currently guides most science education reform activities. This conventional theory of action seeks to influence local school conditions and improve student learning by establishing and implementing policies that focus on standards, curriculum materials, and professional development. The approach emphasizes goals, content, and pedagogy but often neglects the powerful influence of local conditions under which reform is expected to take root. In the second section of the chapter, we consider a different theory of action; one that guides science education reform from the perspective of local school leadership. We outline the community and policy constraints shaping the capacity for reform among local school leaders, and we argue that successful leaders reshape organizations by treating these constraints as affordances for transforming instructional practice. In the final section, we offer suggestions for how reformers can connect their goals with local theories of action to promote science education reform at the organizational level. We hope that these suggestions will enable leaders and policymakers to pursue reform agendas within the real-life constraints of school operation.

COMPONENTS OF THE PREVAILING SCIENCE EDUCATION REFORM THEORY OF ACTION

If we take the massive efforts by the United States to improve science education following the Soviet launching of Sputnik as the starting point, we can safely say that science education has been a national priority for over five decades. In this chapter we briefly review the history of national reform efforts and argue that the prevailing theory of action focuses on innovation at the level of schools, but then treats schools themselves as "black boxes," either excluded from the reform agenda or, at best, dealt with indirectly. The prevailing theory of action has three central pillars: (1) establish *standards* that create a common set of expectations, lending order and coherence to what teachers teach; (2) create standards-based *curricular materials* developed by experts with deep subject-matter knowledge; and (3) provide *professional development* opportunities to support curriculum implementation. This theory of action holds that when these components are implemented at the national, state and local levels, we can expect lasting changes in science teaching and learning.

Standards

Setting national-level content standards has been a central reform strategy for changing local practices. The report of the National Commission on Excellence in Education (NCEE), *A Nation at Risk* (NCEE, 1983) reserved some of its most trenchant criticism for science education, demanding that educators adopt "more rigorous and measurable standards and higher expectations for academic performance and student contact" (p. 3). (For an updated version of the rhetorical critique of science education, see the National Research Council's 2007 report, *Rising Above the Gathering Storm: Energizing And Employing America for a Brighter Future.)* At the time the report was written, the content of science instruction was influenced largely by textbook publishers, institutional inertia, and the force of tradition. Curricular materials were widely regarded as diffuse and outdated, emphasizing breadth over depth in a pattern that one prominent report condemned as "overstuffed and undernourished" (AAAS, 1991, p. xvi).

In the 1980s, reformers began to frame an agenda for improving science instruction that focused on nation-wide standards for high quality science learning. The benchmarks and standards published by the American Association for the Advancement of Science (AAAS, 1993) and the National Research Council (NRC, 1996) facilitated a profound shift in the conception, design, and implementation of science education reform. These documents provided coherence where there had once been chaos. Embraced by many as a "mechanism for school improvement" (Porter, 1994) the AAAS and NRC standards were followed by reform documents such as the AAAS's *Atlas of Science Literacy* (2001), that connected standards to specific goals, learning outcomes, school improvement measures, and teacher development benchmarks to build a "standards-based" roadmap for scientific literacy. Many state departments of education quickly adopted or adapted these national standards for their local and state-wide efforts to reform and standardize the science curriculum (Burry-Stock & Casebeer, 2003; Swanson & Stevenson, 2002). The commitment to standards continues to guide current reform efforts at both the state and national levels (Krajcik, McNeill, & Reiser, 2008).

Curriculum Materials

Developing and disseminating innovative curriculum materials has long been a favorite strategy by which researchers and policymakers have sought to influence classroom teaching and learning (Atkin & Black, 2003; DeBoer, 1991; Welch, 1979). New curricula, usually but not always in the form of textbooks, are a comfortingly familiar form of educational resource that can be easily adapted across a variety of classrooms and districts (Ball & Cohen, 1996; Schneider, Krajcik, & Marx, 2000). The earliest federally funded curriculum projects from the 1960s provided not only textbooks but also laboratory materials and films (see Physical Science Studies Committee, 1960; Biological Sciences Curriculum Study, 1963). Early efforts to develop reform-based curriculum explicitly excluded teachers from the development process (cf. Bruner, 1977). More recent curriculum development efforts entail extensive classroom and subject-matter research, and they aim to situate curricula within current standards and theories of learning (D'Amico, 2005; Krajcik & Reiser, 2004; Schneider & Krajcik, 2002; Singer, Marx, Krajcik, & Chambers, 2000). The best of these research-driven curriculum products are developed and revised in response to school-based field testing, classroom observation, and teacher feedback. Materials created in this manner range in scope from hour-long activities to multiyear programs, and vary in medium from textbooks to new technologies and laboratory activities. (See also DeLucchi & Malone, Chapter 12, this volume, for a discussion of how state and federal policy affects the development of curriculum.)

Shortly after development, most new curricula are simply released "into the wild." It is rare for the developers to stay involved with the dissemination and further development of the materials, which then begin their own independent, market-driven existence. Rowan (2006) suggests that researchers and publishers who develop new curricula face internal pressures to move on to new projects, even if it would benefit teaching and learning for a school to maintain coherent curricula. One reason for this is that external funders are always looking to fund something new, not support something that is already established. With new materials constantly under development, it comes as little surprise that past generations of innovative curricula litter the field of science education reform. Even though their market share pales in comparison to that of the commercial publishers, the sheer volume of research-based curriculum material, particularly in secondary science, overwhelms the capacity of any particular school or teacher to keep up with recent developments. In addition, teachers and administrators have few tools for discerning the value of any given set of materials. Even *locating* earlier generations of resources can be a challenge. Furthermore, commercially promising materials are sometimes acquired and sold by for-profit corporations, which further complicates the process through which curricula are presented to teachers and administrators. As a result, many organizations rely on gatekeeper Internet resources, such as merlot.org, amster.com, and free.ed.gov that aggregate and index materials to provide easier access for practitioners. Although these websites serve an important consolidating function, they offer little guidance to teachers in choosing appropriate materials or implementing the materials they choose.

The commercial market poses further challenges for curriculum material dissemination. Standards-based reform is guided by the assumption that the producers of curriculum materials, in order to survive in a disordered market, will use the various science content standards to filter existing materials and to select innovative curricula to incorporate in new editions. Evaluations of middle and high school textbooks conducted by the staff of Project 2061 (Kulm, Roseman, & Treistman, 1999), however, turned up both startling omissions in content and a variety of extraneous material. The Project 2061 textbook evaluation criteria are available at http://www.project2061.org/publications/textbook/hsbio/report/analysis.htm. The disconnect between the hypothetical "focusing" function of standards-based reform and the reality of textbook content stems from the long-standing practice of layering new content on top of older material. Curriculum developers also must balance their interest in complying with national standards against pressures to deliver content familiar to classroom teachers. In short, even without considering the attenuating effect of divergent implementation, the effect of curriculum-based reform is limited by the complex and chaotic marketplace for new materials.

Professional Development

In-service professional development is the final of three common components of the prevailing theory of action for science education reform. Pre-service teacher education is obviously an important contributor to the preparation of science teachers. However, school-level leaders have little direct effect on teacher preparation programs. Reformers who wish to influence science education as a whole can address issues of teacher preparation (see, e.g., NRC, 2001; NSTA, 2003; Kahle & Woodruff, Chapter 3, this volume), but we feel that those who seek to influence practice within the school need to focus on the resources and practices available to school-level leaders for improving teacher capacity.

Federally funded in-service professional development efforts may even pre-date curriculum development as a strategy to influence science teaching (Welch, 1979). Some of the earliest involvement of NSF in science education was through the support of summer institutes for teachers. As noted above, the inadequate subject-matter preparation of science teachers is frequently identified as a major barrier to the improvement of K–12 science education. Recent statistics reveal that only 72% of K–12 science teachers had a college major and certification in a science-related field, and 20% of high-school science teachers did not have appropriate subject-matter certification (NCES, 2008). Professional development projects, regardless of type, usually are intended to improve teachers' "science content knowledge, process skills, and attitudes toward teaching science" (Radford, 1998, p. 74). According to Supovitz and Turner (2000), this ideally is done by providing teachers with "concrete tasks ... focused on subject-matter knowledge, connected to specific standards for student performance, and embedded in a systemic context" (p. 963). Professional development may be provided as part of a curriculum development project; it also may be focused on a particular pedagogical approach or on the alignment of instruction or assessment with standards. The National Research Council, for example, produced Professional Development Standards to accompany the National Science Education Standards. The NRC's (1996) *Professional Development Standard A* states that the:

> professional development for teachers of science requires learning essential science content through the perspectives and methods of inquiry. Professional development science learning experiences seek to involve teachers in actively investigating phenomena that can be studied scientifically, interpreting results, and making sense of findings. (p. 59)

Science professional development opportunities have resulted in a "menu" approach that results from the diversity of providers in the wider ecology of school improvement. Teachers and schools may choose among options that extend from biomedicine to nanotechnology and from physics to ecology; some topics reflect national standards while others do not, and the content of professional development programs can be delivered in broad strokes or very specifically. The NRC and the Smithsonian Institute, for example, offer four summer workshops to improve "teachers' understanding of science and pedagogy" in order for them to "become more able to engage young minds in the sciences" (National Science Resource Center, 2003, p, 11). At the other extreme, a 2008 NSF Summer

Institute in Applied Biotechnology & Bioinformatics at the University of California-Davis, promoted the integration of new high-technology skills and knowledge "into the traditional high school science classroom. Participants will learn biotechnology and bioinformatics skills and develop curriculum around various disciplines" (University of California-Davis, n.d.). Discerning consumers of professional development can assemble a coherent learning trajectory from this menu of options. For many teachers, however, this diverse marketplace results in a fragmented and inconsistent professional development program. The prevailing theory of action, which relies on providing interested, motivated teachers with the option to acquire intensive experience with new science concepts and practices, fails to take into account the context of everyday practice.

Research on effective professional development for teachers finds that the amount of time teachers spend in professional development is strongly correlated with improved student achievement (Yoon, Duncan, Lee, Scarloss, & Shapley, 2007). (See also Kahle & Woodruff, Chapter 3, this volume, for a further discussion of the impact of professional development on teacher practice.) Effective professional development depends on the incorporation of teacher learning into daily instructional practice. As such, professional development projects are most effective when teachers are able to make explicit connections to their particular school contexts, ideally with the help of sustained, school-based follow up (Bredeson, 2002; Darling-Hammond, Wei, Andree, Richardson, & Orphanos, 2009). In short, effective professional development experiences need to both elicit opportunities for developing new ideas and provide a framework for integrating new ideas into the contexts in which teachers' actually work. As we will discuss in more detail below, the success or failure of these experiences implicitly relies on the support of school leaders.

SUMMARY

Science education reforms have typically been designed and developed in settings far from the local contexts in which they will be implemented. Despite the best efforts of researchers to take "complex practices" and local conditions into account (Confrey, 2006), each new innovation will be transformed by local pressures and competing interests at the state, district, and school levels. The prevailing theory of action for science education reform is guided by a decontextualized view of teaching practice in which teaching can be shaped by standards, curriculum, and professional development with little regard to the contexts in which practice takes place. Key leverage points for reform are identified at the level of policy where standards are determined, in research universities where science education resources and curricular materials are developed, and in the classroom where curriculum are chosen and enacted by teachers.

The individual school itself is absent from this theory of action. Summer institutes and workshops are typically offered at sponsoring university or research institutes (Westerlund, 2002), removing teachers from the school setting for a period of time ranging from hours to days. Although institutes and workshops effectively distance teachers from the day-to-day distractions of the classroom, and they facilitate direct and unfiltered communication between teachers and teacher educators so that teachers can focus single-mindedly on exciting new material, professional development opportunities provided outside of schools can omit considerations of local context; in particular, how local conditions constrain the application of new pedagogies or curricula. Once teachers return to their schools, they must adapt new ideas to the existing culture and expectations of their schools and classrooms.

We argue that this theory of action is untenable because it fails to engage schools in systemic change. Reform policies established and enacted outside the school are unlikely to be successful at the local level without a more careful consideration of the sociocultural contexts of innovation. We suggest that school leaders, who create the spaces for innovation within highly routinized and change-averse institutions, are key to successful science education reform. In the next section, we examine the opportunities that leaders have created for innovative practice in the context of existing school systems, and use concrete examples to illustrate how reform-minded school leadership can help teachers and students make better use of the raw material of science education reform, including the standards, curriculum materials, and professional training opportunities already in existence.

SCHOOL LEADERSHIP AND REFORMED SCIENCE INSTRUCTION

As described above, science education reform typically is planned *outside* of schools for implementation *in* schools. The transition from good ideas about K–12 science teaching and learning to systemic improvements in K–12 science classrooms is the responsibility of school and district leaders. Teachers, by themselves, can and often do, initiate innovative practices for teaching and learning. But, without coordinated organizational support, teacher initiatives can be pushed to the margins of the school instructional program and rendered irrelevant to the overall instructional practices of the school. When teachers take on the tasks of transforming the organizational conditions of teaching and learning, they become *de facto* instructional leaders; but implementing systemic science reform requires that both formal and informal leaders take responsibility for improving the instructional system. Furthermore, the work of instructional leaders must necessarily address the local contexts that influence teaching and learning. From a school leadership perspective, whether science education reformers promote innovation through standards, curricula, or professional development, they can be successful only if they engage the constraints that leaders face in designing a world for improving teaching and learning.

School leaders act as gateway custodians for the ideas and practices driving systemic school improvement (Honig & Hatch, 2004). Leaders are responsible for bringing new faculty into schools and for measuring the effectiveness of teacher practice through the teacher evaluation process. Leaders acquire and allocate resources—including money, time, and people—to support local instructional initiatives. Leaders typically authorize the selection (or the creation) of school- and district-wide curricula and instructional programs. Within schools, leaders use their authority to structure professional interaction among teachers by creating department structures, calling meetings, and establishing committees to work on specific problems or projects. Professional community, widely recognized as a key organizational prerequisite for substantive reform (Bryk & Schneider, 2002; Stoll & Louis, 2007), typically results from faculty interactions that take place in the meetings organized by school leaders.

School leaders are not always seen by reformers as positive contributors to innovation, however. The ways in which leaders control the structures and processes they are responsible for are perhaps more frequently seen as obstacles to change. Leadership agendas often conflict with, and neutralize, reform efforts. Worse, the failure of leaders to manage the school structures and processes in a way that would create the opportunity for improvement in teaching and learning can undermine teachers' efforts and desire to pursue meaningful reforms. The failure to establish conditions for improvement can be rooted in a lack of leadership will and skill. The difficulty in improving the conditions of practice, however, also is a reflection of the highly constrained design spaces within which leaders work. School leaders work in complex *systems of practice* shaped by structures and priorities that are the result of historical decisions about the organization of teaching and learning (Halverson, 2003). Many features of a school's system of practice are beyond the scope of local school leaders' capacity to change. For example, practices concerning age grading, union contracts, and special education provide significant constraints on the range of innovation. Faculty members also come to schools with strong beliefs about how teaching should be organized and with deeply formative prior experiences that influence how they believe they should interact with students. The increasing use of standardized assessments and curricula at both national and state levels constrains and standardizes instruction, not only in the classroom but also across grades and across schools. As these new standardized structures are incorporated into the daily work of schools, they form a resilient system of practice that is remarkably resistant to change.

District and school leaders can create the space for instructional change by addressing or co-opting the external pressures that bear most heavily on their school. Many leaders become so focused on responding to accountability pressures that they exercise their power to create the impression of compliance with policy demands while avoiding significant changes to instructional program (Meyer & Rowan, 1983). (See also Osborne, Chapter 2, this volume.) Other leaders seek to orchestrate substantial instructional changes in some subject areas while leaving other areas unexamined (Center on Education Policy [CEP], 2007). In the language of decision-making, these leaders engage in "satisficing" behaviors (Simon, 1983) that help schools meet accountability requirements but also yield to local pressures to maintain existing practices. As local leaders gauge competing pressures to improve different areas of the instructional program, science education reform seldom emerges as the top priority (even as international comparisons push policy makers to see science education as a national priority).

In the following, we engage in a two-part discussion that first draws out how local contexts blunt reform, and then identifies the leverage points that could be used to reinvigorate reform efforts at the local level. We discuss the different ways in which the science education reform agenda is filtered through the policy pressures that operate at two levels of the K–12 educational system; elementary schools and high schools. Our analysis will demonstrate how differing institutional contexts guide (and qualify) leadership efforts in distinctive ways. At the elementary level, we describe how the science reform agenda has been co-opted by the high-stakes accountability pressure to improve reading and writing. At the secondary level, we describe how leaders in high-poverty districts must make do with a shortage of resources, while leaders in resource-rich districts face pressure from empowered parents to preserve existing practices that they interpret as critical to college admissions. At first glance, it may seem that these pressures simply stifle science education reform. In our subsequent discussion, however, we point to leverage points that science reformers can exploit within the current contexts to further the goals of science education reform.

The Context of Elementary School Leadership

Much of the attention for science reform by policy makers has been justly targeted at early elementary school programs. Science reformers, however, increasingly have become frustrated by school-level resistance to innovative practices. This resistance is the direct result of high-stakes accountability policies. Elementary school leaders and teachers have reshaped mathematics and language arts instruction in response to the high-stakes accountability demands of the No Child Left Behind Act (2001) by increasing the attention given to those subjects. Even though high-stakes tests are now required in science as well, under NCLB those test results are not used in determining a school's adequate yearly progress (AYP), so that most schools have chosen to focus on raising achievement in mathematics and language arts (Marx & Harris, 2006). Schools have increased the allotments of time for mathematics and literacy instruction and reduced the time and resources available for science. A Center on Education Policy (CEP) report found that, from 2001–2006, elementary school instructional time in English and language arts increased by 47%, and mathematics instructional time increased by 37%. About 1/3 of this increase in instructional time came at the expense of science instruction (CEP, 2007).

The increased attention to literacy instruction, in particular, is forcing a change in the nature of science instruction in many places. When elementary schools make a commitment to science education reform, it often takes the form of content-based literacy instruction. Lee and Luykx (2005), for example, felt the need to persuade school leaders and teachers of their science intervention's value by describing how it could improve the students' reading and literacy skills. Although it is useful to be able to read about science with understanding, if that is the primary goal, science instruction could be stripped of its focus on experimentation and inquiry. The design of state science tests typically emphasizes reading comprehension and the use of logical inference skills over specific subject matter knowledge, asking students to answer multiple-choice questions based on their ability to draw inferences from short textual passages. While critics such as Yager (2005) argue that the reduction of science to literacy misses the main point of teaching science, the format of existing state tests suggest to many leaders that elementary science can be adequately addressed as a form of reading comprehension. (See also Sarmant, Saltiel, & Léna, Chapter 6, this volume; and DeLucchi & Malone, Chapter 12, this volume, for a further discussion of how elementary science educators are using the science curriculum to teach basic linguistic skills.)

Examples of Elementary School Reform: Formative Assessment and Professional Communities

Although the pressures of high stakes accountability policies in mathematics and language arts divert attention from science education reform, some elementary school leaders have responded in ways that leave the door open for substantive improvement in science as well as literacy and mathematics education. Two important leadership strategies that characterize local theories of action are: (1) *investment in formative assessment practices* and (2) *the creation of professional communities to share local expertise*. Both strategies contain lessons for reformers seeking to improve science teaching in a high-stakes accountability environment.

Investment in Formative Assessment Practices

It is difficult for schools to use existing standards-based state assessments for formative assessment of student learning. This perspective on formative assessment is substantially different from the version advanced in the classic Black and Wiliam (1998) discussion. Early work on formative assessment emphasized providing pupils with appropriate feedback to guide the learning process. Contemporary discussions of formative feedback position teachers, instead of students, as the crucial learners. Benchmark assessment systems, for example, provide teachers with information on student learning as formative feedback on *teaching* practices. (For more detail on this shift in the use of formative assessment, see Halverson, Prichett, & Watson, 2007.)

Information from state tests is not timely and it is often not aligned with local learning goals or instructional practices. The results of state tests take too long to arrive for teachers to use information to adjust their instructional programs. And, even in the best of circumstances, where the state tests are aligned to the state content standards, specific items may be only loosely related to the school's instructional program. These limitations have led school and district leaders to either purchase benchmark assessment systems from external vendors or develop local formative assessment practices to guide teaching toward tested outcomes (Mandinach & Honey, 2008). Even if these purchased or locally developed tests are well aligned to the school's instructional program, and often they are not, it is unlikely that teachers will gain much benefit from them unless they have the opportunity to share the results of the tests with other teachers in the context of existing lessons, quizzes, and homework, and to discuss with their colleagues the implementation of new instructional strategies and approaches (Newmann & Wehlage, 1993). Such discussion of assessment results also provides an opportunity for the teachers to evaluate the quality and usefulness of the formative assessments and to recommend changes in them. This process of enriching instructional practices with formative assessments gives teachers a path toward collective ownership of the school instructional program.

The Creation of Professional Communities to Share Local Expertise

Assessments produce information about what students have learned, but schools also need professionals who are capable of acting on that information to improve the learning experience for students. In response to that need, many elementary schools have improved student outcomes, particularly in language arts, by cultivating internal instructional expertise in robust professional communities. These communities of local experts propose curricular and instructional initiatives, and then they use coaching and team teaching to leverage the insights sparked by the results of the formative assessments (Blanc, Christman, Hugh, Mitchell, & Travers, 2009; Halverson, 2010). Although some of the research on coaching suggests that new curriculum and instructional initiatives are sometimes not distributed through the school because resources to support coaches are misallocated or co-opted by preexisting instructional priorities (e.g., Mangin & Stoelinga, 2007), other studies conclude that coaching provides a promising strategy to improve professional practice in schools (Showers & Joyce, 1996).

The use of formative assessments, followed by curricular and instructional initiatives that are then distributed through the teaching community by means of coaching and team teaching can create what Bereiter and Scardamalia (1993, p. 106) call *second-order environments*. These second-order environments foster progressive problem-solving activities and push participants to continuously examine and revise their own expertise and generate new, innovative solutions to educational problems. Such learning communities rely on teachers and other professionals who are trained in new practices to enhance existing school expertise and catalyze new opportunities for interested colleagues to acquire useful knowledge and skills. The best professional communities allow teachers to work with coaches and support staff to try out new practices in classrooms, and to use formative assessment data to measure the degree to which those new practices improve student learning. Second-order environments provide a chance for teachers to both learn about new ideas in conversation with colleagues and to sharpen practices through collaborative experimentation. Thus, professional communities provide a path for leaders to distribute existing expertise and catalyze the development of new instructional expertise.

Assessment becomes formative when it sparks the kinds of communication and reflection that successfully transform practice. Simply providing benchmark science testing data to school communities who do not have sufficient expertise may simply result in more data overload. Elementary schools often start with fewer subject matter experts, which puts a premium on the ability of the school to generate the internal capacity for teachers to share information on teaching and learning. However, as school leaders increasingly move toward a theory of action focused on the development of professional communities and grounded in formative measures of student learning, it is possible to see how new approaches to teaching and learning might emerge in science as well as mathematics and the language arts. Science educators can learn from the experience of school leaders in structuring professional communities in mathematics and the language arts. This would mean developing benchmark science assessments that catalyze deep insights for both students and teachers, and fostering in-school professional communities that support teachers as they try out new practices and discuss the results of their innovation. If professional development opportunities allow teachers to engage with innovative content as learners, and link that content with benchmark assessment data, that professional development could help teachers understand patterns in student learning. Furthermore, situating opportunities to learn science in the context of preexisting efforts to improve literacy instruction could help teachers transfer their professional learning strategies to new domains. By aligning their efforts with the theory of action that guides literacy and mathematics education, science educators may be able to make accountability demands work in their favor.

The Context of Secondary School Leadership

The challenges of improving science instruction are different and perhaps more complex in high schools. While many elementary schools have been able to change internal practices to meet the demands of NCLB accountability policies in mathematics and language arts, secondary schools continue to struggle to achieve basic goals such as preventing student dropout and providing adequate preparation for college-bound students. It has been remarkably difficult to reform instructional practices in high schools, even in high-achieving schools that would seem to possess the resources to support reform. As we will argue in the next section, the culture of professional autonomy and public pressure for narrow definitions of success pose obstacles to reform in all secondary schools. The resistance to change in both high- and low-resource contexts points to a crucial role school leaders can play in identifying the key instructional areas for experimentation and innovation. This section will outline the contexts within which secondary school leaders work by tracing how the traditions of professional practice and organization, together with community pressures, reinforce existing models of science instruction. We then describe *Project Lead the Way*, an example of how innovations might be designed to take advantage of the contexts in which secondary school leaders engage in science reform.

Traditions of Professional Practice and Organization

The first obstacle to system-wide reform in secondary science education stems from existing traditions of professionalism. High school teachers have strong traditions of autonomy, and they define their professional roles according to their personal beliefs about what students can and should learn (McLaughlin & Talbert, 2001). The structure of professional interaction in secondary schools reinforces these traditions through academic departmental structures that often act as professional confederations rather than learning organizations (Siskin, 1995). This is not to say that high school teachers are reluctant to embrace new ideas and practices: many of the most exciting high school innovations are developed by teachers who use their autonomy to fundamentally alter instructional traditions. Clifford (2009) described how high school science teachers in two schools used strong collegial relationships and external university and professional organization networks to create the conditions that enabled them to successfully modify their teaching. Still, their reform efforts depended on teachers and schools that were willing to take on leadership tasks. Without such risk-taking leadership, focused on the development of collegial relationships within the school, school administrators and academic departments tend to reinforce teacher autonomy, making it difficult for secondary school leaders to instigate cross-school instructional innovation.

How Student and Parent Expectations Contribute to Curricular Rigidity

The second obstacle to school-wide reform in secondary science education derives from the interplay between student aspirations, parent demands, and college admissions standards. This interplay creates pressures on high school leaders to assign the school's most qualified teachers to the highest achieving students in a traditional science curriculum sequence. Secondary school student populations typically are divided into two kinds of students: those for whom high school is the path to higher education, and those for whom high school is the last stop in the educational process (Sedlak, Wheeler, Pullin, & Cusick, 1986, p. 48). Students in the first group, those for whom high school is a path to college, expect to take a specific sequence of classes (typically involving biology, chemistry, and physics) that articulate well with the admissions expectations of selective colleges. Although the traditional sequence has been supplemented by the addition of Advanced Placement courses, it has remained substantively unchanged for decades.

States reinforce the traditional course sequence by increasing science course requirements for graduation. Traditional coursework fits the expectations and admissions requirements of most state and private university systems. The University of California system, for example, advises students to take biology, chemistry, and physics; the state of Washington requires incoming high school students to take two years of laboratory science, including one year of biology, chemistry, or physics to be considered for admission into state university. (California data can be found at http://www.admissions.ucla.edu/Prospect/adm_fr/fracadrq.htm; Washington data at http://www.hecb.wa.gov/research/issues/documents/MCASOverviewstudents.pdf) Because college admissions programs review transcripts rather than course content, and because alternative titles can give the appearance of a less rigorous curriculum, high schools feel pressure to maintain existing course titles. Finally, parents' perceptions of college admission requirements cause them to demand the kinds of programs that lead to successful college admission (Henderson & Berla, 1997). The press for a "legitimate" and high quality science course sequence brings together faculty members with strong science credentials to teach in the core academic program (Murphy, Beck, Crawford, Hodges, & McGaughy, 2001).

The college-preparatory academic program contributes to a *de facto* tracking system that splits students into science haves and have-nots (Gamoran & Weinstein, 1998; Lucas & Berends, 2002). Because a high percentage of students in the college-preparatory track already meet state minimum competency standards, the high-resource college-preparatory track would seem a natural home for science innovation. Unfortunately, the conservative atmosphere that surrounds the core academic program results in a narrowly defined focus on achievement rather than curriculum reform or innovative instruction. New courses organized around content areas such as nanotechnology, systems biology, information sciences, or engineering often have difficulty gaining acceptance in the college-preparatory track, in part because there is no room in the traditional course sequence for college-bound students, and in part because students have not been prepared by the existing courses to engage in the core concepts of the emerging areas of inquiry in those new courses.

Given that the most experienced teachers are matched with college-bound students in the traditional core sequence of science courses, the most innovative reform-based science programs may need to be implemented outside this sequence with non-college-bound students and less experienced teachers. This presents a challenge for those teachers outside the college-preparatory track, who often have weaker science backgrounds and professional networks. Also, although offering reform-based science education courses outside the college-preparatory curriculum may increase the school's capacity to address the needs of traditionally underserved students, it does not obviate the need for reform in college-preparatory science. Strengthening science education in high schools, both within the college preparatory track and outside of it, requires leaders to create space for innovation.

An Example of Secondary School Innovation: Project Lead the Way

In the elementary school section above, we described an emerging theory-of-action focused on the development of assessment-driven professional communities as a way to help reformers situate science reforms in existing school contexts. Middle schools and junior high schools present yet another set of design challenges for science reform. In many ways, middle and junior high reform rests part way between the issues of elementary and high school contexts; middle school programs face accountability pressures similar to elementary schools, while many middle school faculties share the departmental organization of high schools. The interdisciplinary organization of many middle-level instructional programs, however, provides a unique affordance for the design of science interventions. For a discussion of the developments and challenges of science reform in middle and Junior high schools, see Lee, Songer, and Lee (2006); Ruby (2006); Hewson, Kahle, Scantlebury, and Davies (2001); and Kesidou and Roseman (2001).

In the secondary schools section we now describe a specific intervention to highlight how reformers have situated innovative programs in existing school contexts. To illustrate how innovative program design can support local efforts to operate *within* these constraints, we briefly consider one of the more compelling contemporary examples of comprehensive reform: *Project Lead the Way* (PLTW). This reform has strong roots in career and technical education, outside of the traditional college track, and emphasizes connections across subject-based departments. We do not argue that PLTW is the first or the only program to succeed in secondary schools; instead, we use PLTW to illustrate how a program can be designed to work within the constraints of existing schools and to reveal how secondary school leaders and teachers can create spaces for reformed practice.

First implemented in 1997, PLTW is a nationally recognized high school pre-engineering program that integrates a series of traditional science and math classes with another series of project-based learning courses that require students to apply mathematic, scientific, and technical knowledge to address engineering problems. The PLTW program includes a 2-week professional development program for teachers and a standardized exit exam. Bottoms and Uhn (2007) found that PLTW students scored significantly higher than their peers on a NAEP-referenced test of math and science and that PLTW students were more likely to complete four years of math than their peers. Phelps, Camburn, and Durham (2009) found that PLTW students reported significantly higher levels of intellectual openness than their peers, as indicated by their willingness to discuss open-ended questions and their desire to learn.

Over 3,000 schools in 50 states use the PLTW program. Each PLTW school signs a contract to abide by the conditions for participation. The PLTW contract reveals critical features of a theory of action about how leaders in secondary school instructional programs can support innovation (PLTW, 2007). The PLTW program requires participating schools to engage in a partnership with other districts, colleges, and universities and the private sector. Although the program uses a traditional summer workshop training approach to prepare the participating teachers, it also includes in-service training intended to link teachers with external networks of ideas and professionals focused on PLTW implementation. Participation in broad professional networks external to the school is an important aspect of successful school reform (Huberman, 1995; Lieberman, 2000) because it creates additional opportunities for share ideas. The structure of PLTW requires schools to serve as "model" programs, available for observation and inspection by other participating schools. Inter-school visits replicate some aspects of the professional communities that elementary school leaders use to promote change in literacy and math instruction.

The PLTW program also requires schools to commit significant resources to implementing the program. First, a school must obtain district-level approval for the program. The school must also implement four new courses in engineering and ensure that the program is integrated into the school instructional sequence. Students participating in PLTW must enroll in at least two classes in the school mathematics program. Because PLTW is seen as an alternative to the traditional science sequence, college-prep students may opt out of PLTW enrollment. However, because PLTW provides a viable science course sequence and a link with the existing school mathematics program, students motivated by the engineering perspective on science may begin to break down the *de facto* wall between academic and non-academic tracks. By creating extended professional networks and integrating an engineering perspective into the existing academic program, the PLTW program demonstrates how reforms can produce sweeping changes in science education by tapping into the existing resources of a school community. School leaders who already are committed to the development of professional communities and program integration can build on the foundation of new programs such as PLTW rather than approaching innovation as a distraction from existing priorities.

AFFORDANCES FOR CHANGE

In this final section, we focus on the role of school leaders in encouraging and sustaining innovative practices that improve student achievement. Our discussion thus far has described constraints that limit the range of action for instructional leadership as well as promising theories-of-action that make room for innovative reform in spite of these constraints. Here, we generalize four leverage points for reformers to consider in supporting science reform from a school leadership perspective: connecting teachers with each other, connecting teachers with resources, protecting the early stage of innovation, and building the subject-matter capacity of teachers. For each leverage point, we suggest how reformers *outside* of schools can address the existing constraints of school reform in ways that can support the work of school leaders in enacting change.

Connecting Teachers With Each Other

The development of professional communities of teachers, as described in the elementary school section above should be a central feature of a school-based theory of action for science education reform. Professional communities have several benefits in the context of highly constrained, tradition-bound systems like schools (Bryk & Schneider, 2002; Halverson, 2003; Louis, Kruse, & Bryk, 1995). First, teacher collaboration promises to increase the depth and rigor of reform by creating a shared focus on persistent classroom dilemmas. Research suggests that collaboration enables teachers to test hypotheses about practice and address instructional problems at a deeper level (Krajcik, Blumenfeld, Marx, & Soloway, 1994; Loucks-Horsley & Matsumoto, 1999; Thompson & Zeuli, 1999). Second, collaboration increases the efficiency of reform by enabling teachers to benefit from the expertise and experience of their peers. Teachers who receive help from colleagues who already are implementing new projects report they are significantly more likely to change their own practices (Penuel, Frank, & Krause, 2006; van Driel, Beijaard, & Verloop, 2001). When those more experienced teachers act as "peer coaches" or "teacher leaders," the gains may be substantial (Ruby, 2006). Developing and using teacher leaders requires time (i.e., release from normal classroom responsibilities) and training to achieve results, but the impact will likely be greater than if school and district administrators tried to do that work themselves. Third, collaboration among teachers increases the durability of reform by enabling teachers to share the burden of innovation and by creating social reinforcement structures for positive change. Case study research suggests that reforms are easier to sustain and less vulnerable to external pressures when implemented by groups of teachers (Lee, Songer, & Lee, 2006). Finally, a strong sense of community among teachers is linked to greater student achievement (Ross, Hogaboam-Gray, & Gray, 2004).

How can science education reformers help school-level leaders such as principals and curriculum coordinators create professional communities? A first step is to encourage school leaders to participate in the professional development activities provided to science teachers. At first, leader involvement might have a chilling effect on teachers' willingness to question their knowledge and practices. However, over time, the participation of formal leaders can help to build a professional community around knowledge instead of rank. This would both improve leaders' own understanding of innovation and signal their support for ambitious changes in classroom practice (Gerard, Bowyer, & Linn, 2008). School-level leaders are typically less knowledgeable about science content and science pedagogy than the teachers they supervise, and, as such, are not well positioned to discern the value of a given science education reform initiative or its relation to other instructional initiatives. (According to 2006 data from the Wisconsin Department of Public Instruction, only 5.5% of certified administrators in Wisconsin, for example, have degrees or credentials in any of the sciences.) On the other hand, school-level leaders are ideally positioned to identify teachers who can take a stronger role in leadership practices. While school-level leaders might resist becoming intimately involved with on-going reform efforts for the same reasons that senior managers everywhere tend to avoid becoming too deeply involved in any particular product or initiative, school-level leaders with about a general sense of the direction and value of a reform initiative could help cultivate the organizational conditions for effective adoption.

Reformers from outside the school should promote the use of distributed leadership strategies, in which school-level leaders delegate some responsibility for school-wide instructional reform to teachers. Distributing responsibilities to specialists, coaches, and department chairs empowers teachers to establish or change instructional program priorities (Clifford, 2009; Spillane, 2006). School-level leaders can release teacher leaders from some of their teaching responsibilities in exchange for reform-specific mentorship and management duties. The growth of community depends on collective ownership of reform or professional development projects. Leaders should find ways to involve all teachers in meaningful, reform-related work, even (especially!) those who are initially reluctant to participate. Finally, outside reformers should connect school leaders with professional networks outside of their own schools. Unlike school leaders, who are experts in their local context, outside reformers are ideally positioned to build bridges to professional communities beyond the school walls, both in other schools and in universities and parallel research communities. To accomplish the next level of professional development, communities in which either subject matter or reform expertise is sparse may need to rely on distributed, virtual expertise networks such as on-line discussion groups or virtual university programs. By encouraging visits and collaboration across institutional lines, reformers can help teachers gain a new perspective on their existing instructional practices and develop innovative practices that suit their particular contexts.

Connecting Teachers With Resources

Another central feature of a school leadership-based theory of action for science education reform is linking teachers with curricular and community resources. This strategy is connected to the existing reform-based theory-of-action through curriculum and materials development. Research suggests that school leaders can play a critical role in science education reform by connecting teachers with resources (Spillane, Diamond, Walker, Halverson, & Jita, 2001). Although some resources will be out of reach for financially struggling districts, financial hardship is not an insurmountable barrier to reform. Leaders in resource-poor schools can be shown how the astute use of social capital and access to local, low-cost resources can contribute to successful innovation:

> in investigating the identification and activation of resources for leading science instruction it is imperative to look beyond the particular school to the multiple contexts in which that school is nested. [A]n interagency perspective, as distinct from an exclusive focus on the individual school, is important … to understand the resources for change. [I]t was essential to look beyond the school to the various agencies with which … staff networked in order to forge change in science education. (p. 937)

University research projects often create opportunities for schools to access cutting edge professional development or innovative curriculum projects. School-level leaders in high-poverty schools can be encouraged to develop research-based partnerships that provide resources for science education reform. The richness of the science education resource pool truly becomes an asset for schools that are without ready access to community-based collaborators. In addition, although some well-known curricular innovations are only available for purchase, many other innovative packages can be obtained at low or no cost. Schools wishing to incorporate innovative curriculum materials, for example, could use the well-established (and essentially free) Bottle Biology program as a starting place (Krajcik et al., 1996). Reformers at the district level can create legitimate opportunities for schools to engage in new practices by collecting and distributing information on these resources and on the professional networks that use them.

Reformers could also focus on helping schools access assessment tools and practices. One of the lessons that teachers, school-level leaders, and reformers have drawn from the NLCB era is that internal instructional practices can be usefully restructured to help meet testing goals. Meeting the demands of high-stakes accountability has created a new market for assessment products (Burch, 2009). In all likelihood, testing will continue to drive instructional practice, and reformers will continue to develop and distribute high-quality assessments to influence teaching and learning. The implementation of formative assessment tools, in particular, can strengthen existing professional communities as teachers work together to make sense of assessments in terms of their daily practices (Prichett, 2007).

Some of the curricular innovations referenced in the first section of this chapter, such as The Full Option Science System (FOSS), contain examples of how assessments can be integrated into curriculum materials. (http://www.fossweb.com/) (See DeLucchi & Malone, Chapter 12, this volume, for further discussion of the formative assessment system built into the FOSS materials.) Many innovative science materials are developed hand-in-hand with new learning technologies—often the same technologies that have been used to pioneer performance assessment systems (Mislevy & Knowles, 2002) or video-game development (Gee, 2003). The Calipers project (Quellmalz et al., 2007), for example, demonstrated how technological simulations can allow teachers to engage in formative assessment of student learning; while the Compass project (Puntambekar, 2006) showed how new technologies can be used to assess student collaboration. Although many of these interventions suffer from the reluctance of conservative institutions to try new approaches, because of the current emphasis on accountability through assessment, new markets have emerged in recent years to satisfy increasing school-based demand for formative assessment tools (Burch, 2009). Reform-oriented researchers can take advantage of these new market opportunities by developing next-generation assessment tools that support rather than compete with instructional innovation (see, e.g., Buxner, Harris, & Johnson, 2008).

Protecting and Supporting Innovations During the Early Stages of Implementation

A third feature of a leadership-based theory of action for science education reform involves an understanding of the developmental stages of innovation. Current thinking about instructional reform is dominated by fidelity models that emphasize the consistency with which an innovation is employed or implemented, and we rush to judgment about an innovation's success or failure based on early student outcomes. This binary, and often premature, judgment about success or failure can overlook the incremental effects that a reform effort may have on a school's instructional program. Larry Cuban's 1998 essay about success and failure in comprehensive school reform is particularly instructive. Drawing on several examples from the history of school reform, Cuban describes how premature judgments of success or failure can cripple a reform project. When a reform is judged a failure in its early stages, public approbation and pressure to abandon the project make any further progress difficult. On the other hand, reforms that are seen as early successes may also suffer when school and community attention shifts to "unsolved" problems. To further complicate matters, "success" means different things to practitioners than it does to reformers and concerned parties outside the school. According to Cuban, teachers emphasize the adaptability of a reform to local circumstance over fidelity to the original reform vision, and they prefer a long-lived reform to an intensely but temporarily popular one. School and district-level leaders, under pressure from community members and policy-makers, may have exactly the opposite preferences. These differing ideas of success can lead to conflict.

Taylor (2001) reinforces Cuban's analysis, drawing attention to the particular vulnerability of reforms in their early stages. After listing a number of factors that combine to make reform more difficult, Taylor admonishes reformers within and outside the school to support innovation and experimentation through sustained professional development and community-building. He also notes that it is critical to buffer reforms from inevitable fluctuations in external support. Protecting particular reform projects means committing to a consistent reform focus, including necessary professional development resources, for a period of years rather than months. It means protecting teachers engaged in the early stages of reform from community pressures and suspending judgment on innovative projects until they have time to reach their potential. Reformers who seek to introduce new ideas from outside the school should urge school leaders to recognize the fragility of early innovation and relieve participating teachers of the need to show immediate results. Within the school, it is probably a good idea to set benchmarks and timelines at the beginning of the reform process so that teachers and school-level leaders share a common set of expectations about the progress of reform.

Individual reforms can contribute to or detract from ongoing efforts to improve school instructional capacity. Maintaining the coherence of a school's instructional program is an important aspect of high-quality school leadership (Newmann, Smith, Allensworth, & Bryk, 2001). When different reforms have differing instructional outcomes, agendas, and resources, they find themselves competing for scarce professional development bandwidth in schools, which distracts teachers and leaders from a focus on school-wide goals. To avoid this problem, leaders must select reforms that reinforce and extend prior efforts to build instructional capacity. As we outlined above in the examples of literacy-based professional community and *Project Lead the Way*, grounding new reforms in existing capabilities can create a fertile environment for instructional change. A school-wide focus on instructional reform can run counter to the tradition of teacher-directed innovation, however. Savvy leaders recognize the necessary balance between bottom-up and top-down reforms, and they can reconcile this apparent opposition by promoting the adoption of innovations that stretch existing capacity, but which do so within the bounds defined by school or district-wide goals.

Build the Subject-Specific Competence of Teaching Staff

We have already recommended that reformers help school-level leaders focus on building teacher collaboration, connecting teachers with key resources, and protecting new reforms. To enhance the effectiveness of each of these measures, we recommend that school-level leaders act aggressively to build the science knowledge of their teaching staff, both by hiring new teachers with strong science preparation and by pursuing science-oriented professional development opportunities for veteran teachers. A number of researchers have demonstrated a connection between teachers' content preparation and their teaching practice, as well as a link between content preparation and student achievement (Monk, 1994; Supovitz & Turner, 2000). Others, however, have argued that this link is somewhat tenuous and perhaps should not be a major determinant of policy (see Kahle & Woodruff, Chapter 3, this volume). Therefore, we would not advocate simply importing experienced scientists directly into classrooms. Becoming a teacher involves more than simply applying content expertise; it means developing sophisticated models of pedagogical content knowledge (Shulman, 1986), in part through experience in the concrete practices of teaching and learning. Furthermore, given the dominance of traditional pedagogy in pre-professional preparation for scientists, many innovative approaches to science education will be at odds with the training of practicing scientists. We suggest that leaders should consider the science teaching capacity of staff *collectively*, and deliberately bring together clusters of teachers who are knowledgeable about science but who are also willing to explore together various new ways to communicate science to students and who are able to support each other in new instructional efforts.

There are at least three ways in which school-level leaders can work to improve teachers' science content knowledge and support innovative science instruction. First, the success of professional communities rests on the staff's ability to share and develop their expertise. The scientific expertise of particular teachers is a critical resource for collegial interaction, and the development of new science knowledge and skills can provide a powerful catalyst for professional learning across the entire teaching staff. Second, experience with scientific inquiry in authentic contexts may lend teachers credibility in discussions about the relevance and advantages of a science education reform project. Although school leaders may work to protect teachers from external pressures, teachers may still encounter challenges from parents and community members who stress more traditional science course sequences. Teachers with experience in scientific research may have greater legitimacy in community-wide discussions about the advantages of innovative science instruction. Finally, a community of teachers with strong *collective* science preparation will probably be most capable of choosing and enacting high-quality content-centered reforms (Radford, 1998). Teachers' collective experience with science should enable them to judge the quality of curriculum materials and avoid those that favor visual or technological flash but lack scientific or pedagogical substance. A high level of comfort and confidence with scientific content will also enable teachers to focus on better ways to teach that content and to adapt the reform materials to their school and classroom context. It is crucial to consider the collective competence of the teaching staff in addition to the individual strengths of its members because in a highly collaborative context, it may actually be beneficial for teachers to have widely diverse backgrounds and strengths (Shulman & Shulman, 2004).

Although we caution against overzealous use of alternative certification pathways for reasons outlined above (see also Kahle & Woodruff, Chapter 3, this volume), we do support the judicious use of such strategies to recruit teachers with strong science preparation; in the context of a strong and supportive teacher community. Even though these teachers may lack some important pedagogical skills and classroom-specific preparation, their scientific knowledge and experience will be an asset to their communities even as they themselves benefit from the pedagogical expertise of more traditionally prepared teachers.

One way to build capacity in the science teaching staff is through careful and selective hiring. But even if that is not possible, it is and will continue to be important to build a strong program of science-focused professional development opportunities. Ongoing professional development is important because the positive effects of subject matter preparation appear to diminish over time, as teachers forget their more advanced science training or that training becomes obsolete (Monk, 1994). Because contemporary science reforms often emphasize inquiry skills and the social and epistemological nature of scientific work, we suggest that experience in scientific research is a particularly important piece of teachers' subject preparation, and a particularly exciting strategy for teacher professional development. And, we cannot emphasize enough how these professional development opportunities should focus on the most effective pedagogical strategies for teaching particular science ideas and skills. Knowing more science content without knowing how it is learned by students is simply not enough.

CONCLUSION

This chapter contrasted the theory of action that often guides science education reform with a school leadership-based theory of action that accounts for the constraints presented by local values and practices. We do not oppose reform that focuses on standards, curriculum resources, or out-of-school professional development. These are and will continue to be important parts of the policy agenda for reform in science education. Content standards will continue to provide the learning goals toward which educators can aim their efforts, and they have the potential to organize a coherent system of curriculum, instruction, and assessment. But, without paying attention to the local conditions and the organizational structures that make up a school and shape its interactions with the local community, in other words, the world of school leaders, reform policies are less likely to be successful. There are many innovative and interesting curriculum-centered resources and professional learning opportunities for teachers. The challenge is to choose the resources that are appropriate for the particular local context, create access to internal professional communities and links to external networks such as university and professional organizations that enable innovative practices to take root and bloom, and to create the space to keep these new practices alive for a long enough time to make a difference.

Blumenfeld, Fishman, Krajcik, Marx, and Soloway (2000) argued that reforms are most likely to succeed when they "fit with existing school capabilities, policy and management structures, and organizational culture" (p. 149). Successful reformers need to work with and through school leaders simply because school leaders are best positioned to evaluate the "fit" between a reform project and the local context, and can therefore play an important role in directing teachers toward reforms that are well suited to the overall circumstances of the school.

REFERENCES

American Association for the Advancement of Science. (1991). *Science for all Americans.* New York: Oxford University Press.

American Association for the Advancement of Science. (1993). *Benchmarks for science literacy.* New York: Oxford University Press.

American Association for the Advancement of Science. (2001). *Atlas of science literacy.* Washington, DC: National Science Teachers Association.

American Association for the Advancement of Science. (2007). Large boost to NSF proposed for 2007. Retrieved March 3, 2009 from http://www.aaas.org/spp/rd/nsf07p.htm#tb

Argyris, C., & Schön, D.A. (1974). *Theory in practice: Increasing professional effectiveness*. San Francisco, CA: Jossey-Bass.

Atkin, J. M., & Black, P. (2003). *Inside science education reform: A history of curricular and policy change*. New York, NY: Teachers College Press.

Ball, D. & Cohen, D. K. (1996). Reform by the book: What is – or might be – the role of curricular materials in teacher learning and instructional reform? *Educational Researcher, 25*(9), 6–14.

Bereiter, C., & Scardamalia, M. (1993). *Surpassing ourselves: An inquiry into the nature and implications of expertise*. London: Open Court.

Biological Sciences Curriculum Study. (1963). *High school biology.* Chicago, IL: Rand McNally.

Black, P., & Wiliam, D. (1998). Inside the Black Box: Raising standards through classroom assessment. *Phi Delta Kappan, 80*(2), 139–148.

Blanc, S., Christman, J. B., Liu, R., Mitchell, C., Travers, E., & Bulkley, K. E. (2010) Learning to learn from data: Benchmarks and instructional communities. *Peabody Journal of Education, 85*(2). 205–225.

Blumenfeld, P., Fishman, B. J., Krajcik, J., Marx, R. W., & Soloway, E. (2000). Creating usable innovations in systemic reform: Scaling up technology-embedded project-based science in urban schools. *Educational Psychologist, 35*(3), 149–164.

Bottoms, G., & Uhn, J. (2007) *Project Lead The Way works: A new type of career and technical program*. Atlanta, GA: Southern Regional Education Board.

Bredeson, P. (2002). *Designs for learning: A new architecture for professional development in schools*. Thousand Oaks, CA: Corwin Press.

Bruner, J. S. (1977). *The process of education*. Cambridge, MA: Harvard University Press.

Bryk, A. S., & Schneider, B. (2002). *Trust in schools: A core resource for improvement.* New York, NY: SAGE.

Burch, P. (2009). *Hidden markets: The new education privatization*. London: Routledge, Taylor & Francis.

Burry-Stock, J., & Casebeer, C. (2003, March). *A study of the alignment of national standards, state standards, and science assessment.* Paper presented at the Annual Meeting of the National Association for Research in Science Teaching, Philadelphia, PA. Retrieved March 24, 2009, from http://eric.ed.gov/ERICDocs/data/ ericdocs2sql/content_storage_01/0000019b/80/1a/f1/28.pd

Buxner, S., Harris, C., & Johnson, B. (2008, March). *Creating tightly aligned assessment for primary science in collaboration with an urban school district.* Paper presented at the annual meeting of the National Association for Research in Science Teaching, Baltimore, MD.

Center on Education Policy. (2007). *Choices, changes and challenges: Curriculum and instruction in the NCLB Era*. Washington, DC: Center on Education Policy.

Clifford, M. (2009). *Dissecting local design: Instructional leadership, curriculum, and science education*. Unpublished doctoral dissertation, Madison, University of Wisconsin.

Confrey, J. (2006). The evolution of design studies as methodology. In R.K. Sawyer (Ed.), *Cambridge handbook of the learning sciences* (pp. 131–151). Cambridge, England: Cambridge University Press.

Cuban, L. (1998). How schools change reforms: Redefining reform success and failure. *Teachers College Record, 99*(3), 453–477.

D'Amico, L. (2005, April). *Design-based research: The Center for Learning Technologies in Urban Schools.* Paper presented at the Annual Meeting of the American Educational Research Association, Montréal, Québec, Canada.

Darling-Hammond, L, Wei, R. C., Andree, A., Richardson, N., & Orphanos, S. (2009). *Professional learning in the learning profession: A status report on teacher development in the United States and abroad.* Stanford, CA: The School Redesign Network Retrieved March 19, 2009, from http://www.nsdc.org/news/NSDCstudy2009.pdf

DeBoer, G. (1991). *A history of ideas in science education: Implications for practice*. New York, NY: Teachers College Press.

Gamoran, A., & Weinstein, M. (1998). Differentiation and opportunity in restructured schools. *American Journal of Education, 106,* 385–415.

Gee, J. P. (2003). *What video games have to teach us about learning and literacy*? New York, NY: Palgrave Macmillan.

Gerard, L. F., Bowyer, J. B., & Linn, M. C. (2008). Principal leadership for technology-enhanced learning in science. *Journal of Science Education and Technology, 17*, 1–18.

Halverson, R. (2003) Systems of practice: How leaders use artifacts to create professional community in schools, *Educational Policy and Analysis Archives, 11*(37). Retrieved September 7, 2005, from http://epaa.asu.edu/epaa/v11n37/

Halverson, R., Grigg, J., Prichett, R., & Thomas, C. (2007). The new instructional leadership: Creating data-driven instructional systems in schools. *Journal of School Leadership, 17*(2), 159–194.

Halverson, R. (2010). School formative feedback systems. *Peabody Journal of Education, 85*(2) 130–155.

Halverson, R., Prichett, R. B., & Watson, J. G. (2007). *Formative feedback systems and the new instructional leadership* (WCER Working Paper No. 2007-3). Madison: University of Wisconsin–Madison, Wisconsin Center for Education Research. Retrieved June 15, 2007, from http://www.wcer.wisc.edu /publications/working Papers/papers.php

Henderson, A. T., & Berla, A. (1997). *A new generation of evidence: The family is critical to student achievement*. Washington, DC: National Committee for Citizens in Education.

Hewson, P. W., Kahle, J. B., Scantlebury, K., & Davies, D. (2001). Equitable science education in urban middle schools: Do reform efforts make a difference? *Journal of Research in Science Teaching, 38*, 1130–1144.

Honig, M. I., & Hatch, T. C. (2004). Crafting coherence: How schools strategically manage multiple, external demands. *Educational Researcher, 33*(8), 16–30.

Huberman, M. (1995) Networks that alter teaching: Conceptualizations, exchanges and experiments. *Teachers and Teaching: Theory and Practice, 1*(2), 193–211.

Kesidou, S., & Roseman, J. E. (2001). How well do middle school science programs measure up? Findings from Project 2061's curriculum review. *Journal of Research on Science Teaching, 39*(6), 522–549.

Krajcik, J. S., Blumenfeld, P. C., Marx, R. W., & Soloway, E. (1994). A collaborative model for helping middle grade science teachers learn project-based instruction. *The Elementary School Journal, 94*(5), 483–497.

Krajcik, J., Blumenfeld, P., Marx, R. W., Bass, K. M., Fredricks, J., & Soloway, E. (1996). The development of middle school students' inquiry strategies in project-based science classrooms. In D. C. Edelson & E. A. Domeshek, (Eds.), *Proceedings of the 1996 International Conference on Learning Sciences International Conference on Learning Sciences*. (Evanston, Illinois, July 25–27, 1996). 450–455.

Krajcik, J., & Reiser, B. J. (2004). *IQWST: Investigating and questioning our world through science and technology*. Ann Arbor, MI: University of Michigan.

Krajcik, J., McNeill, K. L., & Reiser, B. (2008). Learning-goals-driven design model: Developing curriculum materials that align with national standards and incorporate project-based pedagogy. *Science Education, 92*(1), 1–32.

Kulm, G., Roseman, J. E., & Treistman, M. (1999). A benchmark-based approach to textbook evaluation. *Science Books & Films, 35*(4), 147–153.

Lee, H., Songer, N. B., & Lee, S. 2006. Developing a sustainable instructional leadership model: A six-year investigation of teachers in one urban middle school. In *Proceedings of the 7th international Conference on Learning Sciences* (Bloomington, Indiana, June 27–July 01, 2006). International Conference on Learning Sciences. *International Society of the Learning Sciences*, 376–382.

Lee, O., & Luykx, A. (2005). Dilemmas in scaling up innovations in science instruction with nonmainstream elementary students. *American Educational Research Journal, 42*(5), 411–438.

Leithwood, K., Seashore Louis, K., Anderson, S. & Wahlstrom, K. (2004). *How leadership influences student learning*. New York, NY: Wallace Foundation. Retrieved September 14, 2005, from http://www.wallacefoundation.org/

Lieberman, A. (2000). Networks as learning communities: Shaping the future of teacher development. *Journal of Teacher Education, 51*(3), 221–227.

Loucks-Horsley, S., & Matsumoto, C. (1999). Research on professional development for teachers of mathematics and science: The state of the scene. *School Science and Mathematics, 99*(5), 258–271.

Louis, K. S., Kruse, S. D., & Bryk, A. S. (1995). Professionalism and community: What is it and why is it important in urban schools? In K. S. Louis & S. D. Kruse (Eds.), *Professionalism and community: Perspectives on reforming urban schools.* Thousand Oaks, CA: SAGE.

Lucas, S. R., & Berends, M. (2002). Sociodemographic diversity, correlated achievement, and de facto tracking. *Sociology of Education, 75*(4), 328–348.

Mandinach, E. B., & Honey, M. (2008). *Data-driven school improvement: Linking data and learning.* New York, NY: Teachers College Press.

Mangin, M., & Stoelinga, S. R. (2008). *Effective teacher leadership: Using research to inform and reform.* New York, NY: Teachers College Press.

Marx, R. W., & Harris, C. J. (2006). No Child Left Behind and science education: Opportunities, challenges, and risks. *Elementary School Journal, 106*(5), 467–477.

Meyer, J. W., & Rowan, B. (1983). The structure of educational organizations. In M. Meyer & W. R. Scott (Eds.), *Organizational environments: Ritual and rationality* (pp. 71–97). San Francisco, CA: Jossey-Bass.

McLaughlin, M., & Talbert, J. (2001). *Professional communities and the work of high school teaching.* Chicago, IL: University of Chicago Press.

Mislevy, R. J., & Knowles, K. T. (2002). *Performance assessments for adult education: Exploring the measurement issues.* Washington, DC: National Academy Press.

Monk, D. H. (1994). Subject area preparation of secondary mathematics and science teachers and student achievement. *Economics of Education Review, 13*(2), 125–145.

Murphy, J., Beck, L. G., Crawford, M., Hodges, A., & McGaughy, C. L. (2001). *The productive high school: Creating personalized academic communities.* Thousand Oaks, CA: Corwin Press.

National Center for Education Statistics. (2008). *Education and Certification Qualifications of Departmentalized Public High School-Level Teachers of Core Subjects* (NCES 2008-338). Retrieved November 6, 2009, from http://nces.ed.gov/fastfacts/display.asp?id=58

National Commission on Excellence in Education. (1983). *A nation at risk: The imperative for educational reform.* Washington, DC: Congressional Research Service.

National Research Council. (1996). *National Science Education Standards.* Washington, DC: National Academy Press.

National Research Council. (2001). *Educating teachers of science, mathematics, and technology: New practices for the new millennium.* Washington, DC: National Academy Press.

National Research Council. (2007). *Rising Above the Gathering Storm: Energizing And Employing America For A Brighter Future.* Washington, DC: The National Academies Press.

National Science Foundation. (2006). *America's Pressing Challenge—Building a Stronger Foundation (NSB-06-02).* Retrieved March 13, 2009, from http://www.nsf.gov/nsb/stem/index.jsp

National Science Foundation. (2007). *A national action plan for addressing the critical needs of the U.S. science, technology, engineering, and mathematics education system (NSB-07-114).* Retrieved February 9, 2009, from http://www.nsf.gov/nsb/documents/2007/stem_action.pdf

National Science Teachers Association. (2003). *Standards for science teacher preparation.* Arlington, VA: National Science Teachers Association. Retrieved August 23, 2009, from http://www.nsta.org/pdfs/NSTAstandards2003.pdf

National Science Resource Center. (2003). *Annual report.* Washington DC: The National Academies/Smithsonian Institution. Retrieved March 11, 2009, from http://www.nsrconline.org/pdf/2003_ar.pdf

Newmann, F. M., & Wehlage, G. (1993). Five standards of authentic instruction. *Educational Leadership, 50*(7), 8–12.

Newmann, F. M., Smith, B., Allensworth, E., & Bryk, A. S. (2001). *School instructional program coherence: Benefits and challenges.* Chicago, IL: Consortium on Chicago School Research. Retrieved July 17, 2003, from http://www.consortium-chicago.org/publications/pdfs/p0d02.pdf

No Child Left Behind Act (2001). Pub. L. No. 107-110,115 Star. 1425.

Oakes, J. (2005). *Keeping track: How schools structure inequality.* New Haven, CT: Yale University Press.

Penuel, W. R., Frank, K. A., & Krause, A. (2006). The distribution of resources and expertise and the implementation of schoolwide reform initiatives. In *Proceedings of the 7th international Conference on Learning Sciences* (Bloomington, Indiana, June 27–July 01, 2006). International Conference on Learning Sciences. International Society of the Learning Sciences, 522–528.

Phelps, A., Camburn, E., & Durham, J. (2009, April). *Engineering the math performance gap.* Madison: University of Wisconsin, Center for Education and Work Research Brief.

Physical Science Study Committee. (1960). *Physics.* Boston, MA: Heath.

Porter, A. (1994). National standards and school improvement in the 1990s: Issues and promise. *American Journal of Education, 102*(4), 421–449.

Prichett, R. (2007). *How school leaders make sense of and use formative feedback systems.* Unpublished doctoral dissertation, University of Wisconsin-Madison School of Education.

Project Lead the Way. (2007). *School District Agreement.* Retrieved February 20, 2009, from http://www.pltw.org/contracts/sample/SAMPLE_Engineering_Technology_SDA_09-10.pdf

Puntambekar, S. (2006). Analyzing collaborative interactions: Divergence, shared understanding and construction of knowledge. *Computers and Education, 47*(3), 332–351.

Quellmalz, E., DeBarger, A., Haertel, G., Schank, P., Buckley, B., Gobert, J., et al. (2007). *Exploring the role of technology-based simulations in science assessment: The Calipers Project.* Presented at American Educational Research Association (AERA), Chicago, IL.

Radford, D. L. (1998). Transferring theory into practice: A model for professional development for science education reform. *Journal of Research in Science Teaching, 35*, 73–88.

Ross, J. A., Hogaboam-Gray, A., & Gray, P. (2004). Prior student achievement, collaborative school processes, and collective teacher efficacy. *Leadership and Policy in Schools, 3*(3), 163–188.

Rowan, B. (2006). The school improvement industry in the United States: Why educational change is both pervasive and ineffectual. In H. D. Meyer & B. Rowan (Eds.), *The new institutionalism in education*. Albany, NY: State University of New York Press.

Ruby, A. (2006). Improving science achievement at high-poverty urban middle schools. *Science Education, 90*, 1005–1027.

Schneider, R. M., & Krajcik, J. (2002). Supporting science teacher learning: The role of educative curriculum materials. *Journal of Science Teacher Education, 13*(3), 221–245.

Schneider, R. M. Krajcik, J., & Marx, R. (2000). The role of educative curriculum materials in reforming science education. In B. J. Fishman & S. F. O'Connor-Divelbiss (Eds.), *Proceedings of the International Conference of the Learning Sciences.* Ann Arbor, MI: University of Michigan, June 14-17, 2000, 54–61.

Sedlak, M. W., Wheeler, C. W., Pullin, D. C., & Cusick, P. A. (1986). *Selling students short: Classroom bargains and academic reform in the American classroom*. New York, NY: Teachers College Press.

Settlage, J., & Meadows, L. (2002). Standards-based reform and its unintended consequences: Implications for science education within America's urban schools. *Journal of Research in Science Teaching, 39*(2), 114–127.

Showers, B., & Joyce, B. (1996). The evolution of peer coaching. *Educational Leadership, 53*(6), 12–16.

Shulman, L. S. (1986). Those who understand: Knowledge growth in teaching. *Educational Researcher, 15*(2), 4–14.

Shulman, L. S. & Shulman, J. H. (2004). How and what teachers learn: A shifting perspective. *Journal of Curriculum Studies, 36*, 257–271.

Simon, H. A. (1983). *Reason in human affairs*. Stanford, CT: Stanford University Press.

Singer, J., Marx, R. W., Krajcik, J., & Chambers, J.C. (2000). Constructing extended inquiry projects: Curriculum materials for science education reform. *Educational Psychologist, 35*(3), 165–178.

Siskin, L. (1995). *The subjects in question: Departmental organization and the high school*. New York, NY: Teachers College Press.

Spillane, J (2006). *Distributed Leadership*. San Francisco, CA: Jossey-Bass.

Spillane, J. P., Diamond, J. B., Walker, L. J., Halverson, R., & Jita, L. (2001). Urban school leadership for elementary science instruction: Identifying and activating resources in an undervalued school subject. *Journal of Research in Science Teaching, 38*, 918–940.

Spillane, J. P., Halverson, R., & Diamond, J. B. (2004). Towards a theory of leadership practice: A distributed perspective. *Journal of Curriculum Studies, 36*(1), 3–34.

Stoll, L., & Louis, K. S. (2007) *Professional learning communities: Divergence, detail and difficulties*. London: Open University Press.

Supovitz, J. A., & Turner, H. M. (2000). The effects of professional development on science teaching practices and classroom culture. *Journal of Research in Science Teaching, 37*(9), 963–980.

Swanson, C. B., & Stevenson, D. L. (2002). Standards-based reform in practice: Evidence on state policy and classroom instruction from the NAEP state assessments. *Educational Evaluation and Policy Analysis, 24*(1), 1–27.

Taylor, J. E. (2001). The struggle to survive: Examining the sustainability of schools' comprehensive school reform efforts. *Journal of Education for Students Placed at Risk, 11*(3-4), 331–352.

Thompson, C. L., & Zeuli, J. S. (1999). The frame and the tapestry: Standards-based reform and professional development. In L. Darling-Hammond & L. Sykes (Eds.), *Teaching as the learning profession: Handbook of policy and practice.* (pp. 341–375). San Francisco, CA: Jossey-Bass.

University of California-Davis (n.d.). *UC-Davis–Biotechnology program.* Retrieved March 20, 2009, from www.biotech.ucdavis.edu/events.cfm#NSF2

van Driel, J. H., Beijaard, D., & Verloop, N. (2001). Professional development and reform in science education: The role of teachers' practical knowledge. *Journal of Research in Science Teaching, 38*, 137–158.

Welch, W.W. (1979). Twenty years of science curriculum development: A look back. *Review of Research in Education* 7, 282–306.

Westerlund, J. F., Garcia, D. M., Koke, J. R., Taylor, T. A., & Mason, D. S. (2002). Summer scientific research for teachers: The experience and its effect. *Journal of Science Teacher Education, 13*(1), 63–83.

Wisconsin Department of Public Instruction. (2006). Data table: Educational qualifications of certified state administrators [CD]. Madison, WI: Department of Public Instruction.

Yager, R. (2005). Science is not written, but can be written about. In E. W. Saul (Ed.), *Crossing borders in literacy and science instruction: Perspectives on theory and practice* (pp. 95–108). Newark, DE: International Reading Association.

Yoon, K. S., Duncan, T., Lee, S. W.Y., Scarloss, B., & Shapley, K. (2007). *Reviewing the evidence on how teacher professional development affects student achievement* (Issues & Answers Report, REL 2007-No. 033). Washington, DC: U.S. Department of Education, Institute of Education Sciences.

ABOUT THE AUTHORS

Rodger W. Bybee holds a PhD in science education from New York University. He recently retired from the biological sciences curriculum study (BSCS) where he was executive director. Prior to that, he was executive director of the Center for Science, Mathematics, and Engineering Education at the National Research Council. His educational interests include teaching science as inquiry, curriculum development, international assessment, and national policies for STEM education. Dr. Bybee has authored, coauthored, or edited many books and chapters. He has published articles in refereed journals including *Journal of Research in Science Teaching*, *Science Education*, and *School Science and Mathematics*. In the 1990s, he chaired the content group for the *National Science Education Standards* and chaired both the science expert group and science forum for the 2006 Program for International Student Assessment (PISA). He has recently given keynote presentations at national meetings and international meetings in Japan, Israel, China, Mexico, Germany, and Finland. His current professional activities include work on common core standards for science education and completing a book on *The Teaching of Science: 21st Century Perspectives*.

Dennis W. Cheek earned a PhD in curriculum & instruction/science education from Pennsylvania State University and a PhD in theology from the University of Durham. He has been a classroom teacher of science and social studies, science department chair, district curriculum developer, state supervisor of nine area career and technical centers, senior administrator in the state education departments of NY and RI, vice president of education at the Ewing Marion Kauffman Foundation, and vice president at the John Templeton Foundation. Dennis is currently a global consultant, visiting scholar at the Center for Contemporary History and Policy at the Chemical Heritage Foundation, and a Senior Fellow at the Foreign Policy Research Institute. He annually lectures in the midcareer doctoral programs in educational leadership and work-based learning at the University of Pennsylvania. He has authored, edited, or contributed to over 800 publications and multimedia products and is a Fellow of the American Association for the Advancement of Science. He was a founding Board member of both the international Campbell Collaboration and the National Creativity Network.

George E. DeBoer is deputy director of Project 2061 of the American Association for the Advancement of Science. He has served as program director at the National Science Foundation and Professor of Educational Studies at Colgate University. He is author of *A History of Ideas in Science Education: Implications for Practice* (Teachers College Press, 1991) as well as numerous articles, book chapters, and reviews. His scholarly interests include science education policy, clarifying the goals of the science curriculum, researching the history of science education, and investigating ways to assess student understanding in science. He holds a PhD in science education from Northwestern University, an MAT in biochemistry and science education from the University of Iowa, and a BA in biology from Hope College. He is a member and fellow of the American Association for the Advancement of Science and the American Educational Research Association, and he is a member of the National Association for Research in Science Teaching, the National Science Teachers Association, and the American Association of University Professors.

Linda De Lucchi is codirector of the Full Option Science System Project (FOSS K–8) and past codirector of the Assessing Science Knowledge Project (ASK) at Lawrence Hall of Science, University of California at Berkeley. For 36 years she has developed instructional materials in science education (FOSS), environmental education (OBIS: Outdoor Biology Instructional Strategies), health education (HAP: Health Activities Project), and special education (SAVI/SELPH: Science Activities for the Visually Impaired/Science Enrichment for Learners with Physical Handicaps). In addition to curriculum development, Ms. De Lucchi has directed numerous teacher preparation projects, and has provided many tens of thousands of teacher-hours of science education inservice at the site level, district level, and national-leadership level throughout the country and abroad (including Israel, the Slovak Republic, the Czech Republic, Japan, and China). Ms. De Lucchi holds a bachelor's degree in zoology from Pomona College and a masters in zoology from the University of California at Davis.

Janice Earle holds a PhD in Education Policy from the University of Maryland. She has been at the National Science Foundation (NSF) since 1991. She currently serves as the Cluster Lead for the Knowledge Building Cluster in NSF's Directorate for Education and Human Resources (EHR) in the Division of Research on Education in Formal and Informal Settings. The Knowledge Building Cluster includes the Research and Evaluation on Education in Science and Engineering (REESE) program, the Foundation's primary education research program, and the CAREER program. Dr. Earle has worked with multiple EHR programs including the Statewide Systemic Initiatives, Instructional Materials, Centers for Learning and Teaching, and Discovery Research K–12. She has served on several interagency initiatives, worked with both the TIMSS and PISA international assessments, served as lead for many funded projects of the National Academy of Sciences, and served on the Planning Committee for the 2009 NAEP Science Assessment.

Noah Feinstein is an assistant professor of science education at the University of Wisconsin-Madison and the coordinator of UW-Madison's Agriscience Education Program. His research focuses on science literacy and public engagement with science. He is particularly interested in how, when, and why science matters to nonscientists. In the past, he has worked with parents of autistic children, attempting to understand how science informed their autism-related learning and advocacy. He currently studies how museums and science centers address issues of equity in science learning, how farmers think about climate change, and how participation in outreach affects the way scientists think about science. He is broadly interested in equity in science education, education for sustainability, climate change education, science and technology studies, autism and developmental disorders, scientist-educator collaborations and informal science learning. He is an affiliate of the Holtz Center for Science and Technology Studies at the University of Wisconsin-Madison.

Richard Halverson is an associate professor of educational leadership and policy analysis at the University of Wisconsin-Madison. He is a former high school teacher and administrator, and earned an MA in philosophy and a PhD in the learning sciences from Northwestern University. Halverson serves as the associate director of the Education Research Challenge Area at the Wisconsin Institutes for Discovery, and as codirector of Games, Learning, and Society at UW-Madison, an internationally known research group that investigates how cutting edge learning technologies can redefine learning in and out of schools. Halverson's research also investigates how schools and school leaders use technologies to create the conditions for instructional improvement. He is coauthor (with Allan Collins) of

Rethinking Education in the Age of Technology: The Digital Revolution and Schooling in America.

Jane Butler Kahle is Condit professor of science education, Emerita, Miami University, Oxford, Ohio. She has been a professor of biological sciences and education and associate dean of the Graduate School at Purdue University and director, Division of Elementary, Secondary, and Informal Education at the National Science Foundation (NSF). She has directed 45 research projects, receiving approximately $55,000,000 in external funds. She was principal investigator of Ohio's Systemic Initiative, funded by NSF and the State of Ohio, and of a research project to assess the outcomes of Ohio's reform of science and mathematics education. She was the founder and original Director of Ohio's Evaluation and Assessment Center for Mathematics and Science Education. Kahle received the Willystine Goodsell Award for her research on women from the American Educational Research Association and the Distinguished Contributions to Science Education Through Research Award from the National Association of Research in Science Teaching. In addition to several presidencies of national professional associations, she was chairperson of the National Research Council's Committee on Science Education K–12. Prior to her appointment at NSF, she was a member of the Advisory Committee for the Directorate of Education and Human Resources and chairperson of NSF's Committee on Equal Opportunities in Science and Engineering. She is the author of 146 refereed papers, 40 chapters in monographs or books, and 5 books. Kahle has served on the editorial review boards of more than 12 scholarly journals.

Pierre Léna is an astrophysicist, an emeritus professor at the *Université Paris Diderot*, and an associate researcher at *Observatoire de Paris*, where he developed his research activities in infrared astronomy, high resolution imaging, including pioneer work on adaptive optics and optical interferometry for the European Very Large Telescope in Chile. For many years he directed the Graduate School Astronomy & Astrophysics in Ile-de-France and supervised numerous PhD projects. His interest in scientific education led him to be the president of the *Institut national de recherche pédagogique* in France (1991–1997) and to get involved in the *La main à la pâte* national, then international, project carried out by the French *Académie des sciences* with numerous collaborations (Book *L'Enfant et la science*, in 2005 with G. Charpak and Y. Quéré and many related publications). A member of the *Académie des sciences* in France, as well as of several other academies, he is currently holding the position of *délégué à l'éducation* of the *Académie des sciences*, hence deeply involved in developing inquiry-based science education in France, in Europe through large cooperative projects, and beyond Europe with the *InterAcademy Panel*.

Sharon J. Lynch holds a PhD in education from Wayne State University. She is professor of curriculum and instruction (specializing in secondary science education) at the George Washington University Graduate School of Education and Human Development, and for the past 2 years she was a program director for the Division for Research on Learning at the National Science Foundation. Her main interest is on the intersection of equity and science education policy and practice. Her most recent large interdisciplinary research focused on the effectiveness and scale-up of innovative middle school science curriculum materials, especially for diverse learners. Lynch is the author of numerous articles and chapters as well as *Equity and Science Education Reform* (2000). She presents papers and organizes symposia at the annual meetings of American Educational Research Association and the National Association for Research in Science Teaching. She is active in encouraging further study of science education policy research.

Larry Malone is codirector of the Full Option Science System (FOSS) Project, and past codirector of the Assessing Science Knowledge Project (ASK), and has been on the staff at the Lawrence Hall of Science, University of California at Berkeley, for 45 years in curriculum development and teacher preparation. He began as materials specialist with the Science Curriculum Improvement Study (SCIS) project. He has been a curriculum developer and instructor for OBIS (Outdoor Biology Instructional Strategies), HAP (Health Activities Project), GEMS (Great Explorations in Math and Science), SAVI/SELPH (Science Activities for the Visually Impaired/Science Enrichment for Learners with Physical Handicaps), and FOSS K–6 and FOSS Middle School Programs. Mr. Malone is a creative materials developer and designer of instructional activities, and serves as the lead writer on the FOSS materials. Mr. Malone holds a bachelor's degree in biology from the University of California at Berkeley and a master's in education from the California State University in San Francisco.

David Meshoulam is a doctoral candidate in the department of Curriculum & Instruction at University of Wisconsin-Madison. He holds a BA in the philosophy of science from Columbia University and an MA in the history of science from UW-Madison. His research interests include the history of education and the role played by history of science in science education. He has presented his work at national conferences, including the American Educational Research Association. He also has several articles in preparation that examine science education in informal settings, ranging from coffee shops to museums. Previous to his graduate work, he taught high-school and middle-school science both in the United States and abroad.

Jodi Peterson is assistant executive director for legislative and public affairs for the National Science Teachers Association. She is responsible for monitoring and identifying policy and legislation affecting science education on Capitol Hill, all outreach activities to Congress, and the association's advocacy initiatives. She also is cochair of the STEM Education Coalition, a group composed of 1,300 advocates from diverse groups representing all sectors of the STEM workforce. The Coalition advocates for increased federal investments in STEM education and strengthening STEM-related federal programs for educators and students. She also oversees all public relations strategies and activities for NSTA programs and is responsible for public relations outreach efforts including monitoring and analyzing key issues pertaining to science education; answering media inquiries from reporters and editors; developing and disseminating all press releases and statements for NSTA; and public relations outreach implementation for selected programs. She is also responsible for coordinating NSTA's Science Matters outreach campaign.

Jonathan Osborne is Shriram family professor of science at Stanford University. Previously he was the chair of science education at King's College London and departmental chair from 2005–08. He taught physics in Inner London for 9 years before joining King's in 1985. He has an extensive record of publications and research grants in science education in the field of elementary science, science education policy (*Beyond 2000: Science Education for the Future*), the teaching of the history of science, argumentation (the Ideas, Evidence and Argument in Science [IDEAS] Project) and informal science education. He was an advisor to the House of Commons Science and Technology Committee for their report on Science Education in 2002. He was also president of the U.S. National Association for Research in Science Teaching (2006–07) and has won this association's award for the best research publication in 2003 and 2004 in the *Journal of Research in Science Teaching*.

Margo Quiriconi joined the Kauffman Foundation in 1991 and currently serves as a director in education. In this role she develops programs and initiatives to advance the Kauffman youth education agenda. Previously, Quiriconi helped to develop numerous Foundation initiatives including Project Early and the math and science education strategy. Before joining the Foundation, Quiriconi worked in a variety of roles at a number of national, state-level, and local organizations focused on improving the well- being of children. She currently serves as a director of the Prime Health Foundation and on the national advisory board of Project Lead the Way. She plays an active role as a volunteer in a number of Kansas City's civic and cultural organizations. Quiriconi has bachelor's degrees from Michigan State University and St. Louis University, and a master's degree from the University of North Carolina at Chapel Hill. Quiriconi is currently pursuing a doctorate degree at Tulane University.

Edith Saltiel is a physicist highly involved in science education. She is an honorary assistant professor at the Paris Diderot University, where she has been all her teaching and research career. She has a PhD in solids physics and, after a thesis in didactics of physics at Paris Diderot University, she created jointly with Jean Louis Malgrange and Laurence Viennot the laboratory of didactics of physics in the 70s. She was codirector of that laboratory from 1979 to 1991. She has been a member since 1996 of the French national program for science education in school, *La main à la pâte,* and was director of the *La main à la pâte* team from 1999 to 2003. She is also a member of the science and technology in primary school working group of the Ministry of Education, a member of the scientific committee of the exhibition, *Mémoire de l'enseignement scientifique à l'école primaire,* and a member of the primary science education assessment committee of the Ministry of Education. She is coauthor of the DVD *Learning Science and Technology in School* (Ministry of Education and Academy of Sciences.

Jean-Pierre Sarmant is a former general inspector of the French Ministry of Education (1994–2004) and professor at the *Lycée Louis-le-Grand, Paris,* Jean-Pierre Sarmant has focused is whole career on the teaching of physics and its improvement. He has written a dictionary of physics as well as several text books of mechanics and electrodynamics for graduate students. His involvement in *La main à la pâte* started in 1999. After submitting a positive report on this new teaching method to the Ministry of Education he became president of a national committee dedicated to its implementation (also known as *le plan de de rénovation l'enseignement des sciences et de la technologie à l'Ecole).* This led him to contribute to the writing of two new national curricula: one for science at primary school in 2002 as a coordinator, and one for natural science and mathematics at junior high school (2002–2004) as a vice president.

Sarah Beth Woodruff holds a PhD in educational leadership from the University of Dayton. She currently is Director of Ohio's Evaluation and Assessment Center for Mathematics and Science Education at Miami University, Oxford, Ohio. As center director, she provides leadership in research design, data analysis, instrument development, and all aspects of evaluation and research for large-scale, externally funded education programs and projects across the nation. Dr. Woodruff has a broad understanding of education, having served as a science teacher, high school principal, and state education official. Her primary research interests include gender and equity issues in science education, science teacher preparation and licensing policy, and teacher professional development related to inquiry teaching and learning. She has coauthored three chapters and recently presented papers at the annual meetings of professional organizations, including the American Educational Research Association, the National Association for Research in Science Teaching, and the American Evaluation Association.

CPSIA information can be obtained at www.ICGtesting.com
Printed in the USA
LVOW08s2258251114

415415LV00003B/55/P